D0747280

Industrial
Design A-Z

Illustration pp. 4/5: *Oyster*, multipurpose storage containers for Tupperware, 1990s

To stay informed about upcoming TASCHEN titles, please request our magazine at
www.taschen.com/magazine or write to TASCHEN America, 6671 Sunset Boulevard,
Suite 1508, USA-Los Angeles, CA 90028, contact-us@taschen.com, Fax: +1-323-463-4442.
We will be happy to send you a free copy of our magazine which is filled with information
about all of our books.

Editing and coordination: Susanne Husemann, Cologne

Design: UNA (London) designers
Cover design: Sense/Net, Andy Disl & Birgit Reber, Cologne

Printed in China
ISBN 3–8228–5057–8

Industrial
Design A-Z

Charlotte & Peter Fiell

TASCHEN

KÖLN LONDON LOS ANGELES MADRID PARIS TOKYO

CONTENTS

For over 200 years, the products of mechanized industrial production have shaped our material culture, influenced world economies and affected the quality of our environment and daily lives. From consumer goods and packaging to transportation systems and production equipment, industrial products encompass an extraordinary range of functions, techniques, attitudes, ideas and values and are a means through which we experience and perceive the world around us.

The nature of industrial products and how they come to be is determined by an ever more complex process of design that is itself subject to many different influences and factors. Not least of these are the constraints imposed by the social, economic, political, cultural, organizational, and commercial contexts within which new products are developed, and the character, thinking and creative abilities of the individual designers or teams of designers, aligned specialists and manufacturers involved in their realization.

Industrial design – the conception and planning of products for multiple reproduction – is a creative and inventive process concerned with the synthesis of such instrumental factors as engineering, technology, materials and aesthetics into machine-producible solutions that balance all user needs and desires within technical and social constraints. Engineering – the application of scientific principles to the design and construction of structures, machines, apparatus or manufacturing processes – is an essential and defining aspect of industrial design. While both disciplines are concerned with finding optimum solutions to specific problems, the primary distinguishing characteristic of industrial design is its concern for aesthetics.

The origins of industrial design can be traced back to the Industrial Revolution which began in Great Britain in the mid-18th century, and which heralded the era of mechanization. Prior to this, objects were craft-produced, whereby both the conception and the manufacture of an object were the work of a single individual. With the development of new industrial manufacturing processes and the division of labour, design (conception and planning) was progressively separated from the act of making. At this early stage, however, design had no intellectual, theoretical or philosophical foundation and was considered just one of the many interrelated aspects of mechanical production. Thus the industrial goods of the years up to the 19th century were created by specialists from the technical, materials and production spheres rather than by an industrial designer. Towards the end of the 19th century, however, manufacturers began to realise that they could gain a critical competitive advantage by improving the constructional integrity and aesthetic appearance of their products. As a consequence, they began to invite specialists from other spheres– most notably, architects – to contribute to the design process.

Industrial design subsequently became a fully-fledged discipline in the early 20th century, when design theory was integrated into industrial methods of production. Among the first professional industrial design practitioners was the German architect Peter Behrens, who was recruited by AEG in 1907 to improve the company's products and corporate identity. Since then, industrial design has become an increasingly important factor in the success of industrial products and the companies that manufacture them.

This survey of industrial design encompasses all aspects of the subject – from heavy industrial products to sports equipment, and medical devices to aerospace vehicles – and is intended as a companion work to our earlier *Design of the 20th Century*. The manufacturers, industrial design consultancies, individual designers and inventors we feature – in entries organized in alphabetical order – are those who have developed some of the most important and influential products and technologies of the last 200 years. Their innovative solutions strike the best possible balance between the intellectual, functional, emotional, aesthetic and ethical expectations of the user/consumer and the influences and factors bearing upon the design process.

Cross-references appear in the text in bold type so as to reveal the many illuminating interrelationships between designers, design consultancies and manufacturers, while a time-line at the back of the book places key industrial design developments within the context of world events.

As more and more countries are drawn into the global free-market economy, so industrial design has become an increasingly vital means of competing on a global scale. By reflecting on the history of industrial design and focusing in particular on its successes, this book hopes to demonstrate how industrial design has consistently sought to de-mystify technology and to deliver it in accessible forms to the greatest number of people.

A further aim of the book is to highlight the phenomenal extent to which manufacturing companies themselves have shaped the history of industrial design. Without their willingness to risk the necessary and sometimes massive investments demanded by the development of new products, there would be very little industrial design. Innovative and commercially-motivated yet socially-minded manufacturing companies are too often the unsung heroes of our material culture.

"Our capacity to go beyond the machine rests upon our power to assimilate the machine. Until we have absorbed the lessons of objectivity, impersonality, neutrality, the lesson of the mechanical realm, we cannot go further in our development toward the more richly organic, the more profoundly human."
Lewis Mumford, *Technics and Civilization*, 1934

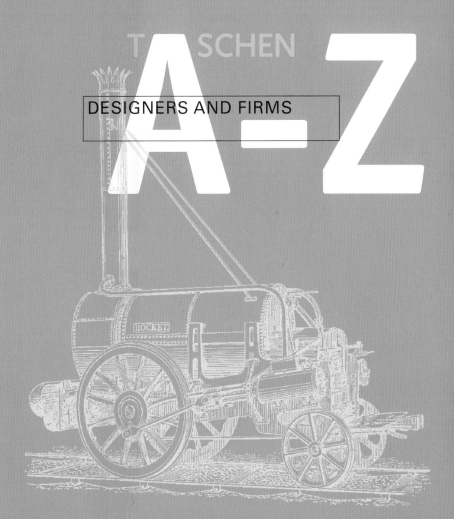

TASCHEN

A-Z

DESIGNERS AND FIRMS

ROCKET

AEG

FOUNDED 1883
BERLIN, GERMANY

Peter Behrens,
Turbine hall for AEG,
1908–1909

→Peter Behrens,
Light fitting, 1907

Peter Behrens, AEG
advertisement, 1907

**ALLGEMEINE
ELEKTRICITÄTS
GESELLSCHAFT**

A·E·G·METALLFADENLAMPE

ZIRKA EIN WATT PRO KERZE

In 1881 Emil Rathenau (1838–1915) visited the "Exposition Internationale d'Electricité" in Paris, where he saw **Thomas Alva Edison**'s light bulb. He was so impressed that he bought the patent licence and in 1883 founded the Deutsche Edison Gesellschaft (German Edison Company for Applied Electricity). The company subsequently changed its name to the Allgemeine Elektricitäts Gesellschaft, or AEG. Its catalogue for the 1900 Paris "Exposition Universelle" was designed by the Jugendstil artist Otto Eckmann (1865–1902), who also designed an Art Nouveau-style logo for the company. In 1907, however, AEG appointed as its artistic consultant the architect and designer **Peter Behrens**, who proceeded to create the first wholly integrated corporate identity for the company. Having re-invented AEG's logo, Behrens went on to design not only a unified range of electrical goods such as kettles, clocks and fans, but also the factory buildings required for their production.

AEG's production engineer, Michael von Dolivo-Dobrowolsky, realised that the key to the successful mass production of high quality goods lay in the standardization of interchangeable components, which would allow them to be used for several products rather than just one. This type of product standardization and the modern methods of manufacture employed by AEG reflected the ideals of the **Deutscher Werkbund**, of which Behrens had been a founding member in 1907. Hermann Muthesius (1861–1927), another central figure in the formation of the Deutscher Werkbund, had urged German companies to

ALLGEMEINE ELEKTRICITÄTS-GESELLSCHAFT, BERLIN

Peter Behrens,
Electric coffee
grinder, c. 1911

↗Peter Behrens,
Clocking-in machine,
c. 1910

establish a national aesthetic of "types" and "standards" that would result in "a unification of general taste". AEG was one of the first companies to embrace and implement Muthesius' ideas and to assemble household products from standardized components, as was the case with Peter Behrens' well-known range of kettles and his electric table fan (1908). With a strong brand identity forged by Behrens' comprehensive design programme, AEG went on to become a leading manufacturer of electrical products. In 1927 the company took part in the Deutscher Werkbund's *Die Wohnung* (The Dwelling) exhibition held in Frankfurt, where it displayed electric fans, water heaters, lamps and kettles, among other products. In the 1960s, AEG became primarily known for its almost puristic high quality white goods. In the 1980s, however, AEG entered a financial crisis and was taken over, first by Daimler-Benz, and in 1994 by the Swedish Company **Electrolux**. Today their concept of high quality and restrained form is very successful and AEG remains a large and prosperous manufacturing company.

Peter Behrens,
Electric desk fan,
1908

A&E DESIGN

FOUNDED 1968
STOCKHOLM, SWEDEN

Hans Ehrich and
Tom Ahlström with
Clean shower and
toilet chair, 1999

→ *Fresh* bath boards
for Etac, 1998

Jordan 1230
Dishwashing brushes
for Jordan, 1974

One of the leading industrial design consultancies in
Sweden, A&E Design was founded in 1968 by Tom
Ahlström (b. 1943) and Hans Ehrich (b. 1942), both of
whom had previously studied design and metalwork at
the Konstfackskolan, Stockholm. From its outset, this
Stockholm-based practice researched the application
of plastics while specializing in design for disability.
A&E Design has since developed numerous aids and
implements for use by the elderly and handicapped –
from a lift-and-ride wheelchair for transporting patients
to the shower, to bath seats and locking armrests for
toilets. The consultancy has also developed other er-
gonomically-led designs including tools for Anza and
dishwashing brushes for Jordan. A&E Design is renowned for innovative product development including me-
ticulous working models. Unusually, around 20% of
the work undertaken at A&E Design is dedicated to the
development of its own products, which "seek an optimal balance between aesthetics, function and market".
In 1982, Ahlström and Ehrich founded a sister design
consultancy, **InterDesign**, that similarly manufactures
models – from mock-ups to fully functional prototypes
– in its own workshop. Although A&E Design is one of
the leading Swedish consultancies, many of its product
designs have attributes more closely associated with
Italian rather than Scandinavian design. Its founders
readily acknowledge the influence of Italian designers,
most notably **Joe Colombo**, and have always maintained
close links with Milan and the design journal *Domus*.
A&E Design has received 14 Excellent Swedish Design
Awards and no less than three Roter Punkt awards
for Highest Design Quality.

Wally Byam

AIRSTREAM

FOUNDED 1934
LOS ANGELES, USA

The history of the iconic Airstream caravan can be traced to plans for a trailer which were submitted to a DIY magazine published by Wally Byam (1896–1962) in the 1920s. Byam received several complaints from readers that the design did not actually work, so he set himself the task of designing his own trailer that would actually perform properly. By lowering the floor between the wheels and raising the roof height, Byam designed the first travelling trailer in which one could stand upright. Byam eventually abandoned publishing in 1929 in order to concentrate on perfecting the revolutionary design of his trailers, in which he applied techniques borrowed from the aircraft industry, including riveted aluminium monocoque structures and streamlining. In 1934, he conceived the name "Airstream" for his trailers, which travelled along "like a stream of air". Two years later, he introduced the *Clipper* built along aircraft principles. Throughout the Depression of the 1930s, Byam's trailers bucked the trend with demand outstripping supply despite their relatively high price. Following the outbreak of the Second World War, however, Byam's growing busi-

Bambi Airstream trailer (re-edition of the 1961/1962 original)

ness was abruptly halted by the US Government's decision in 1941 to
restrict the use of aluminium to weapons manufacturers. Byam spent the
war years working for a Los Angeles aircraft manufacturer, where he undoubtedly gleaned new ideas for his trailers. After the war, Airstream was
reborn and by 1952 increased demand for Byam's high quality trailers had
led to the construction of new manufacturing plants in Jackson Centre, Ohio
and Santa Fe Springs, California. The gleaming Airstream trailer came to
symbolize the American Dream; just like the earlier covered wagons of the
pioneers, these futuristic silver bullets promoted the spirit of freedom and
adventure.

Following Byam's death in 1962, a foundation was created to host annual
caravan meetings across America in order to enhance cross-cultural relation-
ships. 1978 saw the introduction of the first Airstream motor homes, and
today the "Wally Byam Caravan Club, International" boasts some 19,000
member trailers/motor homes. The sleek good-looks of the Airstream trailer
have bred much fanaticism among its enthusiasts and Wally Byam's dream
of some 65 years ago has truly become an international phenomenon. In the
words of Byam, Airstream strives "endlessly to stir the venturesome spirit
that moves you to follow a rainbow to its end ... and thus make your travel
dreams come true."

ALESSI

FOUNDED 1921
OMEGNA, ITALY

Group of Alessi
workers outside the
factory, early 20th
century

→Luigi Massoni &
Carlo Mazzeri,
Model No. 870
cocktail shaker, 1957

Carlo Alessi, *La
Bombé* tea and
coffee service, 1945

Giovanni Alessi founded FAO (Fratelli Alessi Omegna)
in Omegna, Italy, in 1921. According to a sign which it
displayed at trade fairs during the 1920s, FAO was a
"Workshop with foundry specializing in working brass and nickel silver sheet".
From its earliest beginnings, FAO was highly innovative and ready to experiment. From 1929, the company concentrated on producing catering and
domestic tableware in either nickel-plated or silver-plated brass. Around
1935, Giovanni's son Carlo Alessi (b. 1916) was appointed principal designer
to the firm. The ensuing years saw FAO making the transition away from
craft-based methods to industrial production. In 1945 – the year in which
he also became general manager– Carlo conceived his *Bombé* coffee set.
This enduring design, produced in four different sizes, was an unashamedly
industrial product.

Also in 1945, another member of the family, Ettore Alessi, joined the firm as
a technical expert in cold metal pressing – Alessi had started using pressed
stainless steel as a consequence of the general shortage of nickel-silver and
brass after the Second World War. From around 1955, the company began

Guido Venturini,
GVo8 Okkio clothes
brushes, 2000

Alejandro Ruiz,
Parmenide parmesan
cheese graters
(*Memory Container*
collection), 1994

inviting outside designers such as Luigi Massoni (b. 1930) to contribute
to its programme. Many of the resulting designs, which tended to be rather
forward-looking, were consequently mass-produced – 1.5 million *Model
No. 870* cocktail shakers by Massoni, for example, have been sold to date. In
the late 1970s, with Carlo now president, the firm began producing limited-
edition, signed ranges by internationally celebrated designers and architects,
such as **Ettore Sottsass** and **Richard Sapper**. In 1983 Alessi registered a new
trademark, *Alessi Officina*, for products of experimental design that could be
developed outside the constraints of industrial production. That same year,
Carlo's son, Alberto, initiated the *Tea and Coffee Piazza* project in which
eleven architects designed limited edition services. This publicity-motivated
"architecture in miniature" project brought Alessi international recognition
and made the company one of the leading exponents of Post-Modernism
within the decorative arts during the 1980s. In the 1990s Alessi introduced
other product lines, such as *Twergi*, which included characterful wooden
household items designed by Ettore Sottsass and Andrea Branzi (b. 1938)
and a range of affordable yet high-style plastic wares by Stefano Giovannoni
(b. 1954) and Alessandro Mendini (b. 1931). Designs such as these are the
outcome of Alberto Alessi's belief that "design is a global creative discipline
with a strictly artistic and poetic matrix, and not simply one of the many
tools at the service of marketing and technology to ensure improved production methods and better sales".

Super Sport, 1947

ALFA ROMEO

FOUNDED 1910
PORTELLO/MILAN, ITALY

In 1910 the company known as Anonima Lombarda Fabbrica di Automobili (ALFA) was founded in the Portello district of Milan, having emerged from the defunct Società Italiana Automobili Darracq, which had been established in 1906 to produce low-cost cars. The new company decided to espouse a distinctly Milanese image and adopted a red cross and the Visconti dragon for its trademark. Its first venture was a production run of 300 *24 HP Torpedo* cars (1910) designed by Giuseppe Merosi. Three years later, the company produced its first sports car, the *40/60 HP Corsa* (1913), which had a top speed of 147 km per hour. Around this period, Nicola Romeo, a Neapolitan entrepreneur, took over the company and added his surname to its title. He also ensured that cars began rolling off the lines in Portello in significant numbers. During this era, the company produced several remarkably progressive cars, including the tear-drop shaped *40/60 HP Aerodynamica* (1914), which could reach an impressive top speed for its day. The company went into full-scale industrial production with Giuseppe Merosi's *RL* (1922–1923). Described by Alfa Romeo as its first masterpiece, the *RL* was also produced in sports and touring versions. The fast and reliable racing version of the *RL*, known as the *Targa Florio*, with its characteristic "prow-shaped" nose, placed second, third and fifth in the famous Sicilian 1,000 mile road race – the Mille Miglia – and brought the world's attention to the fledging Milanese car industry. At this stage, the Carrozzeria Castagna also designed several car body variations including a cabriolet version of the four-cylinder *RM* and an elegant

↘ *6C 2500 Super Sport*, 1947

Giulietta Spider, 1955

open touring version of the *RM*. In 1923, the designer Vittorio Jano joined the firm and his *P2* won Alfa Romeo's first world title. His *8C-2300*, which won Le Mans in 1931 and 1934, ensured that the marque became synonymous with racing during the early 1930s. Although this was the golden age of Alfa Romeo, the company was nationalized in 1933. Sadly, the Italian Government did not provide the company with sufficient funding and so it moved away from customized prize-winning racing cars in order to concentrate on general use vehicles, although it did continue producing its sleek, high-quality sports cars. After the Second World War, Alfa Romeo mass-produced elegant road cars such as the *6C 2500 Super Sport* (1947) with super-lightweight coachwork. Incorporating the latest aerodynamic research, the *Super Sport* and the similar *Freccia d'Oro* possessed a streamlined beauty and a sense of poise. While the market in the 1950s was demanding more functional and less luxurious models, Alfa Romeo was disinclined

↑ *Spider Duetto*, 1966

Alfa *GTV*, 1995

to completely sacrifice its sporting heritage and decided instead to produce affordable performance cars. By the mid-1950s and early 1960s, the company had increasingly industrialized its production methods and was manufacturing classic models such as the *Giulietta Spider* (1955) styled by **Pininfarina** and the *Giulietta Sprint* (1954) styled by **Bertone**. Under the directorship of Orazio Satta, the company had come up with the *Giulietta* project so as to meet the growing demand for smaller and cheaper private vehicles. The *Giulietta* managed to compress Alfa Romeo's high performance technology into an engine of only 1300 cubic centimetres. In 1962, the slightly larger *Giulia* was launched and the company began targeting the European market with classics such as the *Giulia Sprint* styled by Bertone. Later, Pininfarina styled the streamlined *Duetto* (1966), which with a few minor changes remains in production today. In 1971, **Giorgetto Giugiaro** styled the groundbreaking *Alfasud* that helped to considerably increase the company's market share. From the post-war period until 1986, Alfa Romeo was under the control of the Italian Government, which oversaw its operations through the Istituto per la Ricostruzione (IRI). Even today, Alfa Romeo designs such as the Alfa *156* (1997) and Alfa *166* (1998) acknowledge their legendary sporting ancestry and project a strong and purposeful aesthetic.

Interior of Alfa *166*, 1998

EMILIO AMBASZ

BORN 1943
RESISTENCIA, ARGENTINA

Emilio Ambasz studied architecture at Princeton University, taught for a year at the Hochschule für Gestaltung in Ulm and held a professorship at Princeton until 1969. In 1967 he also co-founded the avant-garde Institute of Architecture and Urban Studies in New York. From 1970 to 1976 he was Curator of Design at New York's Museum of Modern Art, where in 1972 he organized the groundbreaking exhibition "Italy: The New Domestic Landscape – Achievements and Problems of Italian Design". He subsequently founded his own design studio, Emilio Ambasz & Associates, in 1977 and the Emilio Ambasz Design Group in 1981 – both in New York. Between 1981 and 1985, Ambasz was president of the Architectural League and taught extensively at Princeton and other American universities. While Ambasz is widely celebrated for his design teaching and writing, he is also acclaimed

Cummins *Signature 600* engine, 1996–1997

Polyphemus
flashlight, 1985

for several notable seating and lighting designs, such as the highly success-ful seating systems *Vertebra* (1977) with Giancarlo Piretti and *Dorsal* (1981), and the lighting range for Logotec (1981). His architectural projects include the Center for Applied Computer Research and Programming, Las Prome-sas, Mexico (1975), the Grand Rapids Art Museum, Michigan (1975), the Museum of American Folk Art, New York (1980) and the San Antonio Bo-tanical Garden Conservatory, Texas (1982). Ambasz also won first prize and a gold medal in a competition to design the master plan for Expo 1992 in Seville. Since 1980 Ambasz has been chief design consultant to Cummins Engine Co. and has received numerous awards for his lighting and seating designs – his *Polyphemus* flashlight (1985) was nominated for both a Com-passo d'Oro and an IDSA (Industrial Designers Society of America) award in 1987, his *Qualis* seating system (1991) was awarded a Compasso d'Oro in 1991, and his *Soffio* modular lighting system won him the Industrial Design Excellence award by the IDSA. He has stated that the methodological princi-ple guiding his work "is the search for basic principles and prototypical or pilot solutions which can first be formulated into a general method and then applied to solve specific problems". Ambasz believes that design should not just fulfil functional requirements but should also take poetic form in order to satisfy our metaphysical needs. He maintains that designers must learn to reconcile the past and the future in their work and give "poetic form to the pragmatic".

APPLE COMPUTER INC.

FOUNDED 1976
PALO ALTO, USA

Steve Jobs

→ *Apple IIc*, 1984

↘ *Lisa*, 1983

High school friends Steven Wozniak (b. 1950) and Steven Jobs (b. 1955) shared an interest in electronics but were both perceived as "outsiders" by their peers. After graduation, they dropped out of further education and found employment in Silicon Valley – Wozniak at Hewlett-Packard and Jobs at Atari. In 1976 Wozniak, who had dabbled in the development of computers for some time, designed what would eventually become the *Apple I* computer. Jobs, who was more business-orientated than his friend, insisted that this product should be developed and marketed. As a result, Apple Computers was founded in Palo Alto on 1st April 1976 in the hope that the new personal computer, the *Apple I*, would revolutionize computers just as **Ford**'s *Model T* had revolutionized cars. The venture did not take off until a year later, when the *Apple II* – the first ever commercial personal computer – was launched at a local trade fair. This was the first personal computer to have a plastic casing and to include colour graphics. 1978 saw the introduction of the *Apple Disk II*, which at the time was the cheapest and easiest-to-use floppy drive available. Increasing sales drove the company's rapid growth and by 1980, the year in which the *Apple III* was launched, the business was employing a workforce of several thousand. In 1981 Apple hit the first of several rocky periods in its history; market saturation meant fewer sales, **IBM** released its first PC and Wozniak, the creative force behind the business, was injured in a plane crash and only returned to the company for brief periods of time. Undeterred, Jobs began working on the *Apple Macintosh*, which had initially been conceived as a $500 computer but in the end far exceeded that amount. Its subsequent launch in 1984 marked a real breakthrough in personal computing. With its user-friendly interface, high-definition screen and mouse, the *Mac* heralded the true advent of the home computer. Unlike the *Apple II*, the *Apple Mac* featured an integrated monitor and disk drive. Designed

Apple I, 1976

by **Frogdesign**, its **styling** was radically different from its more angular competitors – it had softer, flowing lines that gave it a sleeker aesthetic. Although the smaller, cream-coloured *Mac* initially sold well, users became increasingly frustrated by its small amount of RAM and its lack of hard drive connectivity. Problems also arose through personal differences between Jobs and the then CEO of Apple, John Scully, which eventually resulted in Jobs being ousted in 1985 from the company he had originally created. Bill Gates' introduction of *Windows 1.0*, which had many similarities to the *Mac*'s GUI interface, brought further difficulties. In particular, while Gates had signed a legal statement to the effect that Microsoft would not use *Mac* technology for *Windows 1.0*, it did not extend to future versions of Windows. This resulted in Apple's losing the exclusive rights to its revolutionary, user-friendly icon-based interface. Although the late 1980s were a prosperous time for Apple, by 1990 the market was swamped with cheaper PC clones that could run the newly launched *Windows 3.0* operating system. While watching its market share erode, Apple nevertheless remained the design industry's computer system of choice because of its superior graphics and publishing applications. This was especially true after the launch of the *Powerbook* in 1991 and the *PowerMac* in 1994. By the mid-1990s, Apple was yet again on the brink of collapse with $1 billion of back orders that it could not fulfil because it did not have the parts to make up the products. In 1996,

Apple logo

Jobs returned to Apple and began making significant structural changes to the company, after which Apple sold its products directly over the Internet and a cross-licensing agreement was drawn up with its one-time rival Microsoft. In 1998 Jobs oversaw the launch of the company's new landmark product, the *iMac* – another design from Apple that redefined the personal computer. Importantly, the affordable, fun-coloured and translucent *iMac*, the design of which had been overseen by the brilliant young industrial designer **Jonathan Ive**, tapped into the lucrative educational and home markets to the extent that,

Apple Macintosh I
28 K, 1984

by the autumn of 1998, it had become the best-selling computer in America. As Ive notes: "One of the primary objectives for the design of the *iMac* was to create something accessible, understandable, almost familiar." With the success of the *iMac* and its family of products, including the *iBook* and *G3* (1993), the death knell of the anonymous and essentially alienating grey box was finally tolled. Significantly, the *iMac* must also be the first computer to have stylistically influenced the design of other products – from tasking lamps to desk accessories. With its commitment to design excellence and new paradigm innovation, there is little doubt that Apple Computer Inc. will continue to challenge and redefine the personal computer market.

Detail of Aprilia *SR Stealth*, 1999

→Aprilia *SR Stealth*, 1999

APRILIA

FOUNDED C. 1950
NOVALE, ITALY

The origins of Aprilia can be traced to a family-run company that manufactured motorised bicycles from around 1950. In 1968, Ivano Beggio joined the firm and from the late 1970s, the company committed itself to design and technological innovation. The company also established a networked production system, with smaller companies producing components for assembly at Aprilia's main factory in Scorzé, nearby its headquarters in Novale, just outside Venice. Beggio's love of racing inspired him to invest heavily in the company's motorcycle racing programme. For Aprilia, the racing circuit is an important testing ground for new developments, which are later incorporated into standard road models. In 1991 Aprilia became the first Italian motorcycle manufacturer to produce a scooter with plastic bodywork and a liquid cooling system – the *Amico 50*. A year later the company won its first World Motorcycle Championship and also produced 55,000 motorcycles and scooters. Within six years this number had soared to 300,000, making Aprilia Europe's second largest manufacturer of motorcycles and scooters, with five factories based in Italy. By commissioning designs from **Philippe Starck**, the company has also kept pace with the latest trends in stylistic taste. Its design centre at Novale maintains close links with the Institute of Applied Art and Design in Turin and the Faculty of Engineering at Padua University, where it has jointly set up a postgraduate course in Motorcycling Theory and Technology. Since 1997, Aprilia has also collaborated with the Art Center College of Design in Pasadena, California – one of the foremost styling and design research centres in America. Through these contacts

Aprilia *SR125*, 1999

with young designers, Aprilia has managed to forge a distinctive house style that, although centred on the idea of "creating emotions", never loses sight of the fundamental values of function and durability. Today, with its high-performance and exquisitely styled product range, Aprilia has captured some 60% of the European sportbike market and in 1999 its awesome *1000 cc RSV Mille* motorcycle was voted Motorbike of the Year by the German magazine *Motorrad*.

RON ARAD

BORN 1951
TEL AVIV, ISRAEL

Ron Arad studied at the Jerusalem Academy of Art from 1971 to 1973, and continued his training under Peter Cook at the Architectural Association in London. In 1981, Arad and Caroline Thorman established a design studio, workshop and showroom known as One Off London, which functioned as a forum for Arad's own furniture designs as well as those of other avant-garde British designers. Combining materials associated with High Tech, scaffolding poles and *objets trouvés*, Arad's early furniture resembled "readymade" art works. His *Concrete Stereo* (1984) was consciously distanced from the precision of standardized hi-fi equipment and accepted notions of "good form". Arad's later limited-edition mild steel furniture and lighting designs of the late 1980s, which required extremely costly labour-intensive techniques to produce, brought him widespread publicity and the attention of established manufacturers. Having moved into industrial production during the 1990s, Arad designed several commercially successful products, including his *Book Worm* shelving (1997), of which 1000 kilometres has been produced by **Kartell**. He also created innovative designs that exploited new ma-

Book Worm for
Kartell, 1997

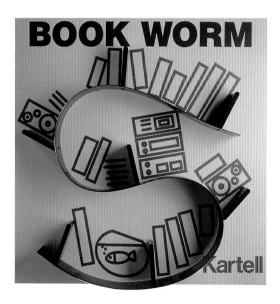

terials and production techniques, such as his *Tom Vac* vacuum-formed aluminium chair (1998), which was later produced in plastic by Vitra, and his *Fantastic Plastic Elastic* chair (1998) for Kartell. In 1997 Arad was appointed Professor of Furniture Design at the Royal College of Art, London, and two years later Professor of Industrial Design. For his sculptural furniture, Arad now uses computer-aided design (CAD) systems, which allow him to translate sketches into three-dimensional models, thus considerably easing the transition from initial concept to finished products.

Packaging design for White Rose, c. 1951

EGMONT ARENS

BORN 1888 CLEVELAND, USA
DIED 1966 NEW YORK, USA

Having worked as a sports editor in 1916 for the *Tribune-Citizen* newspaper in Albuquerque, New Mexico, in 1917 Egmont Arens opened his own bookstore in New York. A year later he began printing newspapers under the imprint "Flying Stag Press" and edited the magazines *Creative Art* and *Playboy* (a journal specializing in modern art, not the Hugh Hefner publication of today). Arens later became editor of *Vanity Fair* magazine and also worked at the advertising agency of Earnest Elmo Calkins, where he began his career as an industrial designer by establishing an industrial styling department. Like other well-known industrial designers who gained prominence in America during the 1930s, Arens styled products for manufacturers so as to make them more alluring to consumers, for example the packaging for the grocery chain A&P (his *8 o'clock coffee* package is still in use). He termed this practice "consumer engineering" and wrote extensively on the relationship between design and marketing, most notably in his book *25 Years in a Package*. Arens' many products in the streamline style, including a range of spun aluminium saucepans, were an attempt to "design" America out of the Great Depression.

Egmont Arens & Theodore Brookhart, Meat slicer, *Model No. 410 Streamliner* for Hobart Manufacturing Company, 1941

RICHARD ARKWRIGHT

BORN 1732 PRESTON, ENGLAND
DIED 1792 CROMFORD, ENGLAND

The famous British inventor Richard Arkwright initially worked as a wig-maker – a vocation that allowed him to travel throughout Britain. His remarkable appetite for learning led him, however, to an interest in spinning technology. In 1764 he began designing and building his first spinning machine, which was patented five years later and became known as Arkwright's *Water Frame*. This water-powered device produced a yarn stronger than that spun by **James Hargreaves**' *Spinning Jenny*, making it suitable for warp weaving. Arkwright subsequently established factories in Nottingham and Cromford that made use of his revolutionary design. A few years later, he expanded his operations by opening several other factories that were able to carry out the complete textile manufacturing process – from carding to spinning. In 1773 Arkwright began manufacturing an all-cotton calico cloth, which revolutionized the British textile industry. As a result of his success, by 1782 Arkwright had a workforce of 5,000 and capital of £200,000. Four years later he was knighted for his spinning innovation that had led Britain to become the leading textile manufacturing nation in the world. Arkwright's *Water Frame* is thus recognized as one of the quintessential designs of the Industrial Revolution.

Patent drawing for the *Water Frame*, 1769

Experimental model of Spinning Machine, c. 1769

ARTEMIDE

FOUNDED 1959
MILAN, ITALY

Emma Schweinberger-
Gismondi, *Dedalo*
umbrella stands, 1966

Vico Magistretti, *Eclisse*
lamp, 1965

Artemide was founded by Ernesto Gismondi (b. 1931) in 1959. Dedicated to the manufacture of furniture and lighting products, it was quickly successful with designs by Vico Magistretti (b. 1920), such as the *Demetrio 45* table (1966), *Stadio* table (1967), and *Selene* and *Gaudí* chairs (1969 & 1970), which were initially manufactured in reinforced plastic and later in injection-moulded ABS. High quality designs such as these assisted in the general acceptance of plastics as noble materials and helped to bring increasing international attention to Italian design. Artemide's many landmark lighting products include Gianfranco Frattini (b. 1926) and **Livio Castiglioni**'s snake-like *Boalum* lamp (1969), Vico Magistretti's *Chimera* floor lamp (1966) and **Richard Sapper**'s *Tizio* task lamp (1972). Always at the forefront of progressive design, Gismondi provided exhibition space in the Artemide showroom for the design group Memphis during the early 1980s. Artemide has also produced lighting designs by **Michele De Lucchi**, **Enzo Mari**, **Ettore Sottsass** and Santiago Calatrava (b. 1951), and its products are represented in over 100 museums worldwide. During recent years, Artemide has manufactured lighting by **Norman Foster** and Ernesto Gismondi, whose *e.light* (2000) was designed to compliment **Apple Computer**'s *iMac* range.

CARL AUBÖCK

BORN 1924 VIENNA, AUSTRIA
DIED 1993 VIENNA, AUSTRIA

Neopan microscope
for C. Reichert
Optische Werke,
1961

Through the activities of his father's workshop Carl Auböck was introduced to metal-casting, leather-working and wood-working at an early age. He later went on to study architecture at the Technische Hochschule in Vienna and subsequently trained at the Massachusetts Institute of Technology in Cambridge, Massachusetts. He also worked as an assistant lecturer under Professor Merinsky at the Institut für Baukunst und Ingenieurwesen in Vienna from 1950 to 1955. After leaving this post, he began working as an independent industrial designer developing a diverse range of products for industrial manufacture: steel cutlery (1957) and steel cookware (1962) for Neuzeughammer Ambosswerk; glassware services for Burg (1961 & 1966); an electric food-warmer for Ultra (1961); ski bindings for Tyrolia (1964 & 1968); school and institutional furniture for Wiesner-Hager (1965); packaging for Radio Köck (1975); cutlery for **Rosenthal** (1974); and the *Neopan* research microscope for C. Reichert Optische Werke (1963). Alongside this work, he designed numerous shops, factories and apartment buildings, mainly in Vienna, and created collections of women's fashions and ski-wear. He also designed glassware and cutlery in non-industrial materials such as brass, horn and wood for his family's workshop. Auböck regarded the role of the designer as that of problem-solver and believed that design practice should be broadly generalized, covering a multiplicity of disciplines, rather than becoming narrowly focused on particular areas of specialization.

AUDI

FOUNDED 1909 ZWICKAU, GERMANY
SINCE 1965 INGOLSTADT, GERMANY

In 1896 August Horch (1868–1951) was hired to supervise the mass production of cars at the factory owned by Karl Benz (1844–1929), who had patented his designs for a vehicle powered by an internal combustion engine ten years earlier. After three years with Benz, in 1899 Horch established his own automobile factory in Cologne, Horch & Cie. His first car, designed with a front engine and a new type of gearbox, proved successful and production capabilities rapidly expanded – by 1908 over 100 cars were being manufactured per annum. In 1909, when Horch left the company, Fritz Seidel was appointed its chief designer and Heinrich Paulmann its technical director. Now no longer associated with the company that bore his name, Horch set up another factory in Zwickau in 1909 and employed August Hermann Lange (1867–1922) as his technical director. The venture began producing cars in 1910 bearing the new Audi marque (the latin word for Horch). Audi 4-cylinder cars driven by Horch triumphed at the Austrian Alpine Runs during the

Wanderer W3
automobile, 1913

→Audi A6 2.7T
Quattro, 1999

↘Audi A4 1.8T Sport,
1999

1910s, winning successive Alpine Trophies from 1912 to 1914 and bringing valuable publicity to the new company. Prior to the First World War, Audi had produced five passenger models and two truck models and had a manufacturing capability of 200 vehicles a year. After the war, German society was on the brink of economic collapse. Its currency had gone into free-fall and there was little if any demand for obsolete cars produced with out of date machinery. Between 1925 and 1929, however, German manufacturers began adopting factory-line assembly systems, which allowed the industry to double its productivity. During the 1920s, smaller cars for ordinary people were manufactured. Audi's first post-war model, the *Type K*, was developed by Hermann Lange. In Germany, this was the first car to be offered with left-hand drive and a central gear shift as standard. In 1932, the Audi and Horch companies, together with the vehicle manufacturer DKW and the automotive division of the Wanderer company, joined forces and formed Auto Union AG of Chemnitz – its logo of four rings symbolizing the four entities. The centralization of these companies brought about greater productivity and increased efficiency. Later in 1936, a Central Experimental Department was formed allowing Auto Union to become the first German manufacturer to undertake crash tests. At the end of the Second World War, Auto Union had lost whole factories – its Audi plant was now in the Soviet Military Zone. By 1949, with the help of bank loans, Auto Union re-emerged at a new production site and began producing vehicles, including those with the Audi marque, from its new assembly plant in the Bavarian town of Ingolstadt. By 1965, however, the "Four Rings" had become a wholly owned subsidiary of **Volkswagen**. A year later, a new range of Audis was produced, including the

Audi 920, 1939

well-equipped Audi *Super 90*. More important models in re-establishing the marque, however, were the Audi *100* (1970), which was developed clandestinely by Ludwig Kraus, and the Audi *80* (1972) by Nuccio **Bertone**. From 1979, the company decided to move Audi up-market and this resulted in the appearance of the high performance Audi *Quattro* in the spring of 1980. This powerful and solidly built car not only epitomized German engineering excellence, but also revolutionized the international rallying scene. These three Audi models served as design blueprints for subsequent models until the introduction of the Audi *A4*, *A6* and *A8* and sporting *S* versions in 1994. Recently, Audi has developed the light weight, safe and economical *A2* – the world's first volume-production aluminium car. Audi sees itself as defining the future with designs such as these, which are born out of its commitment to "cutting edge technology and visionary design". In 1996 Audi announced its sponsorship of the International Audi Design Award, with the objective of creating a forum for the inter-disciplinary exchange of ideas. According to Audi, "progress is not a matter of abolishing yesterday, but of retaining its essence ... to create a better today". Its famous motto, "Vorsprung durch Technik" (Advancement through Technology) succinctly sums up the company's guiding principles.

Audi *TT Coupé Quattro*, 1999

Hansjerg Maier-
Aichen, *Rivista*
magazine rack, 1997

AUTHENTICS

FOUNDED 1980
HOLZGERLINGEN NEAR STUTTGART, GERMANY

Authentics was established in 1980 to manufacture "products of everyday life". It was founded by Hansjerg Maier-Aichen (b. 1940), who studied interior design in Wuppertal and painting at the Staatliche Hochschule der Bildenden Künste, Munich, as a subsidiary of Artipresent GmbH, of which Maier-Aichen was managing director. Using state-of-the-art plastics moulding technology and an essentialist approach to design, Authentics has attempted to raise the status of synthetic materials and to "create sensible and appropriate solutions". Its products – CD-holders, magazine racks, laundry baskets, waste bins, buckets, lamps, soap dispensers and so forth – have been designed by an international array of talented young designers, including Matthew Hilton (b. 1957), Konstantin Grcic (b. 1965) and Sebastian Bergne (b. 1966). Through the quality of their products and the intelligent exploitation of the inherent characteristics of synthetic materials, such as translucency, texture and colour, Authentics have "restored the dignity" of thermoplastics, and in particular polypropylene. While its designs are often copied, Authentics remains at the forefront of the rapidly growing home accessories market, having succeeded in creating a powerful brand identity through its promotion of "minimal, economical and ecological design concepts".

Konstanin Grcic,
2 Hands baskets,
1995 & 1997

DAVID BACHE

BORN 1926 BIRMINGHAM, ENGLAND
DIED 1995 SOLIHULL, ENGLAND

Land Rover *Long
109-inch 10*-seater
station-wagon, 1958

During the late 1940s, David Bache began his apprenticeship as a designer for the motor manufacturer Austin, which merged with Nuffield in 1949. Bache initially worked under the Italian designer Riccardo Burzi, who had designed for Austin since the 1920s and who had previously been employed by Lancia. As an assistant, Bache designed the instrument panel for the Austin *A30*. In 1954 he moved to Rover, where his first vehicles to go into production were the re-styled landmark *Series II* Land Rover (1958) and the Rover *P5* (1959). Bache redesigned the Land Rover from the original farm vehicle by altering its proportions and cut-lines and by adding flash-gaps. The *Series II* has been described as the definitive Land Rover and is a classic British design. In 1963 Bache designed the sleek and sophisticated Rover *2000*, but due to the turmoil brought on by nationalization of the British car industry, some of his best designs, many of which were produced with the assistance of his engineering partner Spen King, were never realized. These included the *P9*, which in its day was compared favourably to the **Porsche** *911*. Bache did, however, see his highly influential *Range* Rover (1970) and *P10* (1975) go into production and achieve much success.

Land Rover *Series II,*
1958

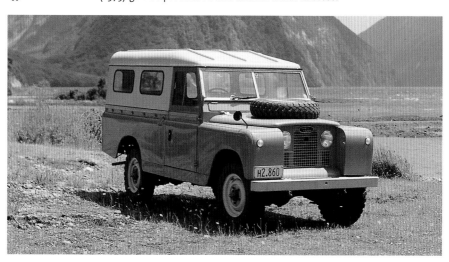

OLOF BÄCKSTRÖM

BORN 1922 FINLAND
DIED FINLAND

O-Series scissors for Fiskars, 1967

Olof Bäckström trained as an electronic engineer, a career he pursued for eleven years before focusing his attention on woodcarving and design in 1954. He initially produced wooden domestic ware, some of his designs winning a silver medal at the XI Milan Triennale in 1957. In 1960 Bäckström designed a range of cutlery for camping that was also awarded a silver medal at the XII Milan Triennale that same year.

In 1958 he became an industrial designer for the Helsinki-based manufacturer **Fiskars**, where his first project was a range of tableware made of melamine. Between 1961 and 1967, Bäckström developed his groundbreaking *O-Series* scissors. Launched in 1967 (a left-handed variation came out in 1971), the scissors had distinctive bright orange ABS handles that ergonomically conformed to the shape of the hand. As a highly proficient woodcarver, Bäckström initially prototyped the design in wood before it was translated into plastic. Fiskars continue to manufacture a modified version of Bäckström's scissors, which demonstrate so clearly that ergonomics can dramatically enhance comfort and performance.

Fiskars scissors closely based on Olof Bäckström's earlier design, 1990s

BANG & OLUFSEN

FOUNDED 1925
QUISTRUP NEAR STRUER, DENMARK

BeoSound 9000 CD
player, 1996

As a child, Peter Bang (1900–1957) experimented with
radio equipment and built a mains receiver that required
neither batteries nor an accumulator. He later trained
as an engineer at the Århus Electrotechnical School and
graduated in 1924. While there, Bang studied alongside
Svend Olufsen (1897–1949), who subsequently conducted radio experiments at his family's estate in Quistrup. While attempting to build a mains
receiver, he turned for help to Bang, who had just returned from America
where he had been researching the latest developments in radio technology.
In November 1925, Bang & Olufsen was officially founded and the two friends
began working on an aggregate to connect battery receivers to the mains
power supply – the *B&O Eliminator*. In 1929 they introduced their own
mains-powered radio known as the *Five Lamper*, which was criticized by the
architect Ole Wanscher for its old-fashioned aesthetic. During the Second
World War, B&O established an illegal radio service for communications
between the British and the Danish Freedom Fighters. Because of the company's links with the resistance, B&O's factory was bombed in 1945. The
factory was quickly rebuilt and production of radios resumed, but again
the company was criticized, this time by the celebrated designer Poul Henningsen (1894–1967), who said its products appeared as "a mass of in-

Jakob Jensen,
Beogram 1200 record
player, 1969

competence". To justify the company's by-line – "The Danish Hallmark of
Quality" – in the mid-1950s B&O began working with some of Denmark's

leading architects and designers,
such as **Sigvard Bernadotte** and
Acton Bjørn. These collaborations
produced spectacular results –
the company's new-generation,
high-quality products were completely integrated and, through
their refined Modern forms, communicated a sophisticated design
rhetoric to a new and highly receptive design-conscious audience.
In the 1960s B&O pioneered the

David Lewis,
Beosystem 2500
music system,
1991

design of integrated stereo record players, "for those who discuss taste and quality before price". Having become a pioneer of audio technology, the company later launched several landmark designs by **Jakob Jensen**, which married superlative quality with sleek good looks. Jensen's "easy touch" layouts and "flush-designed" systems quite literally redefined the design vocabulary for hi-fi equipment. During the 1980s, however, the company suffered badly from Asian competition and eventually formed an alliance with **Philips** in order to survive. In the early 1990s B&O changed its approach to manufacturing, moving strategically from mass production to order-production. There was also an important shift in terms of design, implemented by David Lewis, who had been Jensen's assistant. Lewis' revolutionary *Beosystem 2500* (1991) with its "all in one system" set itself apart from the anonymous stacked black boxes produced by B&O's competitors. Believing that aesthetics were as important to consumers as function and ease of use, Bang & Olufsen skillfully balanced external design with internal technological considerations in order to create distinctive products that powerfully communicated the company's values and vision about technology and concepts. Today, Bang & Olufsen continues its commitment to design and remains a distinguished leader in its field.

OSCAR BARNACK

BORN 1879 LYNOW, GERMANY
DIED 1936 BAD NAUHEIM, GERMANY

Oscar Barnack's design for the *Leica I* of 1925 was nothing short of a photo-graphic revolution. Prior to Barnack's designs for the manufacturer Ernst Leitz Wetzlar (later renamed Leica), 35 mm cameras had been relatively bulky and hence difficult to use. Barnack's first 35 mm Leica camera, the *UR-Leica*, was designed and prototyped in 1913 and was eventually put into production around 1918 as the *Leica A*. Rather than using rolls of perforated cine-film, this groundbreaking camera utilized a rigid film cassette, which allowed it to be portable and light. Both the *Ur-Leica* and the later *Leica A* of 1925 (which had a film winder and controls) were highly convenient cameras that had optically superior *Elmar* lenses developed by Max Berck. The *Leica I*, with its distinctive "hockey stick" element on the front of the camera, was launched to great acclaim at the 1925 Leipzig Fair. It set the standard for subsequent Leica designs and ultimately changed the art of still photography. The *Leica I* was designed to be extremely versatile and could be used to photograph anything from microscopic specimens, por-traits and landscapes to even the outer limits of the visible solar system. Barnack's designs, which were completely functionally conceived, possessed a very pure quintessentially Machine Age aesthetic.

Leica I camera, 1930

BAUHAUS

Josef Albers,
Teaglass, 1926

The Weimar State Bauhaus was born in April 1919 out of the merger of the city's School of Arts & Crafts and Academy of Art into one new interdisciplinary school of craft and design. For **Walter Gropius**, the school's first director, construction was an important social, symbolic and intellectual endeavour and this sentiment pervaded Bauhaus teaching. While based in Weimar, the Bauhaus promoted an inter-disciplinary approach to the arts and was fundamentally craft-based. The curriculum included a one-year pre-liminary course where students were taught the basic principles of design and colour theory. After completing this foundation year, students entered the various workshops and trained in at least one craft. These workshops were intended to support themselves financially through private commis-sions. During the earliest period of the Bauhaus, it was Johannes Itten (1888–1967) who played the most important role within the teaching staff. He taught theories of form, colour and contrast as well as the appreciation of art history. Itten believed that spatial composition was governed by natu-ral laws and students were taught the importance of simple geometric forms such as the circle, square and cone. These elemental shapes came to exemplify Bauhaus design, which was characterized by unadorned sur-

Marianne Brandt,
Tea and coffee
service, 1924
(re-edited by Alessi,
1985)

Marianne Brandt &
Hans Przyrembel,
Light fitting, 1926

Otto Rittweger &
Josef Knau, Tea
infusers with holder,
1924

faces and functional forms. Eventually, conflict arose between Gropius and Itten and the latter left in March 1923, marking the end of the Bauhaus' Expressionist period. Josef Albers (1888–1976) and László Moholy-Nagy (1895–1946) were appointed as Itten's successors and pursued a more industrial approach, with students being taken on factory visits. In 1923 the Bauhaus staged a landmark exhibition highlighting the full scope of its activities. One development seen at this exhibition was the new image that the Bauhaus had forged for itself – the graphics from this period were self-consciously Modern, incorporating sans serif "New Typography". The exhibition, which also featured De Stijl designs, won the school international critical acclaim. This success was short-lived, however, when the right-wing parties won the majority vote in the State elections in Thuringia in 1924, the Bauhaus' budget was slashed by half. In 1925 the Masters voted to dissolve the school in Weimar and Gropius relocated the Bauhaus to Dessau, where the ruling Social Democrats and the liberal mayor were far more receptive to

its ideals. This industrial city offered the Bauhaus the financial support it so desperately required on the understanding that the school would part-fund itself through the manufacture and sale of the designs it produced. The money it put up was sufficient to allow a new purpose-built school to be constructed, and in 1926 the Bauhaus moved into its newly-completed Dessau headquarters designed by Walter Gropius.

The Dessau Bauhaus, with its highly rational pre-fabricated building, marked an important turning-point in the school's evolution away from craft and towards industrialized production. By now, Gropius had become disillusioned with socialism; he believed that Henry **Ford**'s type of industrial capitalism could benefit workers and that, in order to survive, the Bauhaus needed to adopt an industrial approach

→Adolf Meyer &
Walter March
(construction);
Georg Muche
(design), Model
kitchen for the Haus
am Horn, Weimar,
1923

to design. In the conviction that a better society could be created through the application of functionalism, Bauhaus designs were now conceived for industrial production and a machine aesthetic was consciously adopted. **Marcel Breuer**'s revolutionary series of furniture, commencing with his *B3* chair (1925–1927), which was designed for Wassily Kandinsky's staff quarters at the Dessau Bauhaus, employed state-of-the-art tubular metal and utterly transformed the language of chair design and production. This group of furniture was later mass-produced by Standard-Möbel and Gebrüder **Thonet** from around 1928. In November 1925, with the financial support of Adolf Sommerfeld, Gropius realized his long-held ambition of establishing a limited company to promote and retail the school's designs. Bauhaus GmbH duly produced a catalogue, which illustrated Bauhaus products, but sales of these items were far from overwhelming. For the most part, this was no doubt due to the severity of their aesthetic. Furthermore, although they looked machine-manufactured, the majority of Bauhaus GmbH products were in fact unsuitable for industrial production. A few licensing agreements were drawn up between the Bauhaus and outside manufacturers, but these did not bring in the revenues that Gropius had hoped for. Eventually, the Swiss architect Hannes Meyer (1889–1954) took over from Gropius as director of the school, which by now bore the subtitle "Hochschule für Gestaltung" (Institute of Design), reflecting its shift away from the arts and crafts towards functionalism. Meyer, who was a Communist, believed that form had to be governed by cost as well as function, so that products would be not just practical but also affordable for working-class consumers. During his tenure, the Bauhaus' approach to design became more scientific. At this time, the Bauhaus also became more politicized with the school site being used as the focus for the political activities of a group of Marxist students. In 1930 Ludwig Mies van der Rohe (1886–1969) took over as director of the Bauhaus. Under pressure to de-politicize the school for its own survival, he established a new curriculum with the preliminary course becoming non-compulsory. The study of architecture was given greater importance and although the applied art workshops continued, their remit was to supply only products that could be industrially manufactured. In October 1931, however, the National Socialists swept to power in Dessau, too, winning 19 out of 36 seats, and in August 1932 their motion proposing the closure of the school was passed. Briefly re-established by Mies as a private school in Berlin, in July 1933 the Masters themselves voted to dissolve the Bauhaus – formally marking the end of this remarkable institution. As one of the first art schools to pioneer the teaching of industrial design, the Bauhaus had a fundamental impact on subsequent industrial design education and practice.

HANS THEO BAUMANN

BORN 1924
BASEL, SWITZERLAND

Hans Theo Baumann studied painting, sculpture and design in Dresden from 1943 to 1945 and interior design, graphics and stained glass at the Kunstgewerbeschule in Basel from 1946 to 1950. In 1953 Baumann worked under Egon Eiermann (1904–1970) on stained glass panels for the Pforzheimer Kirche and the Gedächtniskirche in Berlin. Two years later he established his own product design office in Schopfheim. Since then he has specialized in the design of ceramics and glassware. Using simple, unadorned geometric forms, which he softens by rounding off edges sculpturally, Baumann's designs combine the visual qualities of refinement and sturdiness. His clients have included, most notably, Arzberg, **Rosenthal**, KPM, Thomas, Daum, Süssmuth and Schönwald. The Baumann Product Design studio was founded in Basel in 1971 and another office established later in Gaillan en Médoc, France. His ceramic and glassware designs, such as the *Brasilia* cof-

Form 5500, Brasilia service for Arzberg, 1975

Delta service for
Arzberg, 1979

fee and tea service (1975) for Arzberg, have brought him significant international recognition. Baumann has also designed textiles, furniture and lighting that express a similar pared-down aesthetic. In 1970, Baumann helped establish a chair for ceramics teaching at the National Institute of Design in Ahmedabad, India, and from 1983 he was a professor at the Hochschule der Künste in Berlin. Baumann's elegant yet robust forms are born out of a deep understanding of materials and their sculptural potential. He believes that "the importance of a design does not spring from the formulas of mathematical calculation – it is solely the result of the ability of the human being to add or take away. It is possible to bring weight, material, purpose, and function into such a relationship with one another that the object which results is beautiful."

TREVOR BAYLIS

BORN 1937
LONDON, ENGLAND

Trevor Baylis grew up in Southall and left school at the age of 15. During his national service in the 1950s he built himself a go-kart in the barracks. After a succession of careers, including that of an underwater stunt artist, and with many years of invention behind him, Baylis developed a clockwork generator that could power an **Apple** laptop computer for 16 minutes before crashing. While watching a BBC television programme on AIDS in Africa in 1991, Baylis had a "eureka moment" and realised that a clockwork-powered radio could bring practical help to third world countries that did not have access to stocks of inexpensive batteries. Having successfully developed and patented the clockwork radio, his requests for manufacturing assistance were turned down by various established companies, including **Philips**,

Freeplay torch, 1999

Marconi and BP as well as the **Design Council** of Great Britain. Baylis' luck changed, however, when his revolutionary design was featured on the BBC television programme, "Tomorrow's World". With the necessary investment subsequently secured and with the assistance of the industrial designer, Andy Davey (b. 1962), Baylis was able to develop the first production version of the *Freeplay* wind-up radio, which was awarded first prize in the Products Category in the 1996 BBC Design Awards. Since then, Baylis has developed a wind-up torch and is currently developing, among other things, a wind-up computer. The manufacturer of the self-powered radios and torches, Freeplay, now sells over 100,000 units per month, which are manufactured in two plants in Cape Town that are run by a disabled workforce. Since the launch of Baylis' ingenious radio in 1994, nearly two million wind-up products have been sold. Baylis hopes to found an Academy of Invention with the proceeds gained from the sale of his products, which will encourage what he describes as "thinking with the hands".

Freeplay radio, 1999

PETER BEHRENS

BORN 1868 HAMBURG, GERMANY
DIED 1940 BERLIN, GERMANY

Peter Behrens studied painting in Hamburg, Düsseldorf and Karlsruhe from 1886 to 1889. In 1890 he married Lilly Krämer and moved to Munich. Initially he worked as a painter, illustrator and calligrapher, and soon after took up arts and crafts. He frequented Bohemian circles in Munich, and was interested in issues relating to the reform of styles of living. His circle of friends included Otto Julius Bierbaum (1865–1910), Richard Dehmel (1863–1910) and Otto Erich Hartleben (1864–1905). In 1892, Behrens the painter co-founded the Munich Secession, whilst his arts and crafts involvement led him to co-establish – with Hermann Obrist (1863–1927), August Endell (1871–1925), Bruno Paul (1874–1968), Richard Riemerschmid (1868–1957) and Bernhard Pankok (1872–1943) – the Vereinigten Werkstätten für Kunst im Handwerk (United Workshops), for serial production of utility objects.

Peter Behrens accepted an invitation from Grand Duke Ernst-Ludwig of Hesse in 1899 to be the second member of the artists' colony then newly established in Darmstadt. There, at the location known as Mathildenhöhe, he built his own house. Self-taught, Behrens designed this as a *Gesamtkunstwerk* (complete-art-work), and from the gardens to the building to the interiors (furniture, lamps, carpets, crockery and glassware, cutlery, towels, décor, paintings etc.) it was created entirely to his designs. The Behrens House marked a significant turning point in the life of the creator. For Peter Behrens it signalled a final farewell to his period in Munich artist circles, and symbolized his relinquishment of Jugendstil and his gravitation toward an austere, sober style of design.

In 1903 he was appointed director of the Düsseldorf College of Arts and Crafts, with a brief to reform it. In this he succeeded brilliantly. Among the new faculty members he brought to Düsseldorf were Josef Bruckmüller, Max Benirschke, Rudolf Bosselt, Fritz Hellmuth Ehmcke and Johannes L. M. Lauweriks.

In 1907 Peter Behrens and ten other individuals – among them Theodor

Kettles illustrated in an AEG catalogue from 1909/1910

ELEKTRISCHE TEE- UND WASSERKESSEL
NACH ENTWÜRFEN VON PROF. PETER BEHRENS

Messing glatt, matt achteckige Form				Kupfer flockig gehämmert achteckige Form				Messing vernickelt, glatt achteckige Form			
Pl. Nr	Inhalt ca. l	Gewicht ca. kg	Preis Mk.	Pl. Nr	Inhalt ca. l	Gewicht ca. kg	Preis Mk.	Pl. Nr	Inhalt ca. l	Gewicht ca. kg	Preis Mk.
3588	0,75	1,75	20,—	3589	0,75	0,75	22,—	3587	0,75	0,75	19,—
3598	1,25	1,0	22,—	3599	1,25	1,0	24,—	3597	1,25	1,0	22,—
3608	1,75	1,1	24,—	3690	1,75	1,1	26,—	3607	1,75	1,1	23,—

ALLGEMEINE ELEKTRICITÄTS-GESELLSCHAFT
ABT. HEIZAPPARATE

ELEKTRISCHE TEE- UND WASSERKESSEL
NACH ENTWÜRFEN VON PROF. PETER BEHRENS

AUSFÜHRUNG:

Leicht auswechselbare Patronen-
heizkörper, komplett, mit An-
schlußstöpsel, 2 m Litze und
Stecker PL Nr 725

Die 0,75 l Kocher haben einen
Stromverbrauch von 275 Watt,
die 1,25 und 1,75 l Kocher einen
Stromverbrauch von 440 Watt

Messing vernickelt, glatt
ovale Form

PL Nr	Inhalt ca. l	Gewicht ca. kg	Preis Mk.
3585	0,75	0,75	19,—
3595	1,25	1,0	21,—
3605	1,75	1,1	23,—

Messing glatt, matt
ovale Form

PL Nr	Inhalt ca. l	Gewicht ca. kg	Preis Mk.
3586	0,75	0,75	19,—
3596	1,25	1,0	22,—
3606	1,75	1,1	24,—

ALLGEMEINE ELEKTRICITÄTS-GESELLSCHAFT
ABT. HEIZAPPARATE

ELEKTRISCHE TEE- UND WASSERKESSEL
NACH ENTWÜRFEN VON PROF. PETER BEHRENS

Messing vernickelt, streifenartig gehämmert
runde Form

PL Nr	Inhalt ca. l	Gewicht ca. kg	Preis Mk.
3581	0,75	0,75	19,—
3591	1,25	1,0	22,—
3601	1,75	1,1	24,—

Kupfer streifenartig gehämmert
runde Form

PL Nr	Inhalt ca. l	Gewicht ca. kg	Preis Mk.
3584	0,75	0,75	20,—
3594	1,25	1,0	24,—
3604	1,75	1,1	26,—

Messing streifenartig gehämmert
runde Form

PL Nr	Inhalt ca. l	Gewicht ca. kg	Preis Mk.
3582	0,75	0,75	19,—
3592	1,25	1,0	24,—
3602	1,75	1,1	25,—

ALLGEMEINE ELEKTRICITÄTS-GESELLSCHAFT
ABT. HEIZAPPARATE

→Electric kettle for
AEG, 1909

Fischer (1862–1938), Josef Hoffmann (1870–1956), Joseph Maria Olbrich (1867–1908), Bruno Paul, Richard Riemerschmid, and Fritz Schumacher (1869–1949) – as well as twelve companies, including Peter Bruckmann & Sons, the publishing house Eugen Diederichs Verlag, Gebr. Klingsor, and the Vereinigten Werkstätten which Behrens himself had co-founded, joined to found the **Deutscher Werkbund** (DWB). Its establishment was influenced by the Arts & Crafts Movement, but from the outset its aims were of a more modern kind, among them a readiness to embrace industry as a factor in creation, the transformation of a class society into an egalitarian mass society, and the re-humanization of the economy, society and culture. More than virtually any other, Peter Behrens embodied both in his own person and in his work for the **AEG** the ideals of the DWB, which indeed earned him the nickname "Mister Werkbund".

The AEG (Allgemeine Elektricitäts-Gesellschaft, or General Electric Company), founded by Emil Rathenau (1838–1915), had taken on Peter Behrens as its artistic consultant in 1907, on the recommendation of Paul Jordan, who controlled almost all of the AEG plants. It was the first time in the history of industry that a major company had taken a step of this kind, and on such a scale it remains unparalleled to this day. Beginning with minor details (changes to arc lamps), Peter Behrens went on to transform the corporate identity of the AEG, from its letter-head to its logo, the design of its

Electric clock for
AEG, 1929

products, its advertising, its exhibition strategy, and indeed many a large factory building. Among these last was his best-known building, the turbine hall flooded with light, hailed as the "cathedral of labour". As artistic consultant, Behrens was neither an employee nor a director of the AEG; he performed his duties as designer and architect on a wholly freelance basis, working in his Neubabelsberg studio near Berlin on commissions from the AEG and other companies and organizations. His other work at this time included the Mannesmann administrative offices in Düsseldorf, the Germany embassy in St. Petersburg, and numerous design jobs,

amongst them the inscription "Dem deutschen Volke" ("For the German people") on the Reichstag building. From 1907 to 1912 the pupils and assistants in his office included **Walter Gropius** (1883–1969), Ludwig Mies van der Rohe (1886–1969), Charles-Edouard Jeanneret (1887–1965; known later as Le Corbusier), Adolf Meyer (1881–1929), Jean Krämer (1886–1943) and many others who were subsequently to decide the future of modernist design and architecture. Mies van der Rohe admitted to taking his famous maxim "Less is more" from Peter Behrens. "His all-comprehending and fundamental interest in the shaping of the entire environment ... held a great attraction for me," Walter Gropius later observed. The Fagus works built in Alfeld by Gropius and Meyer, like the Bauhaus building in Dessau, continued to explore the interplay of structure and form that the turbine hall had engaged with. Here Behrens set an example for the next generation.

In 1921 Peter Behrens was once again invited to the Düsseldorf College of Arts and Crafts, but in 1922 he accepted a chair at the Academy of Fine Arts in Vienna, where he conducted a master class in architecture. From 1936, following the death of Hans Poelzig (1869–1936), Behrens headed the architecture division of the Prussian Academy of the Arts in Berlin.

During the 1920s, both of Peter Behrens' offices, in Neubabelsberg and Vienna, won commissions for important buildings, such as the administrative offices of the Hoechst paint works (which was to be Behrens' only Expressionist building), the storage depot of the Good Hope Colliery at Oberhausen, a terraced house at Stuttgart's Weissenhof colony, the Villa Gans in Königsstein, the synagogue in Zylina, and the first modernist building in England, the New Ways residence, built in Northampton for Wenman Joseph Basset-Lowke. The complex Behrens built in Linz in the early 1930s for Austria Tabak Regie "is without doubt the finest building constructed in a German-speaking country after 1933" (L. Benevolo). However, the Nazis branded Peter Behrens a "cultural Bolshevik" and excluded him from the award of state commissions.

The achievement of Peter Behrens in the first half of the 20th century was pioneering. His thinking was taken out to the wider world by his pupils, particularly by Gropius, Mies van der Rohe and Le Corbusier. Behrens' invention of the concept of corporate identity had a direct influence on later companies such as **Braun**. The development of a comprehensive philosophy of form is now an accepted maxim. The watchword of Peter Behrens in brief, a message even more relevant today than in his own times, was "the humanization of the industrial world".

(Text: Prof. Till Behrens, Frankfurt/Main)

ALEXANDER GRAHAM BELL

BORN 1847 EDINBURGH, SCOTLAND
DIED 1922 BEIN BHREAGH, NOVA SCOTIA, CANADA

The American audiologist Alexander Graham Bell is best remembered as the inventor of the telephone. He was the third generation of his family, who originated from Edinburgh, to be recognized as a leading authority in elocution and speech therapy. His father's publication, *Standard Elocutionist*, was so popular that it went into nearly 200 editions. Trained by his father, Alexander initially worked as an elocution and music teacher in Elgin, County Moray. In 1864 he became a residential master at the Weston House Academy in Elgin, and it was there that he began his first researches into the nature of sound. Four years later he moved to London, where he became his father's assistant. Following the untimely death of his eldest brother from tuberculosis and the ill health of his younger brother, who had also contracted the disease, in 1870 the surviving Bell family, including Alexander, emigrated to Ontario, Canada. A year later, in Boston, Alexander lectured on and demonstrated his father's system of teaching people with impaired hearing how to speak. In 1872 he founded his own school for training teachers of the deaf, and a year later was appointed professor of vocal physiology at Boston University. While there, Bell began experiments with a young mechanic named Thomas Watson, which resulted in the development of

Alexander Graham Bell speaking for the first time on the newly established telephone connection between Chicago and New York, 18th October 1892

an apparatus that could transmit sound electrically. Bell was granted a patent for a telegraph machine that could send multiple messages, and began outlining the specifications for his new invention, the telephone, which was patented in 1876. That same year, Bell set up the **Bell Telephone Company** to commercialize his invention. Almost immediately, this attracted the attention of market rivals such as the Western Union Telephone Company, which attempted unsuccessfully to infringe Bell's patent. In France in 1880, Bell was awarded

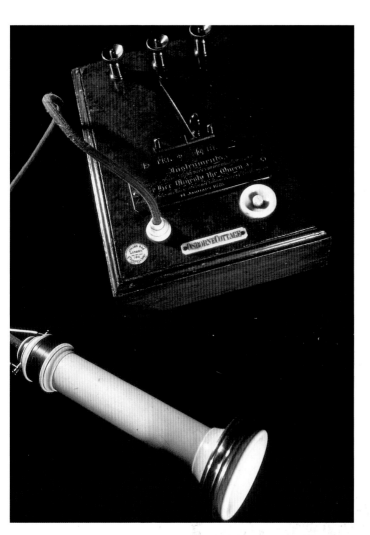

Early Bell telephone
and terminal panel,
1877
This example was used
by Queen Victoria at
Osbourne Cottage in
Southampton to
communicate with
Alexander Graham Bell
in London on 14th
January 1878.

the Volta prize of 50,000 francs. This enabled him to establish the Volta
Laboratory, where with Charles Sumner Tainter and his cousin Chichester
Bell, he developed the "Graphophone" in 1886 – a precursor of **Emil Berliner**'s
Gramophone. After 1900, Bell experimented with man-carrying kites and
undertook research into a number of diverse fields, from sonar detection
to hydrofoil watercraft. Bell's revolutionary designs quite literally changed
the world of communications and entertainment and in so doing heralded
the birth of the Modern Age.

BELL HELICOPTER TEXTRON

FOUNDED 1941
GARDENVILLE, NEW YORK, USA

UH-1D helicopters coming into a forward landing base in Vietnam, 1967

Immediate-response combat unit of the US Army unloading from a Bell "Huey" helicopter in Vietnam, late 1960s

In the early 1930s, a brilliant young inventor named Arthur Young produced and demonstrated a model of a helicopter that could fly. Impressed by the design, Larry Bell, the founder of the Bell Aircraft Company, set up a workshop for Young in Gardenville in 1941. Bell had already earned considerable recognition for its conventional aircraft, including the *P-39 Airacobra* and the *P-59*, America's first jet-powered aeroplane. Bell later went on to develop the first supersonic aircraft, the *X-1*. In 1945, with Young's helicopter about to enter production, Larry Bell incorporated Young's unit into the Bell Aircraft Corporation and moved his main plant to a larger site in Niagara Falls. Bell's first helicopter prototype, the *Model 30*, could reach speeds of 100 mph in level flight. This was succeeded in 1946 by the *Model 47*, which was granted the world's first helicopter licence and heralded the beginning of a brand new industry. That same year, Bell began supplying helicopters to the US military. In 1957 Bell's helicopter division was restructured as the Bell Helicopter Corporation and this was eventually purchased by Textron in 1960. The *Model 47* was widely used during the Korean War, and over the succeeding years other, more advanced military and commercial rotorcraft were developed, including the success-

ful *206 JetRanger* (1960–1965) and the *UH-1D Iroquois*, nicknamed the "Huey" (1960–1963). This legendary helicopter was used extensively in Vietnam for medical evacuations, gun-carrying and anti-tank missile firing and became an iconic symbol of the war. Since 1946 Bell Helicopter has produced more than 34,000 military and commercial helicopters. Today, the company is recognized as the worldwide leader in the design and manufacture of rotary-wing aircraft, and continues to develop highly innovative rotorcraft.

BELL TELEPHONE COMPANY

FOUNDED 1876
NEW YORK, USA

Henry Dreyfuss,
Model 300 telephone
for the Bell Telephone
Laboratories, 1937

Bell Telephone
Company
advertisement, *Look*
magazine, 1948

Alexander Graham Bell established the Bell Telephone Company in New York in 1876 as a means of commercializing his telephone patent of the same year. In 1881 Bell purchased Gray & Barton, the manufacturing company of his competitor, Western Electric. With this acquisition, Bell secured a virtual monopoly of both equipment and transmission technology within the fledgling telephone industry. In 1900 Bell formed the American Telephone & Telegraph Company (AT&T), which in 1925 set up a research division known as Bell Telephone Laboratories. Since then, Bell Laboratories have developed literally thousands of revolutionary inventions and associated product designs, such as the world's first synchronized sound-and-motion picture system (1926), a pioneering electrical relay digital computer (1937) and the transistor (1947). Researchers working for this prestigious research and development facility were awarded Nobel Prizes for physics in 1937, 1947 and 1978 for their innovative contributions to the communications field. The Bell Telephone Company also used the skills of independent design consultants, such as **Henry Dreyfuss**, designer of the historic *Model 300* telephone. AT&T's Bell Laboratories continue to innovate new technology applications and commission design consultancies such as **Frogdesign** to assist in the development of state-of-the-art communication tools such as the *Personal Communicator* (1992).

GOOD NEWS about Long Distance

Today, Long Distance calls go through in about two minutes on the average. Now and then there are delays, but we're handling nine out of ten calls while you hold the line.

We've added many new circuits and switchboards and more are coming along. Our operators know their job.

Our aim is to put your out-of-town calls through faster than ever before. And we're working hard to do it.

BELL TELEPHONE SYSTEM

MARIO BELLINI

BORN 1935
MILAN, ITALY

Mario Bellini studied architecture at the Politecnico di Milano, graduating in
1959. From 1959 to 1962 he was design director at La Rinascente, the influential chain of Italian department stores that did much during the post-war
years to promote industrial design. In 1962 he founded an architectural office with Marco Romano and in 1973 established Studio Bellini in Milan.
As chief design consultant to **Olivetti**, a post he has held since 1963, he designed the *Divisumma 18/28* calculators (1973) and the *Praxis 35* and *Praxis 45*
typewriters (both 1981). From 1969 to 1971 he was president of ADI (Associazione per il Disegno Industriale) and in 1972 he showed a mobile micro-
living environment entitled *Kar-a-Sutra* at the "Italy: The New Domestic
Landscape – Achievements and Problems of Italian Design" exhibition held
at the Museum of Modern Art in New York. This led to his appointment as
a research and design consultant to the car manufacturer **Renault** in 1978.
During the 1970s, Bellini organized workshops to explore the complex rela-
tionships between humans and their man-made environment – a theme
that has informed virtually all his work. From 1986 to 1991 Bellini was editor

Logos 50/60 adding
machine for Olivetti,
1972

→ *Divisumma 18*
adding machine for
Olivetti, 1973

UHF 33 adding
machine for Olivetti,
1981–1982

→ *ET 55P* adding
machine for Olivetti,
1987

← *Lettera 25* typewriter for Olivetti, 1974

TES 401 typewriter for Olivetti, 1978

← *Praxis 35* typewriter for Olivetti, 1981

ET 55P typewriter for Olivetti, 1987

of *Domus* and since 1979 he has been a member of the Scientific Council for the Milan Triennale's design section. He has held professorships at the Istituto Superiore del Disegno Industriale, Venice (1962 to 1965), the Hochschule für angewandte Kunst, Vienna (1982 to 1983) and at the Domus Academy, Milan (1986 to 1991). Bellini has also lectured at many other design colleges including the Royal College of Art, London. His most notable furniture designs include the *Le Bambole* seating system for B&B Italia (1972), the *Cab* seating for Cassina (1977) and the *Figura* office seating programme codesigned with Dieter Thiel for Vitra (1985). He has also designed lighting for **Flos**, **Artemide** and **Erco** and audio equipment for **Yahama** and **Brionvega**. Bellini's designs have won numerous prizes, including seven Compasso d'Oro awards. Bellini believes that "an effective improvement in the quality of our environment may be obtained through the reunification of design standards ...[and].... that such a reunification can be accomplished only by superseding the compartmentalization of sectors and moving towards a more anthropocentric view of man's environment, at least as far as disciplinary and didactic, or cultural, aspects are concerned."

SIGVARD BERNADOTTE

BORN 1907 STOCKHOLM, SWEDEN
DIED 2002 STOCKHOLM, SWEDEN

Facit PI typewriter for Åtvidabergs Industrie, 1958

The son of King Gustavus VI of Sweden, Count Sigvard Bernadotte (later Prince Sigvard Bernadotte) studied at the University of Uppsala (1926–1929), at the Kungliga Konsthögskolan, Stockholm (1929–1931), and at the Staatsschule für angewandte Kunst, Munich, in 1931, prior to joining the Georg Jensen workshop. He was the first designer there to adopt geometric forms rather than naturalistic forms, as his *Bernadotte* flatware of 1939 demonstrates. During the late 1930s he also designed a volumetric metal and Bakelite cocktail shaker for an English manufacturer. After serving as a director of the Georg Jensen silversmithy, in 1949 he established, a Copenhagen-based design consultancy with the Danish designer **Acton Bjørn**. This multi-disciplinary studio later expanded, opening branches in Stockholm and New York, and Bernadotte and Bjørn co-designed several notable industrial products,

Margrethe mixing bowl with salad servers for Rosti, 1954 (co-designed with Acton Bjørn)

including the melamine *Margrethe* stacking bowls for Rosti (1954) and the compact *Facit PI* typewriter (1958). In 1964 Bernadotte independently founded the Bernadotte Design studio, whose output was distinguished by its use of strong geometric forms and interesting colour combinations. Sigvard Bernadotte is an important pioneer of industrial design in Europe and is also the first European designer to have become a member of the American Designers' Institute.

BERTONE

FOUNDED 1912
GRUGLIASCO, ITALY

The Carrozzeria Bertone was founded in 1912 by the
cartwright Giovanni Bertone (1884–1972). It was not
until 1921, however, that the bodywork company received
its first order from SPA to build automobile coachwork
for its car, the *Torpedo*. With the industrialized approach
to automobile manufacture beginning to emerge, coachwork building methods had to follow suit quickly –

*Guilietta Spider for
Alfa Romeo, 1954*

wooden bodies were now bonded to the metal chassis using assembly-line
techniques. Bodywork needed ever-greater strength owing to the increasing
speeds that vehicles could now achieve. In response to this, Bertone constructed a number of hard-top-like canopies to transform models such as
Vincenzo Lancia's *Lambda* into safer and more robust sedan versions. The
following decade, the Bertone company was transformed from a craftsman's
workshop into a coachwork factory. Giovanni Bertone's son, Guiseppe
"Nuccio" Bertone (1914–1997), joined the family business in 1934 and three
years later the company's streamlined bodywork for the **Fiat** *1500* won the
Turin styling competition. As its manager, Nuccio Bertone transformed the
business into one of the world's premier automotive design consultancies.

*Carabo for Alfa
Romeo, 1968*

In 1954 the **Alfa Romeo** *Giulietta Sprint* was launched at the Turin Motor

Sketch of *Genesis*
concept car, 1988

Show and Bertone's body styling for this vehicle became a hallmark of Italian post-war design. During the 1960s Bertone styled cars for Alfa Romeo, Fiat, **BMW**, Aston Martin and **Ferrari**, and received widespread acclaim for its **Lamborghini** *Miura*, launched in 1966. Two years later, Bertone's futuristic styling of the Alfa Romeo *Carabo* signalled the beginning of the so-called "wedge-line" trend in automobile styling that remained fashionable for many years. In 1971 the first prototype of the Bertone-designed Lamborghini *Countach* was unveiled to much praise, but owing to the oil crisis of 1972, the studio was forced to concentrate on more utilitarian vehicles, such as the **Volkswagen** *Polo* (1975). In recognition of his immense contribution to car styling, Nuccio Bertone was recently nominated for the prestigious title of "Car Designer of the Century". Many well-known car designers, including Giovanni Michelotti (1921–1980), **Giorgetto Giugiaro** and Sergio **Pininfarina**, served apprenticeships at the Bertone Style Centre. The forward-looking and often angular formal vocabulary of the Bertone studio, which has distinguished such disparate vehicles as the Ferrari *Dino 308* GT4 (1973), the Lancia *Stratos* (1973) and the **Volvo** *780* Coupé (1985) is born out of the firm's philosophy that design should not only anticipate future trends but should also give style to functional form.

MAX BILL

BORN 1908 WINTERTHUR, SWITZERLAND
DIED 1994 BERLIN, GERMANY

Max Bill trained as a silversmith at the Kunstgewerbeschule in Zurich from 1924 to 1927 and while there was influenced by trends in contemporary fine art such as Cubism and Dada. From 1927 to 1929 he studied art at the Dessau **Bauhaus** and fully embraced the school's functionalist approach to design. Upon completion of his studies, Bill returned to Zurich and worked as a painter, architect and graphic designer. In the 1930s he became the leading exponent of Constructivism within the Swiss School of graphics, and during this period designed graphics for the Wohnbedarf store in Zurich. He established his own architectural practice in 1930 and, as a member of the Swiss Werkbund, designed the Neubühl estate near Zurich in the Modern style (1930–1932). In 1931 Bill adopted Theo van Doesburg's theory of "concrete art", which argued that universality could only be achieved through clarity. From 1932 Bill also worked as a sculptor and became a member of various art organizations, including the Abstraction-Création group in Paris, the Allianz (Association of Modern Swiss Artists), the CIAM (Congrès International d'Architecture Moderne) and the UAM (Union des Artistes Modernes). In 1944 Bill turned his attention to industrial design. His aluminium wall clock of 1957 was manufactured by **Junghans**, for whom he also designed a range of wristwatches characterized by a similarly rational layout and a strong industrial aesthetic. Bill also designed a minimalist stool, the *Ulmer Hocker* (1954), one of his best-known products. Bill was responsible for the founding of the Bundespreis award and the exhibition "Die gute Industrieform", which sought to promote excellence in industrial design. Importantly, he also co-founded the in-

Kitchen clock for
Junghans, 1956

fluential Hochschule für Gestaltung in Ulm in 1951 and was the school's rector and head of its architecture and product design departments from 1951 to 1956. At Ulm, Bill championed Bauhaus-type geometric formalism, believing that products based on mathematical laws had an aesthetic purity and thereby a greater universality of appeal. This approach to industrial design was continued by **Hans Gugelot** when he took over the product design department. Upon leaving Ulm, Bill set up his own Zurich-based studio in 1957 and concentrated on sculpture and painting. He was the chief architect of the "Educating and Creating" pavilion at the Swiss National Exhibition of 1964 and was made an honorary member of the American Institute of Architects that same year. As much a theorist as a designer, Max Bill believed that "good design depends on the harmony established between the form of an object and its use".

Wristwatches for Junghans, 1962 (re-edited by Junghans, 1993)

Biró logo

British patent
for László Biró's
ballpoint pen, 1938

LÁSZLÓ BIRÓ

BORN 1899 BUDAPEST, HUNGARY
DIED 1985 BUENOS AIRES, ARGENTINA

László Jozsef Biró, a Hungarian-born journalist, created the first ballpoint pen, which he patented in 1938. The principle of the ballpoint pen, however, dates back to 1888, when a patent was filed by the American John J. Loud for a device to mark leather. This design had a number of failings and the patent was not commercially exploited. During the 1930s, while working as an editor of a magazine in Hungary, Biró became aware of an ink used for printing that was quick-drying and smudge-free and so decided to develop a pen that could use the same ink. As this thicker ink would not flow from a regular pen nib, Biró devised a point by fitting his pen with a tiny ball bearing at its tip. As the pen moved along the paper, the ball rotated, picking up ink from the ink cartridge and leaving it on the paper. In 1940 Biró, who was a fervent Communist, left Hungary when it entered into an alliance with Nazi Germany and subsequently emigrated to Argentina. Biró became an Argentinian citizen and adopted a new Spanish-sounding name, Ladislao José Biró. By 1942 his ballpoint pen had been perfected as a result of a special ink formulated by his chemist brother, Georg. After filing a fresh patent in 1943, Biró raised US$ 80,000 worth of backing to allow his pen to go into full-scale mass production. In 1944 Henry George Martin, an English accountant who had invested heavily in Biró's company, Eterpen Co., brought the fully patented ballpoints back to England

Fig.1

Fig.2

and offered them to the British and Allied armed forces. Unlike regular fountain pens, Biró's ballpoints were especially useful in unpressurized aircraft
because they did not leak. Over 30,000 ballpoints were manufactured for
the RAF by the Miles Martin Pen Company in Reading, and an even greater
number were produced for the American forces. At the end of the war, the
innovative and robust ballpoint pen went on sale to the general public
around the world. It was initially seen as a status symbol, no doubt due
to its high cost: in America the pen retailed at $12.50. Nevertheless, an in-
credible 10,000 were sold on its first day of sales. In Britain the pen was
priced at 55 shillings ($£$ 2.74) – the equivalent of a secretary's weekly wage
packet. In 1949 Marcel Bich (1914–1994), an expert in plastics machining,
perfected his own much less expensive ballpoint pen, which sold under
the name "BIC Point", a shortened version of his name. Backed by an innovative advertising campaign, the newly founded Société BIC was selling 42
million units annually after only three years. In 1957, Société BIC purchased
Biró Swan, the descendent of the Miles Martin Pen Company. Today, BIC
markets a range of some 20 writing instruments that are direct descendants
of the original Biró pen. Astonishingly, over 15 million BIC pens are sold
every day.

ACTON BJØRN

BORN 1910 COPENHAGEN, DENMARK
DIED 1992 DENMARK

Acton Bjørn studied architecture and town planning and began working as an industrial designer during the Second World War. In 1949 Bjørn and **Sigvard Bernadotte** established Bernadotte & Bjørn, one of the first Scandinavian industrial design consultancies. This Copenhagen-based practice not only designed metalwork and ceramics but also industrial machines. Among their most notable work was office equipment for Facit and plastic kitchenware for Rosti during the 1950s. In 1966 Bjørn won a Danish ID Award for his *Beolit 500* transistor radio for **Bang & Olufsen**. This design was one of the first radios to achieve a functional simplicity comparable to that of a telephone. The same year, he opened an office in Copenhagen specializing in household appliances, office furniture and packaging. He also received recognition for his corporate identity work, executed mainly for banking and insurance companies. Bjørn's designs succeed in balancing practical function with simple sculptural forms.

Magic Wand hand-mixer for ESGE, 1955

Beolit 500 transistor radio, 1965

MISHA BLACK

BORN 1910 BAKU, AZERBAIJAN
DIED 1977 LONDON, ENGLAND

Misha Black studied at Central School of Arts & Crafts in London for a short while, but was essentially self-taught as a designer and architect. In 1930 he founded Studio Z and three years later began working with Milner Gray (1899–1994) at the Bassett-Gray Group of Artists and Writers. In 1935 this became the Industrial Design Partnership, the first inter-disciplinary design consultancy in Britain, whose designs most notably included television sets and radios for Ekco. From 1940 to 1945 he worked as an exhibition designer for the Ministry of Information. In 1943 Black and Milner Gray founded the **Design Research Unit** in London, followed two years later by the Design Research Group. The DRU was the most important design consultancy in post-war Britain and produced numerous industrial designs – from **Kodak** cameras and Watney beer packaging to trains and electric heaters – all of which were distinguished by a clarity of lay-out and a no-nonsense, engineering approach. Black also contributed to major exhibitions, including the 1946 "Britain Can Make It" exhibition. Throughout his career he tirelessly promoted the Modern design cause.

Misha Black &
J. Beresford Evans,
E3001 locomotive
for British Railways,
c. 1956

BLACK & DECKER

FOUNDED 1910
BALTIMORE, USA

Black & Decker was founded in Baltimore in 1910 by Duncan Black (1883–1951) and Alonzo Decker (1884–1956) for the manufacture of specialist machinery. Early products included milk-bottle cap machines, letter-graphs, pocket-sized adding machines, candy-dipping machines and machinery for the U.S. Mint. In 1914 the company filed a key patent for a revolutionary electric drill that had a pistol grip and a trigger-style switch. This first ever portable power tool and its later successors brought the company huge success – within a decade of its foundation, Black & Decker's annual sales exceeded US$ 1 million. In 1935 the company commenced manufacturing in the UK. It was not until 1950 that Black & Decker began making tools for home use. Its first domestic product was the famous "Little Red Drill", which had a 1/4" capacity. In 1964 the company developed the *Cordless Zero Torque* tool for **NASA**'s Gemini Space Project. In 1971 Black & Decker designed the *Apollo Moondrill* for the Apollo 15 Mission – a cordless zero-torque tool that could remove core samples from the lunar surface. One of the company's greatest innovations was the *Workmate*, designed by Ron Hickman (b. 1932) in 1972. This multi-use bench was the blueprint for many subsequent variations. In 1979 Black & Decker launched its highly successful hand-held cordless *Dustbuster* vacuum cleaner, and a decade later introduced its *Multi Tools*, including the multiple-use *Multisander*. The development of multipurpose tools eventually resulted in the launch, in 1998, of the four-in-one *Quattro* tool, which incorporates a jigsaw, drill, screwdriver and sander. Driven as much by state-of-the-art technology as by design innovation, Black & Decker has managed to straddle both the industrial and domestic markets with its extensive range of tools for building, gardening (it launched its first range of lawn-mowers in 1969) and cleaning.

Workmate, the first version was designed by Ron Hickman in 1972

PHH-I drill, c. 1999

BMW

FOUNDED 1916
MUNICH, GERMANY

BMW *3/15 PS*,
1929–1932

Bayerische Motoren Werke was founded as an aircraft
engine manufacturer in 1916. Within just three years
Franz Zeno Diemer had set the world altitude record in
an aircraft powered by BMW engines. In 1923 the company began producing
motorcycles. Five years later, it acquired a car factory in Eisenach and with
it, a licence to manufacture a small car, the *Dixi*, which became known as the
BMW *3/15 PS*. The first car to be designed in-house was the BMW *3/20*,
which was launched in 1932. The company designed several streamlined
cars over the course of the 1930s, most notably the BMW *331* and the classic
BMW *328* sports car. BMW also dominated car racing during those years,
with its aerodynamic *328* roadsters winning their class at the 1938 Mille
Miglia. In 1940 the company began mass-producing the *801* aircraft engine
– some 30,000 were manufactured over the succeeding five-year period –
and also won the Mille Miglia again with a modified "aerodynamic coupé"
version of the *328*. In 1941, car production was suspended and resources
were directed to the design and construction of rockets and aircraft engines,
including one of the world's first jet engines, the *003*. By 1945, however,
BMW's factories in Eisenach and Berlin had been lost and the remaining
plant in Munich was dismantled. After a three-year production ban, BMW
resumed the manufacture of motorcycles in 1948 with the single-cylinder
BMW *R24*. In 1951 the company produced its first post-war car, the *501*, but

BMW *331*, 1930s

BMW *Isetta 600*,
1955

BMW *1602–2002*
models and the
succeeding *3 Series*

it was a commercial failure. Four years later BMW launched the diminutive *Isetta* (an italian licence), which bridged the motorbike and automobile markets. This hybrid design was extremely popular in Germany with around 200,000 examples being produced. Having decided to concentrate on the small car market, BMW launched the *Model 700* in 1959 and the *Model 1500* in 1962, which established the design vocabulary of its later compact touring cars. Despite this strategy, the company faced bankruptcy in the late 1960s. Critically, BMW was only able to emerge from the financial doldrums by introducing a range of high-quality, conventionally styled road cars that drove like sports cars. In 1973 the more aggressively styled BMW *2002* became the first mass-produced turbo-powered car in the world. Two years later the *3 Series* was introduced, followed by the *6 Series* coupé and *7 Series* in 1976 and 1977 respectively. During the 1980s and 1990s the company continued to expand; in 1997 it launched its retro-designed *M Roadster*, and a year later introduced the fifth generation of the highly successful *3 Series*. The achievements of BMW have largely rested on the continuity of its brand values, which are based primarily on a commitment to technological innovation and design.

BMW *3 Series*, 1998

No. 1208 Santos
coffee maker, 1958

BODUM

FOUNDED 1944
COPENHAGEN, DENMARK

The Bodum domestic products company was established in Copenhagen in 1944 by Peter Bodum (1910–1967) to import glassware from Eastern Europe to sell in Denmark. In 1958 it introduced its first branded product, the *Santos* coffee maker designed by Kaas Klaeson. One of the company's most famous products, the *Bistro* coffee maker (1974), was the outcome of a collaboration between Peter Bodum's son, Jørgen (b. 1948), and the industrial designer Carsten Jørgensen (b. 1948). Both believed in simple, effective design without unnecessary adornment and in the use of appropriate Modern materials to create useful everyday products at a reasonable price. In 1983 Pi Design, located in Switzerland, was founded as the Product & Design development company of the Bodum Group. Jørgensen was appointed a director of Pi and made responsible for the design and corporate identity of Bodum worldwide. Since then, Pi has designed a broad range of distinctive and highly rational products for Bodum that eloquently express the company's philosophy, namely that "some everyday things are at their best when they merely constitute a plain function and are simple in shape and proportions". In 1986 Bodum established its first retail outlet in London and later went on to open other shops in Paris, Copenhagen, Lisbon, Oporto, Zurich and Lucerne.

⟍*No. 1508 Bistro*
coffee maker, 1974

No. 1846 Assam tea
maker, 1992

← *No. 1928 Chambord* coffee maker, 1958

→ *No. 5670 Antigua* coffee grinder, 1996

← *No. 5500 Ibis* kettle, 1992

→ *No. 5600 Le Curl* kettle, 1998

BOEING

FOUNDED 1916
SEATTLE, USA

William Boeing (1881–1956) studied engineering at Yale University. In the late 1900s he became fascinated with aircraft and, after a period of some research, concluded that he could build a better bi-plane than any then available. In 1915 he asked the designer, Westervelt, to develop a new aeroplane that he could manufacture. The resulting aircraft was the birch-and-canvas *B&W* twin-float seaplane. During the First World War, Boeing received its first production order from the US Navy for 50 *Model C* seaplanes. Around this time, William Boeing declared: "We are embarked as pioneers upon a new science and industry in which our problems are so new and unusual that it behoves no one to dismiss any novel idea with the statement, It can't be done." Under his careful guidance the company grew and produced a wide range of aircraft – from mail planes and flying boats to military observation planes and torpedo planes. While its competitor, Douglas, was winning international acclaim for its first around-the-world flight in 1924, Boeing was developing state-of-the-art pursuit fighters and was building more fighter aircraft than any other manufacturer. The company's series of fighter aircraft from the 1920s and 1930s included the Boeing *P-26 Peashooter* monoplane, which flew 27 mph faster than any of its bi-plane competitors

Boeing *B52 Stratofortress*, 1948–1952

Boeing *B17 Flying Fortress*, 1934–1935

Boeing *B29 Superfortress*, 1939–1940

Boeing *707*,
1955–1957

and employed innovative production techniques in its construction, including the arc-welding of its fuselage frame. With the introduction of anti-trust
legislation in 1934, William Boeing grew disheartened and left the firm. The
company's new president, Claire Egtvedt, determined that the future lay not
just in the development of new military aircraft but also in the design and
manufacture of large passenger aircraft, such as the long-range four-engine
Clipper seaplane. During the Second World War, Boeing massively increased
bomber production and by 1944 was building an astonishing 362 *B-17 Flying
Fortresses* per month. The enormous *B-17* became legendary for its ability to
remain flying even after sustaining severe damage. In total, Boeing plants
built 6,981 *B-17*s in various models with another 5,745 being built under a
nation-wide collaborative effort by **McDonnell Douglas** and **Lockheed**. In
1942 Boeing began production of the long-range heavy bomber, the *B-29
Superfortress*, which is mainly remembered as being the aircraft from which
the world's first atomic bomb was dropped. During the post-war years
Boeing reduced production but did continue developing military aircraft,
most notably the *B52 Stratofortress*. Produced between 1952 and 1962, this
devastating bomber first saw action in Korea. Remarkably, the last production variation of the aircraft, the *B-52H*, remains in service today – testifying
to the superior design and strength of its airframe. During the 1950s the
president of Boeing, William Allen, moved the company into the commercial

Boeing 747,
1963–1968

Advertisement for
Boeing 707, 1950s

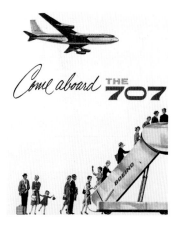

jet market with designs such as the *377 Stratocruiser* –
a luxurious civilian version of the *B-29*. In 1952 the company's net worth was risked so as to begin development
of the *Dash-80* commercial jet. The company eventually
had to bow to pressure from the airlines and re-design
the *Dash-80* so as to match Douglas' *DC-8*. The resulting design was the landmark *707*, which featured more
powerful engines, a wider fuselage, larger wings and a
greater seating capacity than the *DC-8*. The company's
family of commercial jets expanded over the years to include the *727*, the *737* and the epic *747* jumbo jet. One
of the most significant technological achievements in
the history of aviation, the *747* is the world's largest
commercial aeroplane. Its first flight was in 1969 and it
continues to be built at the Boeing factory – the world's
largest building by volume in Everett, Washington –
clearly demonstrating the aircraft's versatility, popularity,
longevity and value. Boeing has also contributed significantly to the American Space Programme, including the
famous Apollo missions.

BOMBARDIER

FOUNDED 1942
VALCOURT, QUEBEC, CANADA

Bombardier *B7*
snowmobile, 1937

Advertisement for
the Bombardier *B12*
snowmobile, 1941

Joseph-Armand Bombardier (1907–1964) was born near Valcourt, Quebec. At the age of 19 he began working as a mechanic in his own garage, and over the next decade researched and experimented with vehicles that could be driven over snow-covered terrain. Between 1926 and 1935 he developed several prototypes and, in 1936, designed and patented a seven-passenger tracked vehicle known as the *B7*. This early snowmobile allowed previously snow-bound areas to be accessed, and by 1940 Bombardier was producing 200 units per annum. In 1941 he developed a larger and more powerful snowmobile, the *B12*, and a year later founded the L'Auto-Neige Bombardier Limitée Company to manufacture these vehicles. A military version of the *B12* was commissioned by the Canadian Government and in 1943 Bombardier developed an armoured snowmobile known as the *Kaki*. After the war, the company diversified its product range to include a device developed by Joseph-Armand Bombardier and his brother, Gérard, for use on tractors. Known as the *TTA* (Tractor Tracking

Bombardier *Ski-Doo*
Mach Z, 1999

Attachment), this traction mechanism improved the performance of tractors on muddy terrain and sold by the thousands around the world. The development in 1953 of an unbreakable warp-proof all-rubber sprocket and the design of a new vulcanizing machine that allowed the production of seamless shock-resistant tracks enabled Bombardier to produce the *Muskeg* multiple-transportation all-terrain vehicle. Between 1957 and 1959 Germain Bombardier, the eldest son of Joseph-Armand, designed and perfected the first ever *Ski-Doo*, a fun sports vehicle that instantly found approval among outdoor enthusiasts. Some 60,000 *Ski-Doos* were being sold annually by the mid-1960s, and by 1972 this figure had risen to 495,000. The oil crisis in the early 1970s, however, dramatically slowed demand for *Ski-Doos* and a year later Bombardier moved strategically into rail transit systems and manufactured the rolling stock for Montreal's newly expanded subway. The company continued to grow through acquisitions – including the aircraft manufacturer Canadair in 1986, the French railway manufacturer ANF-Industrie in 1989, the business jet manufacturer **Learjet** in 1990, and the famous aircraft manufacturer **de Havilland** in 1992. Guided by its motto, "strength in diversity", and its commitment to innovative design solutions, which have resulted in several new vehicle typologies, Bombardier has become one of Canada's largest and most prestigious manufacturing companies.

Robert Bosch

BOSCH

FOUNDED 1866
STUTTGART, GERMANY

Robert Bosch (1861–1942) served an apprenticeship as a precision instrument maker with Wilhelm Maier in Ulm from 1876 to 1879, and spent a year in America, working for Sigmund Bergmann and **Thomas Alva Edison**. On his return to Europe in 1885, he stayed briefly in London, where he worked for **Siemens** Brothers. In November 1886 Bosch opened his own workshop, the Werkstätte für Feinmechanik und Elektrotechnik in Stuttgart, which produced the *Low-Voltage Magneto* for stationary gas engines (1887) and a table-top telephone. By 1898 the company had opened a subsidiary in London, the Compagnie des Magnétos Simms-Bosch, and a year later sales offices were established in Paris and Budapest. In 1901 Robert Bosch's manufacturing operation moved into its first factory, built on the site of the company's existing premises. 1902 saw the launch of the crucial high voltage *Magneto* spark plug, for which the graphic designer Lucian Bernhard (1883–1973) was commissioned to create bold packaging and advertising posters, which helped the company establish its powerful brand identity. In 1909 another factory was built in Feuerbach to increase production capacity. Over the succeeding years, the company diversified its product line by introducing various automotive innovations, including headlamps (1913), the first electric starter motor (1914), car horns (1921), the first electrically powered windscreen wipers (1926) and the first standard diesel injection pump for trucks

Early advertisement for Bosch *Magneto* spark plugs

At Last You Can Get Them,

THE LONG MISSED AND LONG WANTED

Bosch-Magnetos

For all kinds of Combustion Engines.

AGENCY FOR THE UNITED STATES

of the world-famous Bosch-Magnetos, made by Robert Bosch, Stuttgart, Germany, the inventor of the Bosch Ignition Apparatus, and sole manufacturer of all machines sold as Bosch or Simms-Bosch Magnetos,

HAS NOW BEEN OPENED

Bosch ·Ignition is used by all the most important motor car and cycle manufacturers. Some of them are :

UNITED STATES
E. R. THOMAS MOTOR CO.
ROYAL MOTOR CAR CO.
THE GARFORD CO.
OLDS GASOLINE ENGINE WORKS
AMERICAN LOCOMOTIVE AUTOMOBILE CO.
DAIMLER MFG. CO.
ETC.

FRANCE	GERMANY	GREAT BRITAIN
Berliet	Adler	Daimler
Charron, Girardot & V.	Horch	Napier
Clément-Bayard	Neckasulm	Singer
Darracq	N. A. G.	Westinghouse
De Dietrich	Mercedes	Wolseley
Delahaye	Wanderer	Etc., etc.
Léon Bollée	Etc., etc.	**ITALY**
Peugeot		Fiat
Renault Frères	**BELGIUM**	Itala
Richard-Brasier	Fabrique Nationale	Isotta Fraschini
Rochet & Schneider,	Pipe	Rapid
Etc., etc.		Etc., etc.

WRITE, OR BETTER STILL. *CALL* AND OBTAIN PARTICULARS

ROBERT BOSCH, 1947 BROADWAY, New York City
Telephone. 4255 Columbus

(1927). In 1932 Bosch took over the Idealwerke AG and produced televisions
and radios under the label Blaupunkt. Bosch continued its commitment to
research and development throughout the 1930s, introducing the first serial
car radio in 1932 and the first standard diesel injection pump for cars in 1936.
During that decade the company also began manufacturing white goods
and presented its first refrigerator in 1933. Bosch expanded rapidly over the
following decades, opening foreign subsidiaries and numerous overseas
plants. It also continued to launch innovative products, amongst them the
first fuel injection system for cars (1951), the first automatic dishwasher
(1964) the first electronic fuel injection system for cars (1967) and the original ABS anti-blocking system (1978). Renowned for its white goods and electric power tools designed by the German industrial designer **Hans Erich
Slany** (b. 1926), Bosch maintains its tradition of producing well-designed
and well-built products that epitomize the high quality of German industrial

design. Today the Bosch Group is one of Germany's largest companies. Its scientists, engineers, designers and technicians are dedicated to the improvement of the function and reliability of existing products and to the development of new products and systems across all areas of its operations, including electrical and electronic automotive equipment, private and public communications technology, power tools, household appliances and thermotechnology as well as automation technology and packaging machinery.

MARIANNE BRANDT

BORN 1893 CHEMNITZ, GERMANY
DIED 1983 KIRCHBERG, GERMANY

Marianne Brandt studied at the School of Fine Art (Großherzoglich-Sächsi-sche Hochschule für Bildende Kunst) in Weimar from 1911 to 1917. She sub-sequently established her own studio in 1917 and worked as a freelance artist until she enrolled at the Staatliches **Bauhaus** in Weimar in 1923. After the preliminary course, she took an apprenticeship in the metal workshop, which was then directed by László Moholy-Nagy (1895–1946). Around this time she designed a coffee and tea service, which was based on simple geo-metric forms – the body of the diminutive teapot being hemispherical. Even at this early date, Brandt appeared more interested in functional form than in traditional craftsmanship and it is not surprising that her later designs became increasingly utilitarian in nature. After taking her journeyman's exam, Brandt became deputy director of the metal workshop in 1928 and organized projects in collaboration with the lighting manufacturers Körting & Mathiesen AG (Kandem) in Leipzig and Schwintzer & Gräff in Berlin. At the Bauhaus she worked alongside fellow metalworkers **Christian Dell** and Hans Przyrembel (1900–1945) and in 1928 co-designed the *Kandem* lamp with Hin Briedendieck (b. 1904), as part of a class project. Brandt's lamps were some of the most important designs to emanate from the Bauhaus because of their suitability for mass production. Although the Bauhaus' focus moved during the 1920s towards the realization of prototypes that were potentially suited to mass-manufacture, very few designs created at the school were successfully put into production. Brandt's lamps were a notable exception. She worked in the architectural office of **Walter Gropius** in 1929, and from 1930 to 1933 developed new design concepts for the Ruppelwerk factory in Gotha. In 1933 she returned to Chemnitz, where she took up painting and attempted to license some of her products to the Wohnbedarf department store. Brandt taught at the Staatliche Hochschule für Angewandte Kunst in Dresden from 1949 to 1950 and at the Institut für Angewandte Kunst in Berlin-Weißensee from 1951 to 1954. Although Brandt was an accomplished painter and received recognition for her photomon-tages, it is her industrial design work that is of greatest significance. Brandt also holds a special position as one of the very first women to design for industrial production – a field that has historically been dominated almost entirely by men.

702 Kandem bedside table lamp for Körting & Mathiesen, 1928

BRAUN

FOUNDED 1921
FRANKFURT/MAIN, GERMANY

Max Braun, *Model
S 50* electric shaver,
1950

→→Gerd Alfred
Müller & Hans
Gugelot, *SM3*
electric shaver, 1960

Max Braun, *Model
S 50* electric shaver,
1950

In 1921 the engineer Max Braun (1890–1951) established a manufacturing company in Frankfurt to produce
connectors for drive belts and scientific apparatus. In
1923 he began producing components for the newly emerging radio industry. Following the advent of plastic pellets in 1925, he was quick to seize
upon this new material, using homemade presses to manufacture components such as dials and knobs. In 1928 the company moved into a functional modern factory building on Idsteiner Strasse in Frankfurt, and a year
later began producing its own radio sets, which were some of the first to
incorporate the receiver and speaker in a single unit. In 1932 the company
expanded its product range and became one of the first manufacturers to
introduce radio/phonograph combination sets. Braun developed a battery-
powered radio in 1936 and a year later won an award at the Paris "Exposition
Internationale des Arts et Techniques dans la Vie Moderne" for its "exceptional achievements in phonographs".

By 1947 the company was mass-producing radio sets,
albeit still styled as furniture rather than as Modern
electronic equipment. During this period Braun also
began production of its *Manulux* flashlights and in 1950
developed its first electric razor, the *S 50*. This shaver
incorporated an oscillating cutter-block screened by a
thin steel shaver foil – a system that is still used today.
In 1950 Braun also branched into domestic appliances
with the *Multimix*. After the death of Max Braun in 1951,
the firm was headed by his two sons, Artur (b. 1925) and
Erwin (b. 1921), who decided to implement a radical design programme that was both rational and systematic.
In 1953 Erwin identified a marketing opportunity for distinctive radios that were "honest, unobtrusive and practical devices" and embodied a Modern aesthetic. To this
end, **Wilhelm Wagenfeld** and designers associated with
the Hochschule für Gestaltung in Ulm, such as Fritz
Eichler (1911–1991), were commissioned in 1954 to redesign the company's radios and phonographs. This

new Braun line was introduced at the Düsseldorf Radio Fair in 1955 and attracted international acclaim. 1956 saw the establishment of an in-house design department headed by Eichler, who proceeded to formulate a coherent corporate style based on geometric simplicity, utility and a functionalist approach to the design process. The Braun design vocabulary was not only used for products but was also applied to all areas of corporate identity, including packaging, logos and advertising. Eichler also commissioned other designers associated with the Hochschule für Gestaltung, such as Otl Aicher (1922–1991) and **Hans Gugelot,** to design sleek, unornamented products. Notable designs from this period include Eichler and Artur Braun's line of radios and phonographs (1955) and **Dieter Rams** and Hans Gugelot's *Phonosuper SK 4* radio-phonograph (1956), which was nicknamed "Snow White's Coffin". Rams also designed the *Transistor 1* portable radio (1956), the *T 3* and *T 31* pocket radios (1958) and the first component-based hi-fi system, the *Studio 2* (1959) – all of which helped establish Braun's international reputation. In 1955 **Gerd Alfred Müller** (b. 1932) joined the Braun design team and was responsible for some of the company's best-known designs from the late 1950s, including the *KM 3* multi-purpose kitchen mixer (1957), which embodied the austere rationalist aesthetic that became synonymous with German post-war design. In 1961 Dieter Rams was appointed head of the company's design department, and in 1968 overall director of design. Rams was to head the Braun design team for some 40 years, and his pareddown functionalist aesthetic permeated all the products it manufactured, from kitchen equipment to alarm clocks to electric shavers. During his tenure, Braun introduced a series of landmark designs including the *Permanent*

Braun Design Team,
hand mixer *MR 300
Compact*, 1987

lighter (1966), which incorporated an electromagnetic device rather than a traditional friction cylinder, the *ET 22* electronic pocket calculator (1976) and the first radio-controlled clock (1977). In 1967 the Boston-based **Gillette** Company acquired a controlling stake in Braun AG. A year later, the International Braun Awards for design in engineering were established. In 1983 the company was itself awarded the first Corporate Design Award at the Hanover trade fair for its "exemplary conception of product design, information and presentation".

In 1990 Braun discontinued its hi-fi production so as to concentrate on the manufacture of personal grooming products, such as the *Silk-épil EE 1* depilator (1989), the highly successful *Flex Control* line of electric razors (1990) and the *Plak Control D 5* electric toothbrush (1991), as well as a range of hair-dryers. During the 1990s Braun also introduced innovative coffee machines, food processors, hand mixers, irons and alarm clocks. In 1996 Braun launched the *Thermoscan* infrared thermometer, which marked its entry into the personal diagnostic appliance market. Certainly, Braun's success stems from the fact that its products are jointly developed by designers, engineers and marketing experts in accordance with basic design principles. The company uses design innovation to achieve technical and functional innovation and has established a tradition of progressiveness within its

Ludwig Littmann,
K 750 Combi Max
food processor, 1997

design team. The strong aesthetic clarity of its products is the outcome of a logical ordering of elements and the quest for a harmonious and unobtrusive totality. Braun acknowledges that "integrated working methods are ultimately reflected in the obviousness of the product expression", and asserts that "Braun Design is the orientation towards lasting worthwhile values: innovation, distinctive, desirable, functional, clear, honest, aesthetic."

Roland Ullmann,
Flex Integral Colour Selection razors,
1997–1998

CARL BREER

BORN 1883 LOS ANGELES, CALIFORNIA, USA
DIED 1970 DETROIT, MICHIGAN, USA

Carl Breer driving
a Chrysler *Airflow*

→→Carl Breer with a
small-scale wooden
model of the
Chrysler *Airflow*,
1934

Determined to increase the market share of his company, which at that time ranked only tenth in the automotive industry, in the early 1930s Walter Percy **Chrysler** (1875–1940) decided to develop a totally new car. The result was the Chrysler *Airflow*, designed by Carl Breer, chief engineer at Chrysler, in collaboration with his staff engineers Owen Skelton and Fred Zeder. In order to improve both the stability and the comfort of the final design, Breer had studied the concept of streamlining by experimenting with a variety of automobile body shapes in a wind tunnel. His idea was to reverse the lift effect normally associated with an aircraft wing by arriving at an aerodynamic form that would press the car more firmly against the road at high speed. With its swept-back "waterfall" nose and its headlamps and fenders partly absorbed into the body, the *Airflow* was the first automobile production to employ leading-edge aerodynamic styling. With this car, Breer also pioneered unitized body construction and "cab-forward" positioning – an innovative approach to weight distribution. Despite incorporating other significant advances, such as Chrysler's first automatic overdrive transmission, the forward-looking *Airflow* proved too radical in its appearance for the wider public and remained in production for only three years. While Breer's remarkable *Airflow* was a commercial failure, its design was immensely influential for later car designers, especially those at **Ford**, **General Motors** and **Porsche**.

Airflow, 1934

MARCEL BREUER

BORN 1902 PÉCS, HUNGARY
DIED 1981 NEW YORK, USA

In 1920 Marcel Lajos Breuer won a scholarship enabling him to study at the Akademie der Bildenden Künste in Vienna. Dissatisfied with the institution, he remained there for only a brief period before finding work in a Viennese architectural practice. From 1920 to 1923 he studied at the Staatliches **Bauhaus** in Weimar, completing the preliminary course, the carpentry apprenticeship and his journeyman's exam. While still a Bauhaus student, Breuer designed his *African* chair (1921) and his *Slatted* chair (1922–1924). In 1925, when the Bauhaus moved to Dessau, Breuer joined the teaching staff as a "Young Master" and was appointed head of the carpentry workshop. There he designed his first tubular metal chair, the *B3* (1925) – this innovative choice of material having been inspired, so it is said, by the Adler bicycle he had recently purchased. Breuer subsequently designed a whole range of tubular metal furniture, including chairs, tables, stools and cupboards, which were manufactured and distributed by Standard-Möbel, Berlin from 1927, and by **Thonet** from 1928/1929. Tubular metal offered many benefits – affordability, hygiene and an inherent resiliency that provided comfort without the need for springing. Breuer regarded his designs as essential equipment for modern living. At the Bauhaus, Breuer also designed the interiors and furnishings for the school's new premises and for the Masters' houses. His *B3* or *Wassily* chair was originally designed for Wassily Kandinsky's accommodation. Breuer not only created standardized furnishings – in 1926 he also designed a small standardized metal house and, a year later, his *Bambos* house. Breuer continued to teach at the Bauhaus until April 1928, after which he directed his own architectural practice in Berlin from 1928 to 1931. Although his building projects remained unrealized, during this period Breuer continued to design furniture, interiors and department stores. He was commissioned by the **Deutscher Werkbund** to design interiors for the German section at the "Société des Artistes Décorateurs Français" exhibition of 1930. In 1932 he completed his first architectural commission, the Harnischmacher House in Wiesbaden, and designed the Wohnbedarf furniture store in Zurich. Two years later, in conjunction with Emil (1893–1980) and Alfred Roth (b. 1903), he co-designed the Doldertal Houses, a pair of experimental apartment blocks in Zurich for Sigfried Giedion (1888–1968), the founder of the Wohnbedarf company. From 1932 to 1943, Breuer devel-

Folding theatre
seating, *Model B1*,
for Thonet,
1930–1931

oped a range of pliant furniture using a patented method of construction
that incorporated flat bands of steel and aluminium. This range of metal furniture was manufactured by Embru and retailed by Wohnbedarf. In 1935
Breuer emigrated to London and initially worked in partnership with the
architect F. R. S. Yorke (1906–1962). Together they completed several architectural commissions, including houses in Sussex, Hampshire, Berkshire
and Bristol and the Gane Pavilion in Bristol (1936), which combined wood
and local stone (a far cry from the Bauhaus aesthetic of steel and glass).
Breuer and Yorke also designed a "Civic Centre for the Future", which remained unrealized. Between 1935 and 1937, as controller of design at Jack
Pritchard's company, **Isokon**, Breuer produced five plywood furniture designs, which were essentially translations of his earlier metal designs. These
Isokon designs reflected the popularity of the earlier plywood furniture by
Alvar Aalto (1898–1976) that had been exhibited in Britain in 1933. In 1937
Breuer moved to the USA, where **Walter Gropius** had offered him a pro-
fessorship at Harvard University's School of Design in Cambridge, Mas-

sachusetts. He also established an architectural practice in Massachusetts with Gropius, and together they designed the Pennsylvania Pavilion at the 1939 **New York World's Fair** and several private houses, including Gropius' own residence. In 1941 Gropius and Breuer dissolved their partnership and Breuer founded his own architectural practice, which he moved to New York in 1946. During the late 1940s and 1950s, Breuer designed some 70 private houses, mainly in New England, and in 1947 he built a home for himself in New Canaan, Connecticut. The Museum of Modern Art, New York, initiated a touring exhibition of his work in 1947 and the following year invited him to build a low-cost house in the museum's grounds, which would suit the needs of an average American family. He furnished this project with affordable plywood cutout furniture. In 1953 Breuer worked as part of a team on the new UNESCO building in Paris and he also designed the Bijenkorff department store in Rotterdam. In 1956 he founded Marcel Breuer and Associates in New York and around this time made concrete his material of choice. He used this medium in a highly sculptural and innovative way for his design of the monumental Whitney Museum of American Art, New York (1966). Breuer was one of the foremost exponents of the Modern Movement, and the enduring appeal of his highly democratic furniture designs testifies not only to his mastery of aesthetics but to his understanding of industrial methods of production.

Model B18 table (1st version), 1928, and *Model B55* armchair, 1928–1929

BRIONVEGA

FOUNDED 1945
MILAN, ITALY

Achille Castiglione,
RR 126 radio-gramo-
phone, 1965

The consumer electronics company, Brionvega, was founded by the Brion family initially for the production of radio sets. In 1952, after Italy's first television network had been established, the company launched the first all-Italian television sets. A decade later, it manufactured the first Italian fully transistorized television, the portable *Doney 14* (1962), which was designed by **Marco Zanuso** and **Richard Sapper**. The same designers also created the *TS 502* radio (1964) with a hinged ABS plastic housing, the *Algol 11* television with an angled screen and the cuboid *Black ST 201* television (1969) – all of which reflected the company's preference for neo-modern forms and primary colours.

Brionvega's product line was less austere than those of German and Danish companies and this helped the company to appeal to a younger market. **Achille** and **Pier Giacomo Castiglioni** designed several innovative products for Brionvega, including a wire broadcasting receiver and the famous *RR 126* radio-gramophone (1965). **Mario Bellini** also designed a number of products for the company, including the *Totem* radio and record player with integrated speakers (1971), as did Franca Helg (b. 1920) and Franco Albini (1905–1977).

Mario Bellini,
Totem hi-fi, 1971

Marco Zanuso &
Richard Sapper,
Black ST 201
television, 1969

By this time however, the company was undoubtedly finding it difficult to
compete with Japanese imports and at some point in the mid-to-late 1970s
it ceased operating. Like its competitors **Braun** and **Bang & Olufsen**, Brionvega employed a Modern aesthetic for its product's casings, but the **styling**
it promoted was more akin to trends within the contemporary fine arts.

ISAMBARD KINGDOM BRUNEL

BORN 1806 PORTSMOUTH, ENGLAND
DIED 1859 LONDON, ENGLAND

Isambard Kingdom
Brunel, c. 1855

Isambard Kingdom Brunel was the son of the French émigré Sir Marc Isambard Brunel (1769–1849), who invented a wide range of machines for printing, boot-making, stocking knitting and timber processing and who also worked as a civil engineer on tunnels and suspension bridges. Brunel senior was in charge of the building of the Thames Tunnel running between Rotherhithe and Wapping, and in 1825 appointed his 19-year-old son resident engineer for the project. Isambard Kingdom Brunel held this post for three years until he was severely injured in a flooding incident – an event that put the London Tunnel project on hold. While recovering his health, Brunel designed a suspension bridge to span the Avon Gorge, and this design was eventually chosen over that put forward by the renowned Scottish engineer, Thomas Telford (1757–1834). Later, as an engineer, Brunel implemented major improvements to the docks in Bristol and designed other docks at Monkwearmouth, Milford Haven, Plymouth, Brentford and Briton Ferry. He became chief engineer to the Great Western Railway (GWR) in 1833 and subsequently introduced the broad-gauge railway track (spanning seven feet), which made trains more comfortable, roomier and safer because of the engine's lower centre of gravity. Brunel's broad gauge also allowed trains

Great Western
engine, 1838

Clifton Suspension
Bridge, 1830–1863

to travel faster owing to the increased stability. The battle between the adoption of Brunel's gauge over standard gauge raged for many years and it was not until 1892 that the GWR, nicknamed "God's Wonderful Railway", eventually conceded to building only standard gauge. Brunel did, however, introduce more than 1000 miles of railroad in Britain and Ireland. He also built two Italian railway lines, was consulted on the construction of railway lines in Australia and India, and attempted to develop a system of pneumatic propulsion for the South Devon Railway in 1844 – a project that was not fully successful. Brunel is also remembered for his many bridge designs, including one in Maidenhead that remains the flattest arch made of brick in the world. He also worked as a marine engineer and designed three landmark vessels, which were each the largest in the world at their launching – the *Great Western* (1837), the *Great Britain* (1843) and the *Great Eastern* (1858). Brunel was knighted in 1841 for his construction of the Thames Tunnel, which although dogged by financial difficulties and technical setbacks, was a sheer marvel for its day. Brunel's extraordinary originality of thought and ability to put theory into practice against all the odds have ensured his enduring fame as a great pioneer of industrial design and the father of modern civil engineering.

The *Great Eastern*
ready for launching

The *Great Eastern*
steamship, 1858

Detail of Budding's
lawnmower

EDWIN BUDDING

BORN 1796 ENGLAND
DIED 1846 ENGLAND

In 1830 Edwin Budding, an engineer from Stroud,
Gloucestershire, invented the lawnmower. His inspiration for the basic mechanical principle came from
a machine which he saw in a local cloth mill, which used a cutting cylinder
mounted on a bench to cut the pile on textiles to make them smooth after
weaving. Budding realized that he could adapt the concept for the cutting of
grass if the mechanism were mounted in a wheeled frame in such a way that
the blades rotated close to the lawn's surface. The subsequent lawnmower
he designed in 1830 was constructed of cast iron and had spur gears connecting the main roller to the knife blades. The gears allowed the blades to
rotate at a speed twelve times that of the large rear roller. In 1831 Budding
went into partnership with John Ferrabee, who agreed to pay for the patents.
Budding's lawnmower went into production and for many decades afterwards served as the essential prototype for subsequent models produced
by other manufacturers.

The first-ever
lawnmower, 1830

ETTORE BUGATTI

BORN 1881 MILAN, ITALY
DIED 1947 PARIS, FRANCE

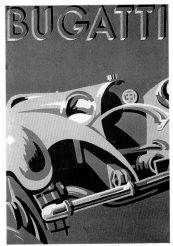

Bugatti advertisement by Gerold, 1932

Ettore Bugatti established an automobile factory in 1909 in Molsheim, Alsace. He specialized in building racing cars such as the *Type 22* and the famous *Type 35* (1924), the latter proceeding to dominate motor racing through the mid-1920s by winning over 2,000 races. During the 1920s the Bugatti workshop also produced a handful (6–8) of the ultra-luxurious *Type 41* road cars. Sometimes referred to as the "Golden Bugatti" or "La Royale", the *Type 41* was possibly the most meticulously constructed car of all time. 1934 saw the introduction of the exceptionally beautiful *Type 57*. The streamlined design of this vehicle – its flush doors in particular – exerted significant influence on the car industry. After Ettore's death in 1947, the company's fortunes declined, but the marque was eventually revived by an Italian entrepreneur in 1987 and three new models were designed, including the *EB 110 GT* super car (1991) – the fastest car in the world at the time – and the luxurious *EB 112* sedan (1993) styled by **Giugiaro** (ItalDesign). In 1995 the company was bought by **Volkswagen**, and 1999 subsequently saw the launch of four powerful and elegant prototypes which reflected Bugatti's strong and remarkable heritage.

Prototype Bugatti *EB 112*, styled by ItalDesign, 1993

Bugatti *Type 35*, 1924

Ettore Bugatti with
the Bugatti *Type 35*

BURG GIEBICHENSTEIN

SCHOOL OF APPLIED ARTS
FOUNDED 1918
HALLE, GERMANY

Stand selling products from the metal workshop of the Kunstgewerbeschule, Burg Giebichenstein, Halle, c. 1927

In 1915 the architect Paul Thiersch (1879–1928) became director of the Handwerkerschule in Halle and transformed the institution by adding courses and establishing two workshops – one for metalworking and the other for enamelling. When the school subsequently moved to new premises in Burg Giebichenstein in 1921/1922, it was renamed the Werkstätten der Stadt Halle, Staatlich-städtische Kunstgewerbeschule Burg Giebichenstein. The institution was one of the first to teach design students the importance of serial and industrial production. In 1923 the Berlin sculptor Karl Müller (1888–1972) was appointed director of the school's metal workshop, and during his tenure the Kunstgewerbeschule became one of the leading German design centres for metalware. Like the **Bauhaus**, the school in Halle made the philosophical transition from hand production to industrial manufacture, but did not depart completely from the idea of craftsmanship. Because the Halle school did not adhere to functionalism quite as strictly as the Bauhaus, the metalware products created there, ranging from tableware to lighting, were less severe and at times more expressive. Similarly, ceramic designs, whether porcelain dinner services or stoneware pots, were more often distinguished by softened elemental forms. Through its promotion of a subtler and more humanistic form of Modernism, the Halle school perhaps struck a better balance between art and industry than did its rival, the Bauhaus.

Marguerite Friedländer-Wildenhain, *Burg Giebichenstein* dinner service for KPM, Berlin, 1930

MIKE BURROWS

BORN 1943
ST. ALBANS, ENGLAND

Mike Burrows developed an interest in cycle design after his car "blew up" in 1976. He founded his own company in 1980 to produce his first bike design and this was soon followed by several recumbent models. Burrows originally conceived the recumbent cycle as a record-breaking high performance vehicle that was intended to go in a straight line only. In 1991 his most advanced monocoque superbike was seen by **Lotus** and adapted for the British cyclist, Chris Boardman, who went on to win a gold medal with it in the 400 metre pursuit event at the 1992 Barcelona Olympic Games. Having accepted an offer of a senior design post with the Giant Cycle Company, Taiwan in 1994, Burrows decided to license the manufacture of his most refined recumbent, the *Windcheetah*, to Advanced Vehicle Design in Altrincham, a manufacturer of radical, high-quality quadricycles. Also known as the "Speedy" because of its 50-mph capability and its 70-mph downhill potential, the *Windcheetah*, like the Boardman bike, was completely revolutionary and allowed "man to power a machine to new limits of speed and distance". The *Windcheetah* utilizes a highly sophisticated cruciform frame and a unique joystick steering system, which provides extraordinary cornering. Lightweight titanium, **aluminium** and kevlar components are used extensively in its construction and the combination of these high performance

Windcheetah T. I. human-powered vehicle, 1995

Mono superbike, 1992

materials, together with wind cheating **aerodynamics**, delivers an outstanding rider experience. As the benchmark for recumbent performance, the *Windcheetah* is a tribute to Mike Burrows' genius.

JAKE BURTON

BORN 1954
NEW YORK, USA

As a child, Jake Burton experimented endlessly with a department store toy known as a *Snurfer* that was manufactured by Sherman Poppen in the mid-1960s. This child's plaything – a short fat ski without bindings or edges – allowed the user to surf on snow. No further development of this proto-snowboard took place over the next ten years, however. In 1977 Burton left the business world of Manhattan and moved to Vermont to become a snowboard "shaper". The same year, he founded the world's first snowboard factory, which in the beginning was a modest garage operation. The first snowboards he produced included the Burton *Backhill*, Burton *Backyard* and *Performer*, which were sold by Jake going from shop to shop. In these early years, snowboarding was very much an underground sport with sledging hills and even golf courses mainly being used. Burton knew that if snowboarders had to walk their way up a hill in order to board down it, the sport would not progress very far. He therefore lobbied local ski hills to allow snowboarders access to the lifts and slopes. Eventually, in the early 1980s, Stratton resort in Vermont acquiesced. As other ski resorts followed suit, Burton developed the *Performer Elite* specifically for hard-packed snow, as his earlier edgeless wooden boards were really only suited to powder snow. Burton's tireless efforts to promote snowboarding eventually ensured its widespread acceptance, but also turned snowboard manufacturing into a highly competitive business. Burton buckled down and focused on product development. In 1992 he moved his factory from Manchester, Vermont, to Burlington. While Jake Burton did not invent the first snowboard – there is an example from the 1920s in the Burton museum – he did invent the sport. Burton's enthusiasm and vision resulted in the creation of a remarkable industry that continues to seriously rival skiing. Not surprisingly, snowboarding has had a significant influence on ski design – the now common parabolic-shaped ski was born directly out of snowboard technology.

Burton *Backhill* & *Backyard* snowboards, 1977

Burton *FL*
snowboard, 1998
(first manufactured
2000)

→Burton *Custom 56*
snowboard, 1998
(first manufactured
2000)

NICK BUTLER

BORN 1942
NORMANTON, ENGLAND

In 1967 Nick Butler founded BIB Design Consultants Ltd. and in 1975 was elected a fellow of the Chartered Society of Designers (CSD). He became a member of the judging panel for the **Design Council** Awards in 1975 and of the Council's Industrial Advisory Committee in 1978. Butler also chaired the British Design Export Group from 1980 to 1981 and was elected a fellow of the Royal Society of Arts in 1983. He served as an external examiner for the Royal College of Art from 1983 to 1986 and has lectured widely on industrial design both in Britain and abroad. In recognition of his contribution to this field, Butler received an OBE (Order of the British Empire) in 1988 and became Master of the RSA's Faculty of Royal Designers to Industry in 1995. During the mid-1990s, Butler consulted to the Bank of England and the Royal Mint on the design of the Euro currency and was a judge for the British Millennium Products programme. His office continues to offer a comprehensive product design service, which includes design audits, feasibility studies, product innovation and development, design engineering, human factors analysis, model-making and prototyping as well as productionization. Butler's product designs, such as the *Durabeam* flashlight (1982) which won a Design Council Award, are simple and intuitive to use and are guided by his belief that "design can make a real difference to how well a product is understood and therefore the efficiency of its usage". BIB has designed personal computers for **Apple**, fax machines for British Telecommunications and Panasonic, watches for Citizen, televisions for Thomson, a range of pens for Dunhill and kitchen equipment for Prestige.

Durabeam
flashlights for
Duracell, 1982

CANON

FOUNDED 1947
TOKYO, JAPAN

Faxphone 8, 1988

The Precision Optical Instruments Laboratory was established in Tokyo in 1933 to conduct research into camera technology. A year later, the company developed a prototype of the *Kwanon*, Japan's first 35mm focal-plane shutter camera, and in 1940 designed Japan's first indirect X-ray camera. Introduced in 1946, the Canon *SII* camera was acclaimed by officers from the Occupation Forces and also by foreign purchasers. In 1947 the firm was renamed the Canon Camera Company and over the following years continued to produce innovations in camera technology, such as the world's first speed-light synchronized flash and shutter camera, the Canon *IVSb* (1952). The Canon *L1* still-camera and the Canon *8T* cine-camera became the first products to receive a "Good Design" designation from the Japanese Ministry of International Trade and Industry in 1957. During the 1960s and 1970s, Canon diversified its product line to include microfilm systems and calculators. In 1965 the company entered the copying machine industry with its introduction of the *Canofax 1000*. Three years later it also launched the first-ever four-track, four-channel recording head. Throughout the 1970s Canon continued developing

Kwanon camera – Japan's first 35mm focal-plane shutter camera, 1934

state-of-the-art cameras, including its top-end SLR, the Canon *F-1* (1971) and the landmark *AE-1* with a built-in microcomputer (1976), which triggered an *AE* SLR boom. In 1975 the company succeeded in developing a laser beam printer (LBP) and in 1976 introduced the world's first non-mydriatic retinal camera, the *CR-45NM*. This was followed in 1981 by the first bubble-jet printer and the new *F-1* SLR. In 1982 Canon launched a colour ink-jet printer and in 1986 the first multi-functional telephone with a built-in fax. Around this time, Canon also began producing digital camcorders, the *Optura* (1997) and the lighter and more compact *Elura* (1999). Over the company's 59-year history, the cumulative production of its cameras has exceeded 100 million units. From handshake-compensating binoculars to digital cameras to ultra-high-quality digital imaging equipment and high-definition television broad-casting lenses, all Canon products have a strong brand identity. Among the company's product design guidelines are: the use of advanced forms, mate-rials and technology; the optimization of comfort with regard to human factors; the dedication to ease of operation; the consideration of the opera-tional environment; and the creation of transcultural solutions.

G-Mark image
stabilizer binoculars,
1995

C.BIO camera
designed by Luigi
Colani, 1982–1983

IXUS camera, 1996

GEORGE CARWARDINE

BORN 1887 ENGLAND
DIED 1948 ENGLAND

Anglepoise lamp,
1932 (produced
from 1933)

Page from a
brochure promoting
the use of the
Anglepoise lamp in
the Birmingham
Hospital Centre,
mid-1930s

George Carwardine, a director of Carwardine Associates, Bath, was an automotive engineer who specialized in the design of suspension systems. In 1932 he patented the design of an articulated table lamp, the *Anglepoise*. This innovative design, which allowed flexible re-positioning, was based on the constant-tension principle of human limbs with the lamp's spring acting in much the same way as muscles. Carwardine balanced a weight against a spring through a linking element so as to allow the weight – and thereby, the arm – to remain stable in a variety of positions. Originally produced by the English spring manufacturer, Herbert Terry of Redditch, from 1933 the *Anglepoise* lamp was an instant success. It found applications in offices, factories and hospitals as well as in domestic environments and was manufactured in large numbers for over 50 years. In 1937 the patent for the lamp was acquired by the Norwegian lighting designer, Jacob Jacobsen (b. 1901), who had been influenced by it when he came to design his own

No. 1288. IN BIRMINGHAM HOSPITAL CENTRE.

FOUR ANGLEPOISE SURGERY MODELS

MODEL 1287.
ANGLEPOISE SURGERY CEILING LAMP. Provides instantly adjustable light for all close work. Always at finger-tip command, yet never in the way. Overall arm-span 3 ft. 3 ins. Choice of chromiumplating or black.

MODEL 1288.
DISSECTING MODEL. A marvellous aid at the operating table, this L-armed Ceiling ANGLEPOISE concentrates a strong shadowless light on any part—from any position. Instantly adjustable, frictionless and silent. Each arm-span is 3 ft. 8 ins.

MODEL 1208C.
SPECIAL ANGLEPOISE TROLLEY MODEL. Invented specially for the surgery. Moves silently and smoothly on large-size rubber castors. The lamp can be instantly adjusted to any position or angle within a 7 ft. sphere.

MODEL 1208A.
WALL BRACKET MODEL. With 8-in. or 12-in. extension as desired. Specially useful for reading, writing, laboratory work, etc., where space is limited. (As an alternative a screw-down base can be supplied.)

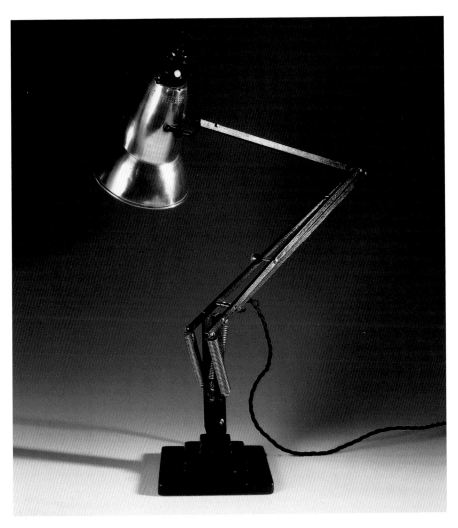

Early example of an *Anglepoise* lamp for Herbert Terry & Sons, 1932 (produced from 1933)

highly successful *Luxo L-1* lamp the same year. Carwardine's *Anglepoise*, which was subsequently manufactured in Norway and retailed under a different name, was highly influential for later generations of task lighting. The fact that this classic design is still manufactured today testifies to its functionality and timeless Modern aesthetic.

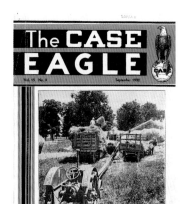

Cover of *The Case Eagle*, September 1932 issue

CASE

FOUNDED 1842
ROCHESTER, WISCONSIN, USA

Jerome Increase Case (1819–1891) read an article in the *Genessee Farmer* about a machine that could be used to "thrash" wheat – a back-breaking chore that had always been done entirely by hand. The time-consuming nature of traditional threshing had hindered the expansion of farms in North America and Case saw the huge potential of a mechanical threshing machine. In 1842 he improved upon a primitive "ground-hog" threshing machine he had designed earlier and founded the J. I. Case Company in Rochester, Wisconsin. Two years later, he built a factory in Racine and became the first manufacturer to build mechanical threshers. His *Sweepstakes Thresher* of 1862 was capable of threshing 200–300 bushels a day, whereas one man could only manage six or seven bushels a day. Case soon became known as the "Threshing Machine King" and in 1865 he adopted the famous eagle trademark. Four years later he produced the first Case steam engine, known as the *Old No. 1*, which boasted 8 hp and which was loved by farmers as they could have it running all day, unlike animal-powered machinery. It was not until the mid-1870s, however, that sales of agricultural steam-powered machines really took off. 1880 saw the launch of the *Agitator* thresher, offering greater capacity and improved efficiency. By 1886 Case was the largest manufacturer of steam engines in the world and in 1892 introduced its first gasoline-powered tractor. By 1913 the company was producing smaller and faster tractors that were the antecedents of today's vehicles, and in 1919 Case added ploughs and tillage tools to its product line, thus enabling it to become a full-line farm equipment manufacturer. Case also dabbled in automobile manufacturing and even fielded a racing team for the first Indianapolis 500 motor race in 1911. In the 1930s, Case exported threshers as far afield as New Zealand and continued to introduce

Tresher/Stacken, 1908

First steam-engine
Old No. 1 with
Eclipse thresher,
1869

Ground/tog thresher,
c. 1842

new products, from the *Case Hammer Mill*, which ground feed for livestock, to the *One-Man* combine. During the Second World War, Case produced more than 1.3 million shells for the American armed forces and manufactured parts for military trucks, Sherman tanks and fighter aircraft. After the war, Case returned to peacetime production of agricultural and construction equipment. Today the company remains a leader in its field, particularly in terms of innovative design and manufacture.

CASIO

FOUNDED 1957
TOKYO, JAPAN

Senty Next G11
Pocket Bell pager,
1996

MG-800 calculator,
c. 1978 – one of
the first calculators
to incorporate a
primitive electronic
game

In 1946 Tadao Kashio, an engineer, set up a company called Kashio Seisakujo to manufacture aircraft parts for the military. Acting on a suggestion from his brother Toshio, Tadao and his three younger brothers also set about developing a calculator. Unlike other calculators, which relied on electrically powered internal gears, the Kashio brothers' design ingeniously adapted the relay elements developed for telephone switching equipment. This new electronic calculator was significantly more compact, since it did not need gears. 1957 saw the founding of Casio Computer Co. and the launch of the all-electric *14-A* model. Ten years later, the company introduced the world's first programmable electronic desktop calculator, the *AL-1000*. At this stage, calculators were still extremely expensive and so Casio decided to develop a design that could be afforded by individuals rather than just corporate consumers. The resulting palm-sized *Casio Mini* (1972) was the world's first personal calculator and helped to make the calculator part of everyday life. Using its pioneering technology, Casio set new standards in miniaturization with the first-ever electronic wristwatch, the *Casiotron* (1974) and the original card-sized calculator, the *Casio Mini Card* (1978). The *Casiotron*, with year, month, day, hour, minute and second display capabilities, was based on the concept that "time is a continuous process of addition". In 1981 the Casio design team decided to develop a wristwatch that was as "strong as a tank but also flexible", and which would last 10 years – the result was the *G-Shock* (1983), which could endure the shock of hitting the ground from a height of 10 metres. Many subsequent *G-Shock* models followed, including a ladies' version, the *Baby-G* (1994), which became an instant "must-have" fashion accessory. Throughout the 1980s and 1990s, Casio continued to specialize in the design and development of state-of-the-art calculators, wristwatches and electronic musical instruments, as well as hybrid products such as the *BM-100WJ* (1989) digital wristwatch with a sensor for forecast-

← BG 100 OCL-7T
Baby G wristwatch,
1995

→ BG 100-3T Baby G
wristwatch, 1995

← AW 500-1E,
G-Mark, G-Shock
wristwatch, 1988

→ DW 003B-9
G-Shock wristwatch,
c. 1991

ing weather, and the *Pathfinder PAT1GP-1* (2000), the world's first wristwatch
with built in GPS for longitude and latitude display. Casio strives to make
original products with innovative capabilities that are easy to operate, versatile and affordable. Research and development of totally new products is a
central theme of its business philosophy.

LIVIO, PIER GIACOMO & ACHILLE CASTIGLIONI

BORN 1911 MILAN, ITALY
DIED 1979 MILAN, ITALY
BORN 1918 MILAN, ITALY

BORN 1913 MILAN, ITALY
DIED 1968 MILAN, ITALY
DIED 2002 MILAN, ITALY

Livio Castiglioni studied architecture at the Politecnico di Milano, graduating in 1936. In 1938 he set up a studio in collaboration with his brother Pier Giacomo and Luigi Caccia Dominioni (b. 1913), designing silver and aluminium cutlery. Their most notable design, however, was the *547* radio (1938–1939) for Phonola. As the first Italian radio to be made in Bakelite, the *547* changed the face of radios, which up till then had mostly been cased in wooden boxes. The *547* was awarded a gold medal at the VII Milan Triennale of 1940, where the Castiglioni office also curated an exhibition of radios. Livio worked as a design consultant for Phonola from 1939 to 1960 and for **Brionvega** from 1960 to 1964. He was president of the ADI (Associazione per il Disegno Industriale) from 1959 to 1960 and designed many audiovisual presentations. He collaborated with his younger brothers on several lighting projects, although his best-known light design, the snake-like *Boalum* (1970), was executed in conjunction with Gianfranco Frattini (b. 1926). Like their elder brother, Pier Giacomo and Achille Castiglioni also graduated from the Politecnico di Milano, in 1937 and 1944 respectively. Achille joined

Livio Castiglioni,
Pier Giacomo
Castiglioni & Luigi
Caccia Dominioni,
547 radio for
Phonola,
1938–1939

Achille & Pier Giacomo Castiglioni, slide projector for Ferrania, 1960

his elder brothers' design studio in Piazza Castello and during the post-war years they undertook town-planning and architectural commissions as well as exhibition and product design. The brothers were extremely active and helped establish the Milan Triennale exhibitions, the Compasso d'Oro awards and the ADI. When Livio left the partnership in 1952, the two younger brothers continued to work together. They designed the "Colori e forme nella casa d'oggi" exhibition at the Villa Olmo in Como, where they first presented their radical "ready-made" designs of 1957 – the *Mezzadro* (Sharecropper's Stool), incorporating a tractor seat, and the *Sgabello per Telefono* (Telephone Stool), which had a bicycle seat. The brothers also produced less contesting designs, such as the Neo-Liberty style *Sanluca* armchair (1959) for the furniture manufacturer Dino Gavina, whose company offices in Milan they designed in 1963. Other notable designs by Pier Giacomo and Achille Castiglioni include the *Tubino* desk lamp (1950), the *Luminator* floor lamp (1955), the *Arco* floor lamp (1962) and the *Taccia* table lamp (1962) – all of which were manufactured in significant numbers. In 1966 they designed the *Allunaggio* seat, which was inspired by the first moon

Achille & Pier
Giacomo Castiglioni,
Toio floor lamp for
Flos, 1962

→Achille Castiglioni
& Pio Manzu,
Parentesi lamp for
Flos, 1970

landing. Their prestigious client list included **Kartell**, Zanotta, Bernini, **Siemens**, Knoll, Poggi, Lancia, Ideal Standard and Bonacina. Apart from furniture and lighting, Pier Giacomo and Achille Castiglioni also designed audio equipment for Brionvega, including the *RR 126* radio-gramophone (1965). After Pier Giacomo's death in 1968, Achille continued to work in industrial design, creating such well-known pieces as the *Lampadina* table lamp (1972) for **Flos**, the sleek cruet set (1980–1984) for **Alessi** and the *Gibigiana* directional table lamp (1980) for Flos.

The Castiglioni brothers exerted much influence on succeeding generations of Italian designers not only through their activities and success as industrial designers, but also through their long-standing involvement in design education. Pier Giacomo taught at the Politecnico di Milano from 1946 to 1968, while Achille was professor of artistic industrial design from 1970 to 1977 and professor of interior architecture and design from 1977 to 1980 at the Politecnico di Torino. In 1981 Achille returned to the Politecnico di Milano as professor of interior design and since 1986 has been professor of industrial design there.

Achille Castiglioni,
design for the *Frisbi*
lamp for Flos, 1978

→Achille Castiglioni,
Gibigiana table lamp
for Flos, 1980

During his long career, which has spanned over half a century, Achille Castiglioni has been honoured with eight Compasso d'Oro awards as well as numerous other design prizes. While the language of design he and his brothers pioneered was grounded in rationalism, it was tempered with ironic humour and sculptural form – an unusual approach to industrial design that has been described as "rational expressionism". This, and the remarkable consistency of quality of their designs, which are both structurally inventive and aesthetically pleasing, make the Castiglioni brothers three of the most important figures in Italian design in the 20th century. Achille Castiglioni believes that "the need for a new and original approach in contemporary design is emerging more and more urgently". "Such an approach," he claims, "can be effected through design activity on two different levels. On the one hand, there is design for mass production; on the other, design for the production of a limited quantity." In both these areas of activity, Achille has not only experimented but also excelled, with his work communicating emotional values while at the same time "straining to achieve adequate results with minimum means".

Benjamin Holt, first "Caterpillar" tractor, patented in 1893

Thirty Caterpillar tractor, c. 1930

CATERPILLAR

FOUNDED 1925
PEORIA, ILLINOIS, USA

Early steam tractor,
c. 1890

During the 1890s Benjamin Holt (1849–1920) and
Daniel Best independently researched and experiment-
ed with steam-driven farm tractors. Both Holt and Best pioneered track-
type and gasoline-powered tractors, and during the First World War Holt's
tracked "Caterpillar" tractor was used extensively by the Allied Forces to pull
artillery and supply wagons. In 1925 the Holt and Best companies finally
merged to become Caterpillar, based in Peoria, Illinois. After years of researching diesel as an efficient source of power for track-type tractors, the
company launched the landmark *Diesel Sixty Tractor* in 1931. During the
1940s, Caterpillar diversified its product line to include motor graders and
electric generators, which were in great demand during the Second World
War, as were its tractors. Caterpillar also produced a special engine for the
M-4 Sherman tank. After the war, Caterpillar introduced numerous new products and in 1950 established its first overseas operation in Britain. Since the
1950s Caterpillar has supplied its powerful *Cat* engines to manufacturers of
trucks, trains, boats and construction equipment. Through electric generation systems, *Cat* engines also supply power to areas inaccessible to utility
power grids, such as offshore oil rigs and developing nations, and provide
emergency power to hospitals, schools etc. During the 1970s Caterpillar developed several innovative products, including what was then the world's
largest and most advanced track-type tractor, the *D10*, as well as the *225* hydraulic excavator and the *3400* family of engines. The company was hit badly
by the recession in the early 1980s but bounced back with revolutionary innovations such as the elevated rear
sprocket design of its track-type
tractors, which increases productivity, simplifies maintenance and
extends drive train life. By 1997
Caterpillar had become the world's
leading manufacturer of construction and mining equipment, diesel
and natural gas engines and industrial gas turbines.

D11R CAT tractor,
late 1990s

CHRYSLER

FOUNDED 1925
DELAWARE, USA

Walter Percy Chrysler
with a model of Carl
Breer's *Airflow*,
1930s

Walter Percy Chrysler (1875–1940) began his working
life as a machinist's apprentice in the railroad industry,
but by the age of 33 had already become the superintendent of motive power for the Chicago Great Western
Railway. In 1910 he moved to the American Locomotive
Company to manage its Pittsburgh works and around
this time purchased his first car, a Locomobile Phaeton.
A couple of years later, Chrysler joined the Buick Motor
Car Company in Flint, Michigan. In 1916 Buick became **General Motors'** first
automotive division, and in 1917 Chrysler was appointed its president and
general manager. He was promoted to vice-president of General Motors in
1919 but retired from the company a year later. He next worked for Willys-
Overland and then the ailing Maxwell Motor Car Company, which he revital-
ized with the development of the Chrysler *Six* (1924) – America's first high-
styled, medium-priced automobile, which set an industry sales record by
selling 32,000 units. The Chrysler Corporation was established in Delaware
in 1925 as the successor to the Maxwell Motor Car Company and Walter
Chrysler was named its chairman. The same year saw the introduction of
the highly successful Chrysler *Four*, with a top-speed of 58 mph, and by 1926
the corporation had risen from 57th to 5th place in industry sales. In 1928

Walter Percy Chrysler
with *Four Series 58*,
1925

Dodge truck, 1940s

Chrysler manufactured the first Plymouth and De Soto models and the company also acquired Dodge Brothers, which was highly regarded for its utility vehicles. Although hit by the Great Depression, Chrysler did not cut back on its research and development programme and pioneered several design and engineering innovations, including the "Floating Power" engine mounting system, which reduced vibration to produce a smoother ride. Although the company hired its first in-house stylist, Ray Dietrich (1894–1980), in 1932, design and **styling** at Chrysler essentially took a back seat to engineering until 1950. A notable exception was the Chrysler *Airflow* of 1934. Designed by the company's chief engineer, **Carl Breer**, and his staff engineers Fred Zeder and Owen Skelton, the *Airflow* introduced leading-edge aerodynamic styling, innovative weight distribution, unitized body construction and the industry's first one-piece curved windscreen. While the *Airflow* was way ahead of its time and consequently a commercial failure, it was immensely influential upon later automotive design.

During the Second World War, Chrysler produced 18,000 32-ton *M-4 Sherman* tanks and around 500,000 Dodge trucks for military use. By 1945 the company had supplied over US$3.4 billion worth of equipment to the Allied Forces. After the war, Chrysler responded to the need for increased car and truck production by building or buying 11 plants between 1947 and 1950. In 1949 the company also appointed its first director of advanced styling, Virgil Max Exner (1909–1973), who had previously worked in **Raymond Loewy**'s

Dodge *Viper*, 1992

Chrysler 392 Hemi
Firepower V-8
engine, 1951

design office. Exner introduced European-like styling to the company and developed the "Forward Look", a style that emphasized movement and speed through the use of curved side windows and tail fins. The Forward Look is perhaps best exemplified by Exner's 1957 Plymouth *Belvedere*. During the 1950s Chrysler also introduced several innovations including power steering, key-operated ignition, electric windows and cushioned dashboards for improved safety. In 1951 it also developed the legendary "Hemi" V-8 engine. With deep hemispherical combustion chambers, the Hemi featured large valves that provided high-volume efficiency and produced 20% more horsepower than standard V-8 engines. The Hemi was discontinued in the mid-1950s but reappeared in 1964 as the ferocious 425 horsepower 426 Hemi.

While it was in production, the 426 Hemi was the perfect power plant for Chrysler's emerging "muscle" cars. Elwood P. Engel (1917–1986) was the designer chiefly responsible for the creation of such beefy and brutishly purposeful cars as the Dodge *Challenger*, the Dodge *Charger*, the Plymouth *Road Runner* and the Plymouth *Barracuda*. From the mid-1960s to the early 1970s, these cars dominated the muscle-car street scene and became

Chrysler *Voyager*,
1999 model

legends in their own time. The oil crisis in the early 1970s put an end to this period of excess in the American car industry, however, and prompted a demand for smaller, more fuel-efficient vehicles. Chrysler found itself facing a financial crisis, and in 1979 the celebrated auto executive Lee Iacocca was brought in to turn the company around. With the well-publicized challenge, "If you can find a better car, buy it", Iacocca managed to save the company with the introduction of the "K-cars" in 1981 – the Dodge *Aries* and the Plymouth *Reliant*. This was followed in 1983 by Chrysler's introduction of the minivan in the form of the Dodge *Caravan* and Plymouth *Voyager*. These vehicles not only created a whole new market – people carriers – but also became the company's best sellers. In 1987 Chrysler re-entered the European market with export vehicles and also purchased the exotic Italian sports car manufacturer, **Lamborghini**. During the 1990s Chrysler launched several notable and financially successful models, including the awesome 700 hp Dodge *Viper* (1992), the **Jeep** *Grand Cherokee* (1993) sports-utility vehicle and a striking evolution of the *Voyager* MPV (1995). By adapting vehicle design to meet changes in living patterns and customer expectations, Chrysler has managed to survive and prosper. In 1998 the company merged with the prestigious German automotive manufacturer, Daimler-Benz AG, and its product range has consequently adapted to the European market.

CITROËN

FOUNDED 1913
PARIS, FRANCE

Flaminio Bertoni,
Citroën *DS19*, 1962

→Pierre Boulanger,
prototype of the
Citroën *2CV*, 1936

↘Citroën *2CV Cl*,
1959

Citroën poster
showing the *Type A*,
1919

André-Gustave Citroën (c. 1878–1935) trained at the École Polytechnique and then worked as an engineer and industrial designer. From 1905 he began manufacturing components for cars and introduced double helical gears to France. In 1913 he founded the Société des Engrenages Citroën in Paris and two years later began manufacturing ordinance – up to 55,000 shells per day – for the French army. At the end of the war, with arms sales falling, Citroën converted his factory to the production of small and inexpensive cars and hired Jules Saloman as his first automotive designer. In 1919 the company launched the first mass-produced car in Europe, the *Type A*. This car significantly advanced the standards of automotive design with its disc wheels, electric lights and on-board starter. It was, however, the slightly later, more compact and much better-selling *5CV* that firmly established Citroën's reputation in the mass market. In 1934 André Citroën almost bankrupted his business by over-investing in tooling for the *Traction Avant* (later launched as the *7A*) and the company was then taken over by Michelin. The *Traction Avant* was an advanced prototypical design that was at least two decades ahead of its time, with front wheel drive and a long wheel base. Citroën's best-ever seller, the *2CV*, was designed by Pierre Boulanger (1886–1950) and first appeared in 1939. Launched in 1948 as a functionalist, no-frills work-horse, the *2CV* was intended to rival **Volkswagen**'s *Beetle*. Between 1948 and 1990, when production ceased, nearly 3,870,000 *2CVs* were built. Citroën's famous and more elegant *DS19*, designed by Flaminio Bertoni (1903–1964) in 1955, was the evolutionary successor of the *Traction Avant*. With its low-slung forward-looking aesthetic, the *DS19* became known as the "Goddess" in France and was, like the *Traction Avant*, an extremely advanced vehicle for its time. Citroën was eventually taken over by Peugeot in 1975, and although it continues to produce very worthy vehicles, the level of design innovation it achieved in the past has yet to be matched.

COALBROOKDALE

BIRTHPLACE OF THE INDUSTRIAL REVOLUTION
SHROPSHIRE, ENGLAND

Christopher Dresser,
Coalbrookdale Co.
cast-iron grate, 1879

←Christopher Dresser,
Coalbrookdale Co.
cast-iron umbrella
stand, c. 1880

Abraham Darby (c. 1678–1717), a British ironmaster, pioneered the smelting of iron ore with coke in the Coalbrookdale area of Shropshire during the early 18th century. Coalbrookdale's close proximity to supplies of natural resources, low-sulphur coal in particular, helped revolutionize the production of iron. In 1709 Darby was the first to produce marketable iron in a coke-fired furnace and his foundry went on to manufacture the first iron boat (1787) and the first railway locomotive (1802) in Britain. Such was the quality of Darby's iron that it could be cast in thin sheets and could thus compete successfully with brass in the manufacture of pots and other hollow ware. The famous cast-iron bridge near Coalbrookdale was also one of the first of its kind. Designed by Thomas Pritchard, the bridge spans 43 metres across the River Severn and was erected between 1777 and 1779 by John Wilkinson and Darby's grandson, Abraham Darby III (1750–1791), partly to advertise the family iron foundry. As a major centre of iron production for over 100 years, Coalbrookdale was renowned in the 19th century for its cast-iron fireplaces and hall furniture, some designed by **Christopher Dresser**. As the birthplace of the **Industrial Revolution**, Coalbrookdale is now a world heritage site.

Philippe Jacques
de Loutherbourg,
*Coalbrookdale by
Night,* 1801

WELLS COATES

BORN 1895 TOKYO, JAPAN
DIED 1958 VANCOUVER, CANADA

Wells Coates studied engineering at the University of British Columbia, Vancouver, and graduated in 1921. He subsequently moved to Britain and from 1922 to 1924 studied for a doctorate in engineering at London University. From 1923 to 1926 he worked as a journalist for the *Daily Express* and for a brief period was one of their correspondents in Paris. During this time he wrote from a humanist perspective, viewing design as a catalyst for social change. In London in 1928, he designed fabrics for the Crysede Textile Company and interiors for the firm's factory in Welwyn Garden City, in which he used plywood elements. From 1931 Coates was a consultant to Jack Pritchard's plywood products company, Isokon – a firm pioneering Modernism in Britain. Coates was also commissioned by Pritchard in 1931 to design the Lawn Road Flats, Hampstead, which are seminal examples of British Modern Movement architecture. In 1933 Coates co-founded MARS (Modern Architecture Research Group) and established design partnerships with Patrick Gwynne in 1932 and David Pleydell-Bouverie in 1933. From 1932 he designed a series of Bakelite radios for the Ekco Radio Company, including his famous circular Ekco *AD 65* (1934), which were conceived for industrial production and were among the first Modern products available to British consumers. After the Second World War Coates worked in Vancouver, designing interiors for **de Havilland** and BOAC aircraft, and in the 1950s producing designs for television cabinets. Just as Wells Coates believed that "the social characteristics of the age determine its art", so his designs epitomize the flowering of British Modernism in the 1930s.

Ekco *AD65* radio for E.K. Cole, 1934

SIR CHRISTOPHER COCKERELL

BORN 1910 CAMBRIDGE, ENGLAND
DIED 1999 SUTTON SCOTNEY, ENGLAND

The concept of an air-cushion vehicle was first proposed
in the 1870s by the British naval architect and engineer,
Sir John Thornycroft (1843–1928). Although Thornycroft
built and patented several models, he failed to solve the
problem of the air cushion becoming "detached" from
the craft. It was not until the 1950s, when Christopher
Cockerell designed a "skirt" that could contain the air cushion, that the air-
cushion principle became viable. In 1959 Cockerell went on to design the
world's first hovercraft, the Westland SRN1. This amphibious vehicle was
capable of carrying three passengers, but required calm water or flat ground
to operate properly. The later, much improved and much larger SRN2 was
designed around a spacious central cabin and, depending on the layout,
could accommodate 56 to 76 passengers and their luggage. In June 1962
American Navy experts observed trials of the SRN2 hovercraft in the Solent,
while British service chiefs were investigating the military potential of this
revolutionary design. Cockerell was knighted in the late 1970s in recognition
of his contribution to vehicle design. Today, enormous hovercraft that can
accommodate over 400 passengers and some 60 cars frequently cross the
English Channel at speeds of up to 60 knots.

Demonstration of
Westland SRN2
Hovercraft at Cowes,
1962

Westland SRN1
Hovercraft, 1959

SIR HENRY COLE

BORN 1808 BATH, ENGLAND
DIED 1882 LONDON, ENGLAND

Henry Cole, teapot
for Summerly Art-
Manufactures,
c. 1846

Contemporary
caricature of Henry
Cole, c. 1850

Sir Henry Cole worked for several years as an official in the Public Records Office in London, where he campaigned for better conservation of archival material. He also devised a scheme for postal reform and studied watercolour painting under David Cox (1783–1859), his works being exhibited at the Royal Academy. He published a series of children's books entitled *Felix Summerly's Home Treasury* in 1841, and two years later published the first Christmas card to a design by J. C. Horsley (1817–1903). In 1846 he became a member of the Society of Arts; that same year, the Society awarded him a medal for a tea service which he had created under the pseudonym of Felix Summerly. Notable for its simplicity of form, this design was manufactured by Herbert **Minton**. As one of the greatest champions of the applied arts, in 1847 Cole founded Summerly's Art-Manufactures, which commissioned designs suitable for industrial production from artists such as John Bell (1811–1895), Daniel Maclise (1806–1870) and Richard Redgrave (1804–1888). Although Cole severed his connection with this venture in 1848, he believed it had "established the compound word 'art-manufactures' in our language". Between 1847 and 1849 Cole organized annual Society of Art exhibitions to promote art-manufactures and in 1851 became chairman of the Society. In this capacity, Cole was largely responsible for the staging of **The Great Exhibition** of 1851 in London and the assembly of a permanent study collection which would eventually form the basis of the Victoria & Albert Museum. He also mounted an exhibition of "Examples of False Principles in Decoration" in an effort to demonstrate the differences between "good design" and "bad design". Another of Cole's contributions to the advancement of industrial design was his reform of design teaching, which placed great emphasis on the creation of simple, functional forms for industrial production.

GINO COLOMBINI

BORN 1915
MILAN, ITALY

4660 wastebin for
Kartell, 1965

4660 wastebins
and *4650* umbrella
stands for Kartell,
1965

From 1933 to 1952 Gino Colombini worked in the architectural office of Franco Albini (1905–1977), where he helped design commercial and domestic buildings as well as furniture. In 1949 Colombini was appointed technical director of **Kartell**, a recently formed plastics manufacturing company. He designed many of the firm's first products, including a kitchen bucket that was awarded the Compasso d'Oro award in 1955. Colombini also designed many other plastic products intended for everyday domestic use, such as a carpet-beater (1957), a milking pail (1958), a lemon squeezer (1958), a children's lunch-box (1958), a washing tub (1957) and various buckets and dustpans (1956–1957). These innovative and highly functional products were among the first to exploit the potential of injection-moulded plastics as materials for high-volume mass production. Together with Leonardo Fioro, Colombini also designed a modular cupboard system for Kartell, which had a tubular metal frame supporting doors and shelves made of shock-resistant polystyrene. This forward-looking design of 1956 predicted the later storage-system furniture of the 1960s and 1970s. Colombini's products, which in most cases were based on the principle of structural, material and functional unity of design, helped Kartell to become one of the leading proponents of Italian design in the 1950s. His product designs revolutionized not only the intrinsic quality but the aesthetic form of everyday objects, and in so doing were extremely influential upon succeeding generations of industrial designers.

JOE COLOMBO

BORN 1930 MILAN, ITALY
DIED 1971 MILAN, ITALY

Cesare "Joe" Colombo was active as a painter before turning his attention to industrial design around 1958. He began experimenting with new materials, including reinforced plastics, as well as novel construction techniques and manufacturing methods. In 1962 Colombo established his own design office in Milan and worked chiefly on architectural and interior design projects. In 1964 he was awarded the IN-Arch prize for his interiors for a hotel in Sardinia (1962–1964), which included ceiling fixtures constructed of Perspex prisms that diffracted light. With his brother Gianni, Colombo developed this idea in his design for the *Acrilica* lamp (1962) for O-Luce. His first design for **Kartell** was the *No. 4801* chair (1963–1964), which comprised three interlocking plywood elements. The fluidity of this design anticipated his later work in plastics, such as the *Universale No. 4860* chair (1965–1967), which was the first adult-sized chair to be manufactured in injection-moulded plastic (ABS). Colombo produced other innovative designs for furniture, including the multi-functional *Boby* trolley system (1968), lighting, glassware, door handles, pipes, alarm clocks and wristwatches. From the beginning of his career, Colombo was interested in domestic systems products, as his early *Combi-Centre* container unit (1963) demonstrates. This interest in systems furnishings gave rise to his *Additional Living System* (1967–1968), *Tube* chair (1969–1970) and *Multi* chair (1970), all of which could be

4860 Universale chair for Kartell, 1965–1967

configured in a variety of ways so as to provide a wide assortment of flexible
sitting positions, thus reflecting his primary goal in design – adaptability.
His most forward-looking designs, however, were his integrated micro-envi-
ronments. These included his *Visiona 1* habitat of the future shown at Bayer's
Visiona exhibition in 1969, which comprised a space age "Barbarella-like" interior where furnishings transmuted into structural elements and vice versa.
Traditional furniture items were replaced with functional units, such as the
Night Cell and *Central Living* blocks and the *Kitchen Box*, so as to create a
multi-functional living environment. For his own apartment, Colombo de-
signed the *Roto-living* and *Cabriolet Bed* units (both 1969), and these were
followed by his *Total Furnishing Unit* (1971), which was a highly influential ex-
ample of "Uniblock" design. Shown at the 1972 "Italy: The New Domestic
Landscape" exhibition at the Museum of Modern Art, New York, the *Total
Furnishing Unit* was proposed as a completely integrated machine for living.
Colombo also designed products for O-Luce, Kartell, Bieffeplast, **Alessi**,
Flexform and Boffi.

SAMUEL COLT

BORN 1814 HARTFORD, CONNECTICUT, USA
DIED 1862 HARTFORD, CONNECTICUT, USA

As a young seaman bound for Calcutta, Samuel Colt carved a wooden model that later became the most famous handgun ever designed. It took Colt several years, however, to develop his idea into a working design. His revolutionary multi-shot revolver, which featured a cartridge cylinder that revolved when the hammer was cocked, was also the first handgun to successfully use a percussion action. In 1835 he patented the design and the revolver was put into production by the Patent Arms Manufacturing Company, which he established in Paterson in 1836. Colt subsequently designed three types of revolver – pocket, belt and holster – as well as two rifles. In 1842 his factory closed because of insufficient orders, but production resumed five years later when the US Government ordered 1,000 revolvers for use in the Mexican War. In 1855 Colt opened a factory in Hartford which became the world's largest privately-owned armoury. Influenced by **Eli Whitney**, who had earlier pioneered the mass production of firearms, Colt developed a production-line system of manufacturing based on the standardization of components for interchangeability – 80 % of which were turned out by machine. Colt reputedly declared that "there is nothing that can't be produced by machine", and industrialized methods of production certainly enabled him to produce a remarkable 150 weapons a day by 1856. His revolvers were widely used during the American Civil War, and at the time of his death Colt had manufactured over 400,000 firearms. In the 1880s the Colt *Peacemaker* became legendary in the Wild West, while the Colt .45 semi-automatic pistol was American standard issue during the First and Second World Wars and

Samuel Colt's patent firearms factory, Hartford

remained the official sidearm of the American armed forces until 1984. Together with the National Institute of Justice, Colt is currently developing "smart gun" technology which will allow guns to be fired only by their registered keepers.

→Revolver, c. 1849

↘Advertisements
for the Colt auto-
matic pistol

Advertisement for
Colt's *Official Police*
.38 calibre revolver

COLT .. **The Dependence of Great National Institutions**

For 70 years, the infallibility of Brink's Express Service has been assured by Colt Fire Arms

Your treasures, too, Safeguarded

THIS Colt Official Police .38 calibre Revolver is a fine example of the unflagging care which, for 93 years, has signalized all Colt Fire Arms manufacture.

Its production entails more than a thousand separate steps, consisting of 564 machining operations, 124 hand operations and 322 visual and gauge inspections, culminating in Colt's expert target test.

Every part of each Colt Revolver and Automatic Pistol is forged from selected, flawless steels of superior tensile strength, machined to micrometer exactness; finished, fitted and assembled by hand.

The Colt Positive Lock embodied in Colt Revolvers makes accidental discharge impossible. It is upon these exclusive Colt features that nations, institutions and households place implicit trust.

Colt's Service Department will help solve shooting problems; suggest the Arm best suited to your needs; assist you in forming or joining a Revolver Club.

Colt's new Catalog No. 26 is the most complete Fire Arms Manual ever issued. Write for it;

COLT'S PATENT FIRE ARMS MFG. CO.
SMALL ARMS DIVISION
HARTFORD, CONN., U.S.A.
Phil. B. Bekeart Co., Pacific Coast Representative, 717 Market St., San Francisco

COLT .. *The* ARM OF LAW *and* ORDER

Now Completely Insured!

Of the several forms of insurance not the least valuable is the protection rendered by a Colt because it *prevents* loss.

Other forms can only partly restore *after* loss has occurred. Make *your* insurance complete with a Colt Revolver or Automatic Pistol.

Send for interesting booklet, "Romance of a Colt"

COLT'S PATENT FIRE ARMS MFG. CO.
Hartford, Connecticut, U. S. A.
Pacific Coast Representative
Phil. B. Bekeart Co., 717 Market St., San Francisco, Calif.

THE ARM OF LAW AND ORDER

CORNING

FOUNDED 1868
CORNING, NEW YORK, USA

Amory Houghton I

In 1864 Amory Houghton I (1813–1882) bought the Brooklyn Flint Glass Company, which he moved in 1868 to Corning, New York. Specializing in pressed glass products, the company was incorporated as Corning Glass Works in 1875. Two years later, Corning scientists attempted to improve the quality of optical lenses, which marked the company's first foray into scientific research. In 1880 Corning began manufacturing blanks for **Thomas Alva Edison**'s incandescent lamps, and in 1908 the Corning research laboratory was established.

The company subsequently developed heat-resistant borosilicate glass, which it named Pyrex, and in 1915 it added Pyrex-branded cookware to its product line. Owing to the popularity of its Pyrex wares, Corning expanded and in 1918 purchased the Steuben Glass Works. Corning continued to develop complex laboratory and scientific apparatus, including large observatory telescope lenses, and in 1935 began commercial production of glass fibres. In 1942 the company started producing cathode ray tubes for radar systems, and during the Second World War completed 174 research projects, including one for "ribbon glass" used in electrical components. Corning commenced automatic production of television bulbs in 1947, followed two years later by the centrifugal casting of television funnels. Colour television bulbs were produced from 1953, and in 1956 the highly successful *Corningware* range was launched – marking the first commercial application of *Pyroceram* glass ceramics.

Television panel and funnel, 1947

In 1980 a range of glass saucepans was introduced under the name of *Visions Cookware*, and since then Corning has pioneered glassware designs for both domestic and laboratory use, while conducting research into other glass applications. In 1994 Corning was awarded the United States National Medal for Technology for its development of six "life-

Cornflower casserole dish from Corning-ware range, 1958

changing" products, including the glass bulb for Thomas Alva Edison's electric lamp, the first mass-produced television tubes and its invention of optical fibres in the 1970s.

Silver Streak iron, 1942–1946 (Corning manufactured the glass shroud element only)

THOMAS CRAPPER

BORN 1836 THORNE, YORKSHIRE, ENGLAND
DIED 1910 LONDON, ENGLAND

Shower fitting,
c. 1902 (detail)

The development of modern sanitation ranks amongst the greatest triumphs of the Industrial Age and one of the areas of design that has benefited humankind the most. One of the most famous figures in this often-overlooked field was Thomas Crapper. Having served an apprenticeship with a plumber in Chelsea, Crapper founded his own plumbing business there in 1861 and began researching and developing new sanitary products. Such was his eventual reputation that in the 1880s his firm was employed to install around 30 lavatories at Sandringham House, the residence of the royal family. While Crapper filed a total of six patents relating to ventilated drains and lavatory mechanisms, he did not invent the flush WC (water closet). He did, however, manufacture a toilet with a syphonic flush known as *Crapper's Valveless Waste Water Preventer*, and his designs for sanitary engineering products were among the most advanced in their time. His manhole covers can still be found all over the south of England; there are even several in Westminster Abbey.

Pages from the
Thomas Crapper &
Company catalogue,
c. 1902

GUSTAV DALÉN

BORN 1869 STENSTROP, SWEDEN
DIED 1937 STOCKHOLM, SWEDEN

In 1906 Dr. Gustav Dalén became chief engineer at the Svenska Aktiebola-get Gas Accumulator Company, a supplier of acetylene gas. Here Dalén undertook much research on gases and turbines, which resulted in the improvement of hot-air turbine engines and the de Laval steam turbine. Dalén became managing director of the company in 1909 and subsequently invented Agamassan – a substance that absorbs acetylene allowing the gas to be concentrated without fear of explosion. In 1912 he won the Nobel Prize for Physics for his invention of the solventil, an automatic sun valve which uses sunlight to regulate a gas light-source so that it is turned off during daylight hours. This innovative design was later used internationally for the lighting of unmanned lighthouses and buoys. Tragically, Dalén was blinded in 1913 by an explosion that occurred during one of his experiments. While he was recuperating, he became aware of the problems his wife was encoun-

The
AGA COOKER
sets a new standard
in ECONOMY

Burning day and night the normal fuel consumption is 1¼ TONS per ANNUM. The stove requires stoking only once in 24 hours, but is always ready for cooking. The hot spots and ovens are automatically regulated to remain at the correct cooking temperatures.

Reliability is absolute

BELL'S GUARANTEE THE ACCURACY OF THESE CLAIMS and confidently invite your personal investigation at the Ideal Home Exhibition, Stand No. 105, Ground Floor.

Bell's
Asbestos
and Engineering
Supplies Limited

Bestobell Works,
Slough, Bucks.

The two-oven *Aga* cooker, 1990s

tering with her antiquated and highly temperamental cast-iron cooking range, which needed constant supervision to keep the temperatures from fluctuating. By applying the concept of heat storage, in 1922 Dalén invented a cooker that combined a small and efficient heat source, two large hotplates and two generous ovens into one robust and compact unit. Known as the *Aga* (taking its name from the initials of his company), Dalén's design achieved a high accumulation of heat that could be delivered in precisely controlled quantities, with a minimum of fuel, within a well-insulated environment. The heat generated was radiant, so that every cubic inch within each oven was utilized. The heat was also thermostatically controlled, which meant the elimination of all knobs and dials. Since the ovens and hotplates maintained different temperatures depending on their distance from the heat source, the *Aga* could roast, bake, steam, simmer, fry, stew, grill, boil and toast – all at once. *Agas* were first imported into the UK in 1929 and are now solely manufactured by Aga-Rayburn in England. All the principal castings for Aga cookers are made at the historic **Coalbrookdale** Foundry in Shropshire – birthplace of the Industrial Revolution. The remarkable Aga revolutionized the design of cookers; even today it has many loyal users who consider it the ultimate oven.

DANESE

FOUNDED 1957
MILAN, ITALY

Enzo Mari, *Timor* calendar, 1967

Enzo Mari, *Formosa* perpetual calendar, 1963

The design-led Milanese manufacturing company Produzioni Danese was founded by Bruno Danese and Jacqueline Vodoz in 1957 to produce tabletop consumer products manufactured in plastics. Seeking "perfect" objects, they commissioned freelance designers to create products with a high design integrity. These included the minimalist melamine and anodized aluminium *Cubo* ashtray (1957) by Bruno Munari (1907–1998) and the *In Attesa* wastebin by **Enzo Mari**, which had an inclined form to increase the probability of scoring a "goal" when throwing rubbish into it. In 1959 Edizioni Danese issued the results of its research into the "phenomena of visual perception" and later in the 1960s received considerable recognition for its combination of functionalism with sleek Italian styling, as demonstrated in Mari's *Formosa* (1963) and *Timor* (1967) everlasting calendars. The majority of Danese products have been designed by Enzo Mari, but other designers who have contributed to the company's range of objects for the home and office include Bruno Munari, Kuno Prey (b. 1958), Angelo Mangiarotti (b. 1921) and Marco Ferreri (b. 1958). In 1994 Danese was purchased by the prestigious Italian furniture manufacturer Alias, and since then has continued its non-conformist tradition of design research in co-operation with the Domus Academy Research Centre, Milan. Among the Danese range are "objects of beauty and spirit, aware of being the guardians of a noble history and tradition in Italian design", which testify to the company's long-held belief that design can be the "bearer of lasting value and innovation".

HUMPHREY DAVY

BORN 1778 PENZANCE, ENGLAND
DIED 1829 GENEVA, SWITZERLAND

In 1796 Humphrey Davy was apprenticed to an apothecary and surgeon with a view to pursuing a career in medicine. He began his scientific studies in earnest in 1797, establishing his own laboratory, where he undertook experiments with nitrous oxide (laughing gas). A year later, he became the chemical superintendent of the Pneumatic Institution in Clifton, where he researched the therapeutic nature of various gases, especially compositions of nitrogen. He also made several major scientific breakthroughs, including

Various examples
of the Davy Lamp,
1815

the isolation of both sodium and potassium and the discoveries that the production of electricity in simple electrolytic cells results from chemical action and that chemical combination occurs between substances of opposite charge. It was, however, his design of a miner's safety lamp that brought Davy the greatest public recognition. Miners had traditionally worked by candlelight or by light from other types of naked flame that could ignite the gas known as "firedamp" in deep mines, with disastrous results. Almost 100 lives were lost in this way in the early 19th century, prompting the Society for Preventing Accidents in Coal Mines to approach Davy to design a safer lamp. His lamp comprised a wick connected to an oil reservoir and a two-layer metal gauze chimney to surround and confine the flame and conduct the heat of the flame away. Davy was knighted in 1818 and awarded the Royal Society's Royal Medal in 1826.

ROBIN DAY

BORN 1915
HIGH WYCOMBE, BUCKINGHAMSHIRE, ENGLAND

In 1935 Robin Day won a scholarship to the Royal College of Art, London, and graduated in 1938. In 1948 he opened a joint design office with his wife, the textile designer Lucienne Conradi (b. 1917). Day worked as a freelance exhibition, graphic and industrial designer, and, at the 1948 "International Competition for Low-Cost Furniture Design" held at the Museum of Modern Art, New York, won first prize with Clive Latimer (b. 1915) for the design of wooden and tubular metal storage units. Shortly afterwards, he was commissioned by Hille International to design Modern furniture for the 1949 "British Industry Fair". In 1950 Day designed Hille's corporate identity and became its chief designer, while also continuing to consult to other companies, most notably John Lewis. Between 1962 and 1963, he developed an injection-moulded stacking chair, known as the *Polyprop*. This revolutionary design was among the first pieces of furniture to fully exploit the mass-manufacturing potential of thermoplastics. Obviously inspired by the earlier fibreglass-shell range of chairs by **Charles and Ray Eames**, the *Polyprop* was more successful in terms of sales because its higher volume method of production meant that it could be manufactured at a lower unit cost. Developed as a seating system with two standard shell types (armchair and side chair) and a wide variety of interchangeable bases, the *Polyprop* has multiple applications, from schools and church halls to football stadiums and airports. Since 1963 over 14 million *Polyprop* chairs have been sold, making it one of the most democratic products of the 20th century. Day believes that "the quality of design [is] important to the well-being of humanity".

Polyprop armchair for Hille International, 1962–1963

DE HAVILLAND

FOUNDED 1920
LONDON, ENGLAND

DHC-6 Twin Otter,
1965

Geoffrey de Havilland (1882–1965) was one of the great aviation pioneers. In 1920 he founded the de Havilland Aircraft Company, which manufactured the *Moth* (1924–1925) and later the famous twin-engine *Mosquito* fighter-bomber (1939–1940). De Havilland also manufactured the *Vampire* jet fighter (1942–1943) and developed the first civil jet transport with the *Comet*, which first flew in 1949 and began regular transatlantic services in 1958, the same year as the **Boeing** *707*. In 1947 a de Havilland subsidiary in Canada, originally founded in 1928, began building the *DHC-2 Beaver*. This legendary bush plane played ed a vital role in the development of remote regions in Northern Canada as well as other inaccessible areas, from Antarctica to equatorial Africa. In 1987 the *Beaver*, which was frequently configured with twin floats, was selected as one of Canada's top ten engineering achievements. Of the 1,692 models built between 1947 and 1967, approximately 1,000 are still in operation – a remarkable tribute to the strength of its design. In 1965 de Havilland introduced the *DHC-6 Twin Otter*, a larger utility bush plane that also went on to carve out a distinct niche in aviation with its exceptional performance and reliability. Since 1997 de Havilland has been a subsidiary of **Bombardier** Inc.

DHC-2 Beaver, 1947

MICHELE DE LUCCHI

BORN 1951
FERRARA, ITALY

Michele De Lucchi studied at the Liceo Scientifico Enrico Fermi in Padua and later trained as an architect under Adolfo Natalini (b. 1941) at Florence University, graduating in 1975. In 1973, together with Piero Brombini, Pier Paola Bortolami, Boris Pastrovicchio and Valerio Tridenti, he founded the architecture and design group Cavart, which promoted a radical design agenda through happenings, publications and seminars. He also collaborated with Superstudio, **Ettore Sottsass** and Gaetano Pesce (b. 1939) and taught architecture at Florence University from 1975 to 1977. He then moved to Milan and began working as a design consultant to the **Kartell** in-house design studio, Centrokappa. De Lucchi later assisted Ettore Sottsass with the planning of the first Memphis exhibition. In 1979 he designed several Post-Modern prototypes of domestic electrical appliances for Studio Alchimia and became a consultant to **Olivetti**. He co-founded Memphis in 1981 and was responsible for the introduction of geometric motifs on the plastic laminates used by the co-operative. In 1986 he founded Solid, a Milan-based design group, and also began teaching at the Domus Academy, Milan. In the early 1990s he established the De Lucchi Group and has since worked widely in

Tolomeo wall light for Artemide, 1995

Japan and Germany. Increasingly, his designs have become more suited to industrial production and his clients have included **Artemide**, Kartell, Bieffeplast, Mandarina Duck and Pelikan. He has won numerous design prizes, including Good Design awards (Japan), Die Gute Form and Deutsche Auswahl awards (Germany) and a Compasso d'Oro (Italy). De Lucchi regards design as a means of communication, a sentiment that has informed his work throughout his career – from contesting young radical to established international industrial designer.

JC DECAUX

FOUNDED 1964
LYONS, FRANCE

Jean-Claude Decaux

In 1955 Jean-Claude Decaux (b. 1937) started up a company specializing in motorway billboard advertising, but after the French government slapped high taxes on this form of publicity in a law passed in 1964, he turned his attentions to urban street advertising. His innovative concept was to provide bus shelters, which would be wholly funded by high-quality advertising, free-of-charge to city councils.

His first bus shelter design was tested in Lyons with the mayor's approval and soon found its way into other French cities, such as Grenoble, Angers and Poitiers. Decaux initially found it difficult to convince advertisers of the benefits of this smaller-format publicity medium, but in the early 1970s his idea finally caught on and the company expanded rapidly, with the *Abribus Standard* shelter (1970) becoming a common sight in French towns and cities.

Jean-Claude Decaux,
SPEA public lavatory,
1980

Decaux also designed street information panels known as *MUPI* (Mobilier urbain pour l'information) in 1972 and *PISA* (Point d'information service animé) in 1976. He offered local authorities free municipal information displays in return for permission to erect *MUPI*'s and bus shelters. In 1980 Jean-Claude Decaux designed an automatic entry public lavatory known as the *SPEA* (Sanitaire public à entretien automatique) and, a year later, introduced the first illuminated news boards. Using the highest quality materials and instituting a rigorous maintenance program, JC Decaux became a leader in the design and manufacture of street furniture, exporting its products worldwide. During the 1980s and 1990s, the company commissioned leading architects and designers such as **Mario Bellini, Philippe Starck, Norman Foster, Knud Holscher,** Martin Székély (b. 1956) and **Ferdinand Alexander Porsche** to design bus shelters, public toilets, information boards, signage and street kiosks. Impressively, the company manages to progress from the initial design stage of a product's development to final installation in just twelve months. Mindful of the aesthetic impact of street furniture on our shared environment, Decaux is particularly adroit at producing large ranges of designs that are stylistically suited to different cultural settings.

Jean-Claude Decaux,
Abribus Standard
shelter, 1970

JOHN DEERE

BORN 1804 RUTLAND, VERMONT, USA
DIED 1886 MOLINE, ILLINOIS, USA

→ *D* tractor, 1923, the first Waterloo tractor to bear the name of John Deere (produced 1924 to 1953)

John Deere served a four-year apprenticeship as a blacksmith in Vermont, before working as a journeyman. He soon gained considerable fame for his careful workmanship and ingenuity. His highly polished hay forks and shovels were in especially high demand throughout Western Vermont. Tempted by tales of a better life in the West, in 1836 Deere moved to Grand Detour, Illinois, where he entered into a business partnership with Major Leonard Andrus. Deere quickly discovered that the cast-iron ploughs originally intended for the light sandy soils of New England could not cope with the clod-like earth of the Mid-West that the pioneers were trying to farm. Having to frequently repair their damaged cast-iron ploughs, Deere became convinced that the problem could be overcome by the addition of a highly polished mouldboard that, if shaped properly, would scour itself clean while turning the furrow slice. In 1837 Deere designed the first commercially successful self-scouring steel plough and began manufacturing the "self-polishers" to sell to local farms. This landmark design, which transformed farming in the West, was followed by another 50 improved models over the next two years. In 1843 Deere managed to arrange the importation of steel from England, and by 1846 his business was manufacturing around 1,000 ploughs per annum. He subsequently established a factory in Moline, Illinois, which by 1857 was producing 10,000 ploughs a year. Deere constantly changed and

Self-scouring steel plough, 1837

improved the design of his ploughs, declaring: "If we don't improve our product, somebody else will." The company was incorporated as Deere & Company in 1868 and the following year John Deere's son, Charles, succeeded him as president. In 1918 the company purchased the Waterloo Gasoline Traction Engine Company in Waterloo, Iowa, and tractors became an important part of the John Deere line. Deere & Company introduced its first tractor, the *Model D*, in 1923, and over the succeeding decades the company diversified its product-line to include cultivators, harvesters, balers and other engine-powered agricultural machinery, as well as construction equipment such as backhoes and loaders. Deere & Company is now the largest farm equipment manufacturer in North America.

↑ *GP* tractor, 1928 (produced until 1935)

MR tractor, 1992

CHRISTIAN DELL

BORN 1893 HANAU, GERMANY
DIED 1974 WIESBADEN, GERMANY

During his apprenticeship as a silversmith at Schleißner & Söhne in Hanau, Christian Dell also studied at the Königliche Preußische Zeichenakademie. From 1911 to 1912 he worked as a silversmith in Dresden before training under Henry van de Velde (1863–1957) at the Kunstgewerbeschule in Weimar from 1912 to 1913. After the First World War he worked as a journeyman and later as a master silversmith for Hestermann & Ernst in Munich. In 1920 he joined the silver workshop of Emil Lettré (1876–1954) in Berlin, establishing his own silver studio a year later in Hanau. From 1922 to 1925 he headed the metal workshop at the Weimar **Bauhaus**, and then taught for seven years at the Frankfurter Kunstschule, where he designed silverware that was made in the school's workshop. During this period, Dell began producing designs that were suitable for industrial production. His lights, such as the adjustable

↘ *Das Frankfurter
Register 1* brochure,
1928

Rondella desk lamp
for Rondella,
1927–1928

Rondella desk lamp (1927–1928) and his *Idell* range, which was later copied by Helo, were mass-produced by Rondella and Kaiser respectively. Dell designed over 500 lights in total and his *Idell* range remained in production for over 60 years. In the 1930s Dell began experimenting with plastics and in 1939 established his own jewellery business in Wiesbaden.

DESIGN COUNCIL

FOUNDED 1944
LONDON, ENGLAND

The first issue of *Design* magazine, 1949

In 1944 the British Board of Trade announced the founding of the Council of Industrial Design "to promote by all practicable means the improvement of design in products of British industry". Two years later, the CoID organized the "Britain Can Make It" exhibition to re-invigorate British design after the war so as to boost foreign exports. This successful exhibition attracted over one and half million visitors and did much to raise the profile of design in Britain. In 1949 the CoID launched *Design* magazine to promote "good design" by addressing professional designers, manufacturers, company buyers and managers. The council was later responsible for the selection and arrangement of the 10,000 exhibits at the Festival of Britain in 1951. In 1956 the CoID's Design Centre was opened in the Haymarket, London, to provide "a shopping guide to well-designed British goods for those who want to buy well and save time". This idea was advanced a year later by the appearance of the Design Council Awards and, in 1958, by the introduction of the well-known black and white kitemark. In 1972 the CoID was re-named the Design Council and in 1976 was granted a royal charter. The Design Council was restructured in 1994 and subsequently helped the government modify the national curriculum, so as to place more emphasis on the teaching of design and technology in schools.

Design Council kitemark, first introduced in 1958

DESIGN RESEARCH UNIT

FOUNDED 1943
LONDON, ENGLAND

Corporate identity
for British Rail, 1965

Exhibition sales
caravan for Olivetti,
c. 1956

In 1922 Milner Gray and Charles and Henry Bassett founded the Bassett-Gray Group of Artists and Writers, one of the first professional design practices in Britain. Collaborating with artists such as Graham Sutherland (1903–1980), Bassett-Gray represented "a body of artists who design for industrial and commercial purposes". In 1934 Bassett-Gray was joined by **Misha Black**. The following year it was reorganized and renamed the Industrial Design Partnership, even though at this time the practice was still primarily focused on graphic, packaging and exhibition design. Following the outbreak of the Second World War, the IDP was shut down and both Gray and Black joined the Ministry of Information. Realizing that there would be a huge demand for consumer goods once wartime restrictions had been lifted, in 1942 Marcus Brumwell (1901–1983) and the design theorist Herbert Read (1893–1968) proposed the idea of forming a collective of designers and architects who could produce designs for large-scale industrial production after the war. Milner Gray, who was by then head of exhibition design at the Ministry of Information, was consulted on the organization of such a group. The Design Research Unit was subsequently established in January 1943 and from the outset emphasized industrial product design. Both Black and Gray joined the DRU, the former originating the majority of its product designs – from televisions and electric heaters to cameras and trains – while Gray concentrated chiefly on packaging and signage. Dorothy Goslett, who helped administrate the office, published a book in 1961 entitled *Professional Practice for Designers*, which was in essence a guide to running a design consultancy. "Designers," she wrote, "have to deal directly with business men who are seldom likely to be sentimental about creative work. ... The relationship between designer and client is therefore strictly a business one. The designer's codes of professional conduct are the foundations on which this relationship is built." As the highest-profile design office in post-war Britain, the DRU did much to elevate the professionalism of industrial design practice and provided "an effective design service for industry" by giving concrete expression to Modern design theory.

Ilford *Pixie* camera,
c. 1960s

DONALD DESKEY

BORN 1894 BLUE EARTH, MINNESOTA, USA
DIED 1989 VERO BEACH, FLORIDA, USA

Packaging design for
Procter & Gamble,
c. 1951

Donald Deskey studied architecture at the University of California, Berkeley, as well as fine art at the Arts' Student League, New York, and the Institute of Chicago's School of Art. He later trained in Paris at the École de la Grande Chaumière, the Académie Colarossi and the Atelier Léger. After the First World War he worked as a graphic designer for a Chicago advertising agency, but a visit to the 1925 Paris "Exposition Internationale des Arts Décoratifs et Industriels" prompted him to turn his attention to three-dimensional design. He subsequently produced a number of screens for Saks Fifth Avenue in 1926 and a year later designed window displays for this New York department store as well as for Franklin Simon. Also in 1927, he established the Deskey-Vollmer partnership with Philip Vollmer, which concentrated on exclusive metal furniture and lighting, but this venture was dissolved in the early 1930s. During the late 1920s, he developed a stained-wood laminate

Filing cabinet for
Globe-Wernicke,
c. 1951

known as Weldtex, while in 1931 he designed Moderne interiors for the Radio City Music Hall – his most prestigious and influential commission, which exemplified the American Art Deco style.

During the 1930s he executed a variety of industrial designs, such as washing machines, vending machines and even a rack for bowling balls, which were exhibited at the Metropolitan Museum of Art in 1934. From the late 1930s until 1975, he was the principal of Donald Deskey Associates and designed everything from graphics and packaging for Procter & Gamble to printing presses for American Type Founders. His design work was characterized by streamlined forms and innovative planning. His sleekly styled filing cabinet for the Globe-Wernicke Company, for instance, had several novel features including a combined drawer pull/identification panel and drawer fronts with contoured edges. While chiefly remembered as one of the greatest proponents of the Art Deco style in America, Deskey was also an important pioneer of industrial design consulting.

Deutscher Werkbund emblem

DEUTSCHER WERKBUND

FOUNDED 1907
MUNICH, GERMANY

Founded in October 1907, the Deutscher Werkbund attempted from its outset to reconcile artistic endeavour with industrial mass production. Its founding members not only included designers such as Richard Riemerschmid (1868–1957), Bruno Paul (1874–1968), **Peter Behrens** and Josef Maria Olbrich (1867–1908), but also a dozen established manufacturers, including Peter Bruckmann & Söhne and Poeschel & Trepte, and design workshops such as the Wiener Werkstätte and the Munich-based Vereinigte Werkstätten für Kunst im Handwerk. Peter Bruckmann (1865–1927) was appointed the association's first president and within a year its membership had risen to around 500. From 1912 the Werkbund began publishing its own yearbook, which included illustrations of and articles on its members' designs, such as factories by **Walter Gropius** and Peter Behrens and cars by Ernst Naumann. The yearbook also listed members' addresses and areas of specialization in an attempt to promote collaboration between art and industry. In 1914 the Werkbund organized a landmark exhibition in Cologne, which included within its grounds Walter Gropius' steel and glass model factory, Bruno Taut's Glass Pavilion and Henry van de Velde's Werkbund Theatre. A year later, the Werkbund's membership had swollen to almost 2,000. The increasing divergence between handcraftsmanship and industrial production fuelled heated debate within the Werkbund, however, with members such as Hermann Muthesius (1861–1927) and Naumann urging for standardization, while others such as van de Velde, Gropius and Taut argued for individualism. This conflict almost led to the disbanding of the association. The widespread need for consumer products after the devastation of the First World War, however, led Gropius to accept the necessity for standardization and industrial

Fritz Hellmut Ehmcke, tobacco box for the Deutscher Werkbund exhibition in Cologne, 1912–1914

AEG stand in the main hall of the 1927 Deutscher Werkbund "Die Wohnung" exhibition in Stuttgart

production, although other members, such as Hans Poelzig (1869–1939), continued to resist change. From 1921 to 1926 Riemerschmid was president of the Deutscher Werkbund and during his tenure the functionalists' approach to design was advanced. In 1924 the Werkbund organized the exhibition *Form ohne Ornament* (*Form without Ornament*). The catalogue illustrated industrially-produced designs and promoted throughout its text the virtues of plain undecorated surfaces and, ultimately, functionalism. In 1927 the Werkbund staged a unique exhibition in Stuttgart, entitled "Die Wohnung" (The Dwelling), which was organized by Ludwig Mies van der Rohe (1886–1969). Although the focus of the exhibition was the Weißenhofsiedlung, a housing estate project for which the most progressive architects throughout Europe were invited to design buildings, "Die Wohnung" also acted as an important showcase for industrial design, with stands exhibiting the latest products from companies such as **AEG**. The modern tubular metal furniture designed by Mies van der Rohe, Mart Stam (1899–1986), **Marcel Breuer**, Le Corbusier (1887–1965) and others, which was used to furnish the interiors of the specially commissioned houses, was also widely publicized and revealed the increasing internationalism of the Modern Movement. Through this landmark exhibition, the Deutscher Werkbund succeeded in achieving a greater acceptance of Modernism both at home and abroad. The Werkbund was eventually disbanded in 1934, and though re-established in 1947, was by then a spent force. Through its activities and in particular its development of the concept of "types" – standardized designs that could be easily assembled from industrially produced components – the Deutscher Werkbund heralded the advent of large-scale industrial production and in so doing had an enormous impact on the evolution of German industrial design.

NIELS DIFFRIENT

BORN 1928
STAR, MISSISSIPPI, USA

Niels Diffrient studied aeronautical engineering at the Cass Technical High School, Detroit, and later trained at the Cranbrook Academy of Art, Bloomfield Hills, Michigan and at Wayne State University, Detroit. He worked in **Marco Zanuso**'s Milan-based design studio from 1954 to 1955 and from 1946 to 1951 in the office of Eero Saarinen (1910–1961). In 1952 he joined **Henry Dreyfuss** Associates, becoming a partner in 1956. While there, he helped to compile data on anthropometrics which was eventually published in the three influential volumes entitled *Humanscale*. He also designed aircraft interiors as well as computers and X-ray equipment. In 1981 he founded Niels Diffrient Product Design in Ridgefield, Connecticut and has since designed ergonomically-conceived office systems furniture for Knoll and for Sunar-Hausmann. With his mastery of ergonomics, Diffrient is able to promote a highly resolved physical interaction between object and user. Regarding design as a "practical art", Diffrient attempts to provide superlative physical function in the most aesthetically pleasing way possible.

Diffrient chair, office seating system for Knoll International, 1979–1980

CHRISTOPHER DRESSER

BORN 1834 GLASGOW, SCOTLAND
DIED 1904 MULHOUSE/ALSACE, FRANCE

→Electroplated bowl
for Elkington & Co.,
1885 (this design
was roughly
sketched in 1864
and illustrated in C.
Dresser's *Principles
of Design* published
in 1873)

Christopher Dresser studied under the botanist John Lindley at the Government School of Design in London from 1847 to 1854 and subsequently lectured there for 14 years. Dresser's advocacy of "Art Botany" – the stylized yet scientifically based depiction of nature – helped replace the overblown and false naturalism common to the High Victorian style with a more formalized type of ornamentation. In 1857 he was appointed professor of "Botany Applied to the Fine Arts" at the School of Design, South Kensington. He also published various articles in the *Art Journal* and three books, which helped establish his reputation and contributed to his receiving an honorary doctorate from the University of Jena.

In 1860 he abandoned botany and embarked on a career in design. He established his own studio, which from the beginning was enormously successful. Dresser supplied designs for metalware, ceramics, glass, tiles, textiles, wallpapers and cast-iron furnishings to at least 30 of the most eminent British manufacturers, including Elkington & Co., **Minton** & Co., **Coalbrookdale** Co., Benham & Froud, James Dixon & Sons, Hukin & Heath and Linthorpe Art Pottery. His electroplated metalware designs are especially noteworthy for their geometric simplicity, which prefigured the formal vocabulary of the Modern Movement. These wares demonstrate what Dresser described as the "breadth of treatment, simplicity of execution, and boldness of design" of Japanese applied art. While Dresser was one of the first Europeans to promote the qualities of the art of Japan, his forward-looking designs also reflected his belief in the efficacy of industrial production and his pursuit of "Truth, Beauty, Power". As a leading design reformer and theorist of the 19th century, Dresser was one of the first professional practitioners of industrial design.

→Tea service for
Watcombe Terracotta
Company, c. 1870

his office. His large corporate clientele included AT&T, American Airlines, **Polaroid**, **Hoover** and **RCA**. His designs were characterized by the use of sweeping sculptural forms, and as such exemplified streamlining in American design. Like **Raymond Loewy**, Norman Bel Geddes and **Walter Dorwin Teague**, Dreyfuss re-styled many products for manufacturers so as to increase consumer demand through stylistic rather that technical innovation. Some of his designs bore a facsimile of his signature – an early example of designer labelling. Dreyfuss was a founder member of the Society of Industrial Design and the first president of the Industrial Designers Society of America. He was also a long-term faculty member of the engineering department at the California Institute of Technology. Dreyfuss' greatest contribution to industrial design practice, however, was his research into **anthropometrics** – the find-

← *Convair* car/plane
for Consolidated
Vultee Aircraft, 1947

← *Convair* car/plane
for Consolidated
Vultee Aircraft, 1947

ings of which were published in his influential books, *Designing for People* (1955) and *The Measure of Man* (1960). He also published a sourcebook of international symbols, which acknowledged the communicative power of symbols over words.

20th Century Limited locomotive for New York Central Railroad, 1938

DUCATI

FOUNDED 1926
BOLOGNA, ITALY

Ducati bicycle with
Cucciolo motor,
c. 1946

→Ducati *MY 916
Biposto* motorbike,
1998

→ *M900 Monster*
motorbike, 1993

↘*750 GT* motorbike,
1971 – the first hy-
persport production
motorbike

Ducati *Mariana 125 –*
first design by
Taglioni, c. 1955

The Società Radio Brevetti Ducati initially manufactur-ed radio components that were based on patents filed by Adriano Ducati (1903–1991). The company grew and in 1935 built its first factory. After the Second World War, the Ducati brothers developed a small auxiliary bicycle motor known as the *Cucciolo* (1946), and when it later acquired a frame, the first Ducati motorcycle was born. In the early 1950s Ducati began producing motorbikes such as the futuristic 175 cc *Cruiser* (1952). In 1955 the company was joined by the brilliant engineer-designer Fabio Taglioni (b. 1920), who had already designed several motorbikes that were notable for their originality. At Taglioni's insistence, the high-performance designs he produced for Ducati were tested in long-distance races such as the Milan-Taranto and the Giro d'Italia. Taglioni began developing the desmodromic system in 1955, but it was not used in a production bike until 1968, when the *450 Mark 3D* appeared. Ducati's *Scramblers* found much success in America, and with the late 1960s maxibike boom, Taglioni designed the twin-cylinder desmodromic *750*, which won the 1972 Imola 200 Miglia. Developed from this winning design, the *750 Super Sport* was later taken up to a 900 cc dis-placement and won the 1978 Formula 1 TT. During the 1980s – the Super-bike era – Ducati expanded its market by introducing new high-performance models with distinctive styling. In 1993 Miguel Galluzzi (b. 1959) created the legendary *Monster* and a year later the *916* was born. Combining "functional-ity and harmony of appearance, logic and emotion", it was subsequently nominated "Motorcycle of the Year".

JAMES DYSON

BORN 1947
NORFOLK, ENGLAND

DCo1 cleaner, 1993

While training in furniture and interior design at the Royal College of Art, London (graduation 1970), James Dyson designed the *Sea Truck* for the inventor Jeremy Fry, which won a **Design Council** Award and the Duke of Edinburgh's special prize in 1975. After working as a designer for the Marine Division of Rotork for four years, Dyson left to develop the *Ballbarrow*. This innovative and better-performing barrow, which used a ball instead of a wheel, became a highly successful product. Dyson did not profit greatly from the *Ballbarrow*, however, because he sold his interest in it to fund the development of a bagless vacuum cleaner. The concept of a bagless vacuum cleaner first captivated Dyson in 1978 when he realized that, when in use, even new bags quickly become clogged with a fine layer of dust, which dramatically and increasingly reduces suction. Between 1979 and 1984, Dyson built 5,127 prototypes of his revolutionary vacuum cleaner, which used centrifugal force to lift dust and separate it from the air, just like the much larger cyclone towers used in saw-mills and spray-paint shops to remove hazardous particles from the atmosphere. By using a cyclonic method of suction and eliminating the bag, Dyson was able to create a vacuum cleaner that provided a high level of constant suction. In 1983 his first prototype, the pink and lilac *G-Force*, was featured on the cover of *Design* magazine and in an article entitled "Form Follows Fluff". When attempting to commercialize his cleaner, however, Dyson discovered that the established vacuum cleaner manufacturers did not want to invest in his invention as it spelt the end of the lucrative planned obsolescence of the replacement bag market. In 1985 Dyson took the *G-Force* to Japan and found a licensee to make it. Six years later, the cleaner won the International Design Fair prize in Japan. Having also eventually managed to find licensees in North America, Dyson was able to establish his own research centre and

DC02 cylinder
cleaner, 1995

factory in Chippenham in 1993 with the royalties generated from overseas sales. Launched the same year, the famous yellow and grey Dyson *DC01* (*Dual Cyclone*) became the best-selling vacuum cleaner in Great Britain within two years, despite its relatively high price. Every aspect of Dyson's design was driven by logic, such as the transparent collecting container, which allows one to see when it is full. The *DC01*'s aesthetic also distinguished it from other cleaners on the market. As Dyson put it: "we wanted it to look like a piece of **NASA** technology. Its superior performance had to be visible. It had to look the business." For its marketing, Dyson adopted a "less is more" approach with bold no-nonsense advertisements that directly communicated the design's benefits with simple slogans such as "say goodbye to the bag" and "100% suction, 100% of the time". In 1995 Dyson launched a cylinder model – the Dyson *DC02* – and opened a larger manufacturing facility in Malmesbury. The following year, the *DC01* was selling at the phenomenal rate of 32,000 units per month, five times that of its nearest competitor, demonstrating clearly that the public were more than prepared to pay a premium price for a better-performing design. In 1996 the exhibition "Doing a Dyson" was staged by the Design Museum in London, tracing the development of the Dyson cleaner, and the following year Dyson became the first British company to be awarded a European Design Award.

CHARLES & RAY EAMES

BORN 1907 ST LOUIS, MISSOURI, USA
DIED 1978 ST LOUIS, MISSOURI, USA

BORN 1912 SACRAMENTO, CALIFORNIA, USA
DIED 1988 LOS ANGELES, CALIFORNIA, USA

Charles and Ray Eames did more to change the public perception of Modern design than just about anyone else in the 20th century. With their mission "To get the most of the best to the greatest number of people for the least", the Eameses had an enormous impact on industrial design practice, producing throughout their careers highly successful and extraordinarily innovative work that exemplified the timeless ideals of good design. The achievements of Charles and Ray Eames continue to be revered today, as they are increasingly recognized as having single-handedly changed the course of design. After training and teaching at the Cranbrook Academy of Art in Bloomfield Hills, Michigan, Charles Eames and Eero Saarinen (1910–1961)

PAW swivel chair for
Herman Miller, 1950

– assisted by Ray Kaiser – submitted entries to the "Organic Design in Home Furnishings" competition held at the Museum of Modern Art, New York, in 1940. Their prize-winning seating designs incorporated two state-of-the-art manufacturing techniques – the compound moulding of plywood into complex curves and the bonding of wood to metal by means of electronic cycle-welding, a process that had recently been pioneered by **Chrysler.** Through the foam-lined monocoque seat shells of these chairs, Eames and Saarinen also advanced the revolutionary idea of continuous contact and support, thus heralding a totally new direction in Modern furniture design. Although wartime restrictions on materials and processes meant this highly innovative seating programme was not manufactured, it informed all of the Eameses subsequent chair designs. In 1942 the Eameses were commissioned by the US Navy to design plywood leg and arm splints and a plywood litter. They established their own company to manufacture these products, which were their first to be mass-produced. In 1945 the Eameses designed a series of compound-moulded plywood children's furniture, which was produced by the Evans Products Company. This was soon followed by their famous series of plywood chairs (1945–1946), which were initially manufactured by Evans and then later by **Herman Miller**. Employing a system of standardized interchangeable components, these revolutionary chairs were the first pieces of furniture with compound-moulded elements to be truly mass-produced. In 1948 the couple proposed a series of moulded fibreglass chairs, which won second prize at the Museum of Modern Art's "International Competition for Low-Cost Furniture Design". Their designs led on to the innovative seating programme which they developed with Herman Miller, Zenith Plastics of Gardena, California, and the engineering department of the University of California, Los Angeles, between 1948 and 1950, and which was some of the earliest unlined plastic seat furniture to be mass-produced. Based on the concept of a universal seat shell that could be used in conjunction with a variety of interchangeable bases to provide numerous variations, this product range, which was manufactured by Herman Miller, was extremely influential. The Eameses slightly later *ESU* (Eames Storage Units) series of 1950 – a modular system of standardized steel frames with either plastic-coated plywood or lacquered Masonite panels, sliding door and drawer options – offered similar functional flexibility and could be configured into room dividers, cupboards, shelves and desks. The *ESU* series bears a passing resemblance to the Eameses' own house (1949), which was largely constructed of pre-fabricated components. The Eames house was designed as part of the Case Study House program sponsored by the journal *Arts & Architecture*, and as such speculated on the idea of product architecture – buildings as industrially produced consumer products. Through their work, which also included film-making, photography and exhibition design, the Eameses skilfully and eloquently communicated the values of appropriateness, social morality, egalitarianism, optimism, informality and essentialism. While balancing the poetic with the pragmatic, the Eameses above all demonstrated in practice how cutting-edge materials and technology can be harnessed to create high-value yet accessible Modern products.

Harley Earl, Cadillac
62 Sedan de Ville,
1956, photographed
with a Douglas *A4
Skyhawk* which was
put into production
the same year

HARLEY EARL

BORN 1893 LOS ANGELES, CALIFORNIA, USA
DIED 1969 WEST PALM BEACH, FLORIDA, USA

Harley Earl's father established the Earl Carriage Works
in 1908, but two years later renamed the venture the
Earl Automobile Works to reflect the transition from
horse-drawn to motorized vehicles. His business involved customizing cars
for film stars and building chariots for use in films. In 1914 Harley Earl began studying at Stanford University, but this was soon abandoned so that
he could train as a designer at his father's auto works. After a period of time,
he began designing luxurious cars for the Hollywood glitterati, including the
soon-to-be-disgraced actor Roscoe "Fatty" Arbuckle and the legendary director Cecil B. De Mille. In 1919 the business was purchased by Cadillac's West
Coast distributor, Don Lee, and the works were subsequently devoted entirely to car customization. Around this time, Harvey Earl devised an innovative modelling technique using clay that promoted greater sculptural expres-
siveness, which he used as a styling aid and which later became standard
practice throughout the car industry. Whereas, in the early years of the car
industry, manufacturers only had to worry about their ability to meet demand, by the early 1920s, with productivity no longer an issue, the market
was growing more competitive. The chairman of **General Motors**, Alfred

La Salle, with Harley
Earl (seated) and
L. T. Fisher, 1927

Cadillac *Eldorado*,
1959

Sloan, realised that aesthetics would play an increasingly important role, and in 1925 he invited Earl to Detroit. Earl's first brief for GM was to bring "something" to the new La Salle brand, which was pitched in-between the well-heeled Buick and luxurious Cadillac models. The resulting 1927 La Salle was the first mass-produced car to be designed by a stylist rather than by an engineer. Sloan realized that the company would profit if it could produce new cars each year that differed from the previous year's model. This idea of annual cosmetic changes to promote stylistic obsolescence led to the creation of General Motors' Art & Color Section in 1928. Earl became supervisor of this unprecedented styling department, which was re-named the Style Section in 1937. He oversaw the development of the very first concept car, the 1938 Buick *Y-Job*, which had the first two-tone paintwork, wrap-round windscreen and "fishtail" rear fender. Earl was also responsible for convincing GM to build a sports car, which resulted in the classic 1953 Chevrolet *Corvette*. Earl's styling, however, became increasingly extreme – with cars becoming lower, longer and more chrome-laden – culminating in the extraordinary 1959 Cadillac *Eldorado* with rocket-ship tail fins. Earl's brand of exaggerated styling was highly influential upon the automotive industry, and his contribution is still remembered at GM in the immortal lines: "Our father who art in styling ... Harley be thy name!"

ECCO DESIGN

FOUNDED 1989
NEW YORK, USA

Beepwear watch/
pager for Motorola/
Timex, 1999

For Women Only
hairbrushes for
Goody, 1997

Ecco Design was founded by Eric Chan in 1989 and has since expanded into a creative team that specializes in the design of consumer products, furniture and computers as well as interactive multimedia, telecommunications, sporting and medical equipment. It has won numerous prizes, including five Good Design awards. The companies it has worked for include **Herman Miller**, Colgate, Bausch & Lomb, Timex and Motorola. Ecco Design states that: "As designers, we challenge ourselves to utilize today's technology to create products that bring harmony between man, nature and society in an increasingly global and multicultural market. We are most interested in translating complicated technology into products that go beyond mere function, that are friendly, understandable, and sensi-

LS550 numeric pager for Motorola, 1998

tive to people's needs." Indeed, Ecco Design's products are often distinguished by a tactility that expresses the inner qualities of the materials it has used, for example, the over-sized rubber handles of the *Grip 'Ems* children's toothbrushes for Colgate (1999) and the softly formed translucent plastic handles of the *For Women Only* hairbrushes for Goody (1997), winner of *ID Magazine*'s Design Distinction award. In 1993 Ecco Design developed the *SolarBlind*, a prototypical venetian blind that absorbs solar energy during the day and emits light in the evening. Ecco Design believes that "the designer's task is to mediate between people and object, poetry and logic, technology and nature".

THOMAS ALVA EDISON

BORN 1847 MILAN, OHIO, USA
DIED 1931 WEST ORANGE, NEW JERSEY, USA

Edison's original carbon filament lamp manufactured by the Corning Glass Works, 1879

→Advertising display light-bulb box made by the Edison Lamp Works, General Electric, 1923

Drawing showing the elements of a light bulb, c. 1880

Thomas Alva Edison pioneered some of the greatest and widest-ranging inventions of the 20th century, yet as a child his education was hampered by dyslexia and severe hearing problems. He was eventually removed from school and taught instead by his mother, who was a former schoolteacher. She gave him an elementary science book which outlined experiments that could be done at home. At the age of ten, Edison was allowed to set up his own science laboratory in the basement of his home. At the age of twelve, he began working as a trainboy for the Grand Trunk Railway and moved his laboratory into a baggage car so that he could do experiments during the layover periods in Detroit. The laboratory soon caught fire, however, and Edison lost his job. After this, Edison sold newspapers along the railroad and reputedly saved the life of a station official's son. The father was so grateful to Edison that he taught him how to use a telegraph. He later work-

ed as a telegrapher in Toronto and, having to send a signal every hour, Edison devised a transmitter and receiver that could automatically telegraph a message even when he was asleep. He later moved to New York and began working on other inventions including a telegraph machine that could send multiple messages, an electric vote recorder and an improved stock market ticker, the rights to which he sold to the Gold & Stock Telegraphy Company for a massive $40,000 in 1870. By now wealthy, Edison established a workshop in Newark, New Jersey, to manufacture stock tickers, high-speed printing telegraphs and an improved version of Christopher Sholes' typewriter that was the first model suitable for commercial use.

There's a Right Lamp
For Every Purpose

Thomas Alva Edison

In 1876 he built an "invention factory" in Menlo Park, New Jersey, and subsequently spent the majority of his time experimenting in this new science laboratory that had 60 employees. He worked at the laboratory for a decade and often had as many as 40 projects running at once. During this period, Edison was applying for up to 400 patents per annum. The numerous inventions conceived at Menlo Park and subsequently transformed into landmark products included the wireless telegraph (1875), the memograph (1876), an improved carbon-button telephone transmitter (1877), the phonograph (1877–1878), the incandescent electric light bulb (1878–1880) and the wireless induction telegraph (1885). Of these, the phonograph and the electric light bulb must be considered his greatest triumphs. The phonograph evolved from his research into telegraphy and was the first-ever recording machine. By developing a device that indented a strip of paper to record Morse Code, Edison observed that when the strip was run quickly, it emitted distinctive noises. The concept was adapted to sound recording by using a stylus that vibrated when exposed to sound, which in turn indented tin foil wrapped around a rotating drum. His incandescent electric light bulb was the result of many years of research and some 1,200 experiments. Using carbonized filaments made from burnt cotton thread, Edison's light bulbs could burn for up to 48 hours and were first used on the steamship *Columbia*. In 1878 he founded the Edison Electric Light Company to develop and eventually market this life-changing invention – the firm was incorporated as the Edison General Electric Company in 1892 and later became known as **General Electric**. In 1882 Edison established the first "electric light-power station", which enabled New York to become the first city to be illuminated by electricity. Another far-reaching discovery made by one of Edison's engineers in 1883 led to the development of the electron tube, which helped establish electronics as a completely new branch of science. Having founded an enormous new laboratory for the "business of inventing" in West Orange in 1887, which employed over 5,000 workers, Edison

Poster advertising the Edison phonograph

Edison phonograph, c. 1880

went on to design improved versions of his phonograph, the first-ever motion picture camera (1891), the kinetograph and projecting kinetoscope for still and moving films (1897), the bipolar dynamo (1899) and the reversible galvanic battery (1900). Over his remarkable life, a total of 1093 patents were issued to Edison inventions. "All progress, all success, springs from thinking", he declared – but his phenomenal achievements were also the result of dogged perseverance and his ability to learn from his innumerable failures. As Edison put it so succinctly: "Genius is 1% inspiration, and 99% perspiration."

Edison bipolar dynamo generator manufactured by the Edison Machine Works, 1899

EKCO PRODUCTS COMPANY

FOUNDED 1888
CHICAGO, ILLINOIS, USA

Peter Müller-Munk,
Ekcoware saucepans,
c. 1950

From the early years of the 20th century, the manufacture of kitchen tools became increasingly mechanized and competitive, with quality often becoming the first casualty in the battle to make and sell goods more cheaply. From the 1920s Ekco mass-produced inexpensive kitchen tools of limited durability that differed little from those of their competitors. In 1935, however, the company brought out some kitchen spoons with holes in their handles that enabled them to be hung up – an innovative feature that helped Ekco gain market share but that was soon copied widely. Around 1945 Ekco's president decided to develop a line of high-quality kitchen tools with superior durability. Since there was no company "designer", the firm's tool-room foreman and product development engineers were given the responsibility of designing this new product range. While the resulting stainless steel and plastic *Flint* range introduced in 1946 was significantly more expensive than competitor products, it sold very well. The success of the *Flint* range demonstrated the public's willingness to pay a premium price for "good design in everyday objects" and established Ekco as a market leader in the 1950s. Richard Latham of **Raymond Loewy** Associates, who designed the packaging for *Flint*, assisted Ekco with project planning from the 1950s.

Flint kitchen tools,
first introduced in
1946

KENJI EKUAN

BORN 1929
TOKYO, JAPAN

→Soy sauce
dispenser for
Kikkoman, 1961

Dubbed "the **Raymond Loewy** of Japan", Kenji Ekuan is Japan's most prominent industrial designer. He initially trained as a Buddhist priest before
studying design at the National University of Fine Arts & Music, graduating
in 1955. With fellow students he founded the **GK** (Groupe Koike) **Design
Group** in 1953, which initially concentrated on industrial design. A scholarship from the Japan External Trade Organization (JETRO) enabled him to
study industrial design at the Art Center College of Design in Pasadena,
California, for one year. On his return to Tokyo in 1957 he became president
of GK Industrial Associates. As well as designing motorcycles and audio
equipment for Yamaha for over 40 years, GK has also created many other
products, including packaging for Kikkoman, bicycles for Maruishi and cameras for **Olympus**. GK later expanded its remit to include graphic design,
signage and urban planning. With his many publications on design practice,
Ekuan has emphasized the need to humanize technology, democratize design and "seek the soul in material things". As both a world-class designer
and a Buddhist monk, Ekuan has always pursued a harmonious connection
between the material and spiritual worlds, and has been fascinated by the
"communicative quality of design as a tangible medium".

Narita Express train
for the East Japan
Railways Company,
1991

Alvar Lenning,
Assistant food
processor for
Electrolux, 1940

ELECTROLUX

FOUNDED 1919
STOCKHOLM, SWEDEN

The Electrolux Group is the world's largest producer
of household appliances, with over 55 million products
being sold worldwide each year. The origins of the company can be traced back to Axel Wenner-Gren's invention of a very early
household vacuum cleaner, the *Lux 1*, which was launched by AB Lux in
1912. In 1919 Lux merged with another Stockholm-based company, Elektro-
mekanista, to form Electrolux, and Wenner-Gren was appointed general
manager of this new entity. Three years later, Swedish engineering students
Baltzar von Platen and Carl Munters invented a cooling machine which converted heat to cold by applying the absorption process, and which could
be powered by electricity, kerosene or gas. This highly innovative refrigerator was subsequently manufactured in limited quantities by their company,
AB Arctic. In 1925 Electrolux purchased the company, increased production
of the refrigerator and over the next decade grew rapidly, establishing manufacturing plants in France, Britain and America. In the late 1930s Electrolux
recognized the competitive advantage offered by high-style design and com-

Ralph Lysell,
Design for the *L460*
refrigerator for
Electrolux, 1948

Oxygen vacuum cleaner, 1999

missioned **Raymond Loewy** to design several streamlined products. 1940 saw the launch of Alvar Lenning's revolutionary labour-saving streamlined *Assistant* food processor, which could mix, grind, grate, squeeze and chop. Later in the 1940s and early 1950s, Swedish designers **Sixten Sason** and Ralph Lysell (b. 1907) produced other streamlined designs for Electrolux, including Sason's famous bullet-shaped *Model 248* vacuum cleaner (1943). In 1963 Electrolux established its own in-house design department under the leadership of Hugo Lindström. Over the following decades, Electrolux increased its share of the white goods market and household appliance sector with various company acquisitions, including Eureka, Husqvarna, **Zanussi** and White Consolidated Inc., which owned brands such as Frigidaire, Kelvinator and White Westinghouse. One of Electrolux's design directors, Michael Green, notes: "In today's marketplace, appliance manufacturers that don't make a serious commitment to design very quickly turn into commodity producers ... this company includes design in the Values and Vision statement."

ERCO

FOUNDED 1934
LÜDENSCHEID, GERMANY

Heidi & Dieter Witte,
7310 spotlight,
1969–1970

Erco's origins can be traced back to a lighting factory established by Arnold Reininghaus in 1934. In 1968 the factory's executive director, Klaus Jürgen Maack, came up with the concept of "Light, not luminaries" that defined the future direction of the company. Viewing light as the fourth dimension of architecture – in that light interprets and defines space and brings rooms to life – the firm pioneered revolutionary track light systems in the late 1960s. Erco designs, such as the *7310* spotlight (1969–1970) by Heidi and Dieter Witte, are typically distinguished by functionalism, standardization and suitability for mass production. Initially mainly used in contract settings, Erco track systems, spotlights and flush-mounted downlights had a strong sense of technical structuralism and were highly influential upon the designs of other lighting manufacturers. Many well-known designers and consultancies have designed lighting for Erco, including **Mario Bellini**, **Knud Holscher**, **Frogdesign**, **Norman Foster** and **Franco Clivio** (b. 1942). Erco is a leader in the field of architectural lighting with its state-of-the-art products, which are installed in such buildings as the Reichstag in Berlin and the Guggenheim Museum, Bilbao.

Mario Bellini, *Eclipse*
spotlight, 1986

ERGONOMI DESIGN GRUPPEN

FOUNDED 1979
STOCKHOLM, SWEDEN

Speedglas welding
mask for Hörnells,
1982

In 1979 two Swedish design groups both founded in
the mid-1960s, Designgruppen and Ergonomi Design,
amalgamated to form the Ergonomi Design Gruppen,
based in Bromma. The 14-member group is dedicated to the research and
development of safe, reliable and efficient designs based on ergonomic
principles. It studies user-system problems by building full-scale models to
evaluate designs in experimental situations. Ergonomi's most notable design is the *Eat and Drink* combination cutlery, drinking vessels and plates
manufactured by RSFU Rehab for use by the disabled. Other award-winning
products include the *Genotropin Pen* for Pharmacia & Upjohn, an ergonomic
coffee pot for SAS, the *Ergo* range of hand tools for Sandvik, a baby carrier
for BabyBjörn and an ergonomically resolved stapler for Esco. In 1972 two
of the group's founders, Maria Benktzon (b. 1946) and Sven-Eric Juhlin
(b. 1940), undertook research into muscular ability and its relationship to
the actions of gripping and holding, and have since specialized in design for
disability. Ergonomi Design Gruppen is guided by the belief that: "Design
is not only appearance. It is form, function and economy. Conscientious
product development can increase sales, reduce production costs, open
new markets and improve quality profile. Design becomes a strategic tool

E Series knives
for Tupperware,
1995–1996

for success." The office is also renowned for its design of machines which
reduce the risk of repetitive strain injury and accidents.

ERICSSON *Review* 4 1956

L. M. ERICSSON

FOUNDED 1876
STOCKHOLM, SWEDEN

Cover of the
Ericsson Review,
1956

Model 88 telephone,
1909

Lars Magnus Ericsson (1846–1926) founded the L. M. Ericsson Company in Stockholm in 1876. The business initially concentrated on the repair of telegraph equipment, but by 1878 it was producing telephones that were based on an earlier design by **Alexander Graham Bell**. Soon after, L. M. Ericsson began manufacturing telephones of its own design, which exported throughout Europe, and which were the first to feature a combined transmitting/receiving handset. In 1909 Ericsson launched a cradle telephone, which was enormously influential throughout Europe. Seeking to update this landmark design, in 1930 the company commissioned the artist Jean Heiberg (1884–1976) and the Norwegian engineer Johan Christian Bjerknes to design a telephone with a Bakelite casing. Although not the first telephone to use plastics in its construction, this sculptural design was highly influential and inspired **Henry Dreyfuss'** later *Bell 300* model (1930–1933). Between 1940 and 1954, Hugo Blomberg (1897–1994), Ralph Lysell (1907–1987) and Gösta Thames (b. 1916) designed and developed another groundbreaking model for Ericsson, the *Ericofon* (launched in 1956). This unusual and forward-looking telephone had a completely unified form that integrated the earpiece, mouthpiece and dial. The design of the *Ericofon*, which incorporated new lightweight materials such as plastic, rubber and nylon, was made possible by the increasing miniaturization of technology. Fun yet stylish, it was the most popular one-piece telephone for over three decades. Today, L. M. Ericsson continues its commitment to innovative

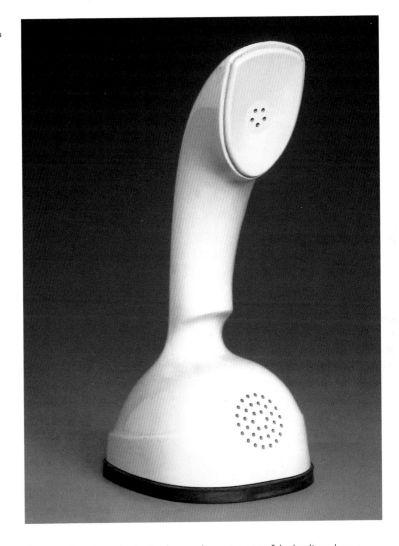

design and cutting-edge technology and remains one of the leading players in the telecommunications industry, with over 100,000 employees. The company is the world leader in mobile telephone systems, connecting nearly 40% of the world's mobile callers. Ericsson is at the forefront of the WAP (Wireless Application Protocol) revolution – the bridge between mobile communication and the Internet – with landmark products such as the *R320*, the company's first mobile phone with a WAP browser.

HARTMUT ESSLINGER

BORN 1944
ALTENSTEIG, GERMANY

Hartmut Esslinger trained as an electrical engineer at the University of Stuttgart before studying industrial design at the Fachhochschule in Schwäbisch Gmünd. In 1969 he established his own industrial design consultancy, Esslinger Design, in Altensteig. His first client was the electronics company Wega Radio. Wega's subsequent acquisition by **Sony** in 1975 introduced Esslinger to the lucrative consumer electronic market that was booming in Japan. His sleek designs, such his *Concept 51K* hi-fi system for Wega (1975), brought him widespread recognition and were remarkable for their Modern dematerialist aesthetic. His work initially appeared to be influenced by the functionalist approach to design promoted by **Max Bill** and **Hans Gugelot** at the Hochschule für Gestaltung in Ulm. It was, however, also driven by a rejection of "the prevailing Modern sentiments of 60s' Germany" and sought instead a "vision of what design could still yet become as opposed to what

Esslinger Design,
Tri-Bel showerhead
for Hansgrohe,
1973

it has already settled for being". In 1982 Esslinger renamed his consultancy **Frogdesign** (frog = Federal Republic of Germany) and opened an office in Campbell, California, so as to meet the design needs of the burgeoning computer industry in Silicon Valley. In 1984 the consultancy designed the off-white *Apple Macintosh* for **Apple Computer** and in so doing redefined the aesthetic parameters of personal computers. Two years later Esslinger opened a sister office in Tokyo, and during the 1990s established branches in technology hot-spots such as the "Silicon Prairie" of Austin, Texas, "Silicon Alley" in New York, "Media Gulch" in San Francisco and "Silicon River" in Düsseldorf. Frogdesign has also designed cameras, synthesizers, binoculars, communications equipment and office seating for such companies as **RCA**, **Kodak**, **Polaroid**, Motorola, Seiko, **Sony**, **Olympus**, AT&T, **AEG**, König und Neurath, **Erco**, Villeroy & Boch, **Rosenthal** and **Yamaha**. Esslinger attempts to humanize technology with sculptural forms or visual references to create more user-friendly products. He has always had an international outlook and recognizes the "benefit of bringing Asian teamwork, European discipline, German precision, American optimism and Californian craziness together" to produce a very unique and culturally melding creative enterprise.

HARTMUT ESSLINGER . 207

EUCLID

FOUNDED 1931
EUCLID, OHIO, USA

First large-capacity
self-powered hauler
for off-road use,
1933

In 1933 Euclid Road Machinery introduced its first hauling unit that could undertake more work at a lower cost than conventionally designed trucks. The reputation of Euclid's equipment was significantly enhanced when it was used in the construction of the Hoover Dam (1930–1936). In 1934 Euclid introduced the first-ever 20-ton coal hauler and by the 1940s the name "EUC" had become a generic term for all off-road dump trucks. During the Second World War, Euclid produced equipment for the US military and afterwards began assembly-line production of its off-highway equipment, which could now manage up to 40-ton loads. Realizing that speed and capacity was limited by the horsepower of available engines, in 1949 Euclid devised the "Twin Power" concept using two diesel engines mounted side by side. This system permitted the hauling of much heavier loads on steeper gradients at faster speeds. In 1968 Euclid introduced the *R105* – the first ever off-highway hauler with a load capacity of more than 100 tons. 1996 saw the introduction of the *R260* mining hauler, which can carry an astonishing 260 tons.

R260 mining hauler,
1996

FERRARI

FOUNDED 1929
MODENA, ITALY

Ferrari logo

In 1920 Enzo Ferrari (1898–1988) joined **Alfa Romeo** and over the next 20 years went from test-driver to racing driver to director of the Alfa Racing Division. In 1929 he founded the Scuderia Ferrari in Modena with the purpose of helping members compete in motor races. In 1940 the Scuderia became an independent company, the Auto Avio Costruzioni Ferrari. The workshop moved to Maranello in 1943 and at the end of the war changed its name to Ferrari. The first Ferrari racing car was the *125 GT Sport*, which won the Rome Grand Prix in 1947. In 1969 Enzo Ferrari sold 50% of the company to **Fiat** (this investment increased to 90% in 1988), but it was nevertheless able to maintain a strong autonomy. In 1972 Ferrari built the Fiorano track close to its factory so as to test its Formula 1 and GT cars. The experience gained on the racetrack has always had a direct influence on Ferrari road cars. By combining extraordinary performance with beautiful, visually seductive lines, Enzo Ferrari began a tradition of creating, as he put it, "expressions of superlative engineering" and "cars that make themselves desired". Over a period of 40 years, models such as the legendary *250 GT* (1960), the *Testarossa* (1984) and the *550 Maranello* (1996) have been exquisitely styled by **Pininfarina** and have come to epitomize the best of Italian design.

Pininfarina, Ferrari
Testarossa, 1984

Pininfarina, Ferrari
550 Maranello, 1996

Pininfarina, Ferrari *Mythos*,
1989 (concept car)

Poster advertising
the Fiat 500, c. 1936

FIAT

FOUNDED 1899
TURIN, ITALY

Fiat is Italy's largest automotive company and one of the world's largest industrial groups, with diversified production ranging from buses and trains to farm tractors and aircraft. Founded in 1899 as the Fabbrica Italiana Automobili Torino by a group of investors including Giovanni Agnelli (1866–1945), Fiat received instant recognition for its luxury cars as well as its racing triumphs. The company grew rapidly in terms of both sales and production, and built its first American factory in 1909 in Poughkeepsie, New York. Later, in an attempt to increase production, Agnelli decided to build the largest automotive plant in Europe, which opened in Lingotto in 1922. Implementing assembly-line methods of production, Fiat used this factory to transform the car from a relatively exclusive item into a much more accessible product, a philosophical approach to manufacturing that the company has adhered to throughout its history. Its most successful pre-war car was the compact Fiat 500, or "Topolino" (Little Mouse) as it became known, which was designed by **Dante Giacosa**. Launched in 1936, the 500 was the smallest mass-produced car in the world and over 519,000 models were produced between 1936 and 1955. Its rounded lines broke with the tradition of box-shaped forms, while its light metal body enabled extremely fluid production.

Dante Giacosa,
Fiat 500, 1936

→ Fiat *Nuova 500A*,
1957

During the post-war period, Fiat's objective was to produce the sort of cars that the large American manufacturers were not making – cars with smaller engines that everyone could afford. This led directly to models such as the Fiat *600* (1955) and the Fiat *Nuova 500* (1957). These inexpensive and diminutive vehicles, together with their variations, were manufactured in their millions and did much to democratize car ownership in Italy. They also became potent symbols of Italy's economic miracle. In 1978 Fiat introduced the world's first flexible robotic assembly lines in its plants in Rivalta and Cassino, and a year later merged with Lancia, **Ferrari**, Autobianchi and Abarth to form Fiat Auto SpA. During the 1980s and 1990s, the company continued producing "people's cars" such as the Fiat *Panda* (1980), the Fiat *Uno* (1983), which was styled by **Giorgetto Giugiaro** and became Fiat's best-ever selling car, and the intermediate sized Fiat *Punto* (1993). In 1999 Fiat launched the versatile 6-seater *Multipla*, a groundbreaking mid-range MPV with non-conformist styling that was voted Car of the Year 2000 by the BBC. Highly innovative designs such as these are developed in Fiat's Advanced Design centre, which employs a youthful team devoted specifically to generating new car concepts.

↘ *Fiat Uno*, 1983,
styled by Giorgetto
Guigiaro,

Fiat *Multipla*, 1999

FISKARS

FOUNDED 1649
FISKARS, FINLAND

Olavi Lindén,
Clippers secateurs,
1996

As the oldest industrial company in Finland, Fiskars can trace its origins to an ironworks established by Peter Thorwöste in southern Finland in 1649. From the 1820s, Fiskars was forging high-quality tableware, kitchen knives and scissors, but it was not until 1967 that the company came to the world's attention when it launched its famous orange-handled *O-series* scissors (1963). Developed by the engineer and wood-carver **Olof Bäckström**, the prototype for the ergonomically conceived ABS handles of these scissors was carved from wood. Licensed worldwide, the scissors were also manufactured as pinking shears and in a left-handed version. Over the succeeding years this benchmark ergonomic design, now known as the *Classic*, has been developed into a family of 18 scissors. More recently, under the guidance of Olavi Lindén (b. 1946), the Fiskars design team has developed a range of garden tools that has been awarded numerous design prizes. Known as *Clippers* (1996), these tools reflect Fiskars' "commitment to quality and systematic application of ergonomic principle, which result in products that are easy, safe and pleasurable to use".

Olavi Lindén, *Handy* axe and *Clippers* secateurs, 1994 & 1996

Philippe Starck,
Miss Sissi table lamp,
1991

FLOS

FOUNDED 1962
MERANO, ITALY

Achille & Pier Giacomo
Castiglione, *Luminator*
floor lamp, 1962

In the late 1950s a spray-on self-skinning plastic coating was developed in the United States. This new material, known as Cocoon, was imported into Italy by Arturo Eisenkeil, who began seeking applications for it. In 1962 he founded the lighting company Flos in partnership with the avant-garde furniture manufacturers Dino Gavina (b. 1932) and Cesare Cassina (1908–1981). That same year, he launched the company's first range of lamps, which were designed by **Achille** and **Pier Giacomo Castiglioni**. Soon afterwards Flos produced its first Cocoon lamps, designed by the Castiglioni brothers and by Tobia Scarpa (b. 1935). From the outset, Flos distinguished itself from its competitors not only by commissioning avant-garde designers but through its strong corporate identity. To this end, the graphic designer Max Huber (b. 1919) designed its first catalogues, while the Castiglioni brothers designed interiors for its Milan showroom in 1968. In 1974 Flos acquired the lighting company Arteluce, which had been established by Gino Sarfatti (1912–1985). Since then, the two companies have produced lighting products by an international array of designers, including **Mario Bellini**, Matteo Thun (b. 1952), Marc Newson (b. 1963), **Jasper Morrison**, Antonio Citterio (b. 1950) and Konstantin Grcic (b. 1965). In 1988 Flos launched its first lamp designed by **Philippe Starck**, the *Arà*, which was followed by *Miss Sissi* (1991). In 1994 the innovative rubber-like *Drop* wall lamp (1993) designed by Marc Sadler (b. 1946) was awarded a Compasso d'Oro, as was the company for its long-commitment to design innovation. Many of the lamps, especially those by Achille Castiglione, are classic examples of Italian industrial design. The fact that they continue to sell well is a tribute to Flos's commitment to technical and aesthetic excellence.

Henry Ford I

FORD

FOUNDED 1903
DETROIT, MICHIGAN, USA

While Henry Ford (1863–1947) did not invent the automobile, it was his vision that made the car accessible to literally millions of people. His primary goal was to "build a motor car for the great multitude ... it will be so low in price that no man ... will be unable to own one." In the 1860s, two events occurred that enabled his dream to be fulfilled – the invention of the open-hearth process in 1864, which heralded the birth of modern steel manufacture, and the oil industry's laying of a pipeline in the Allegheny River valley in 1865, which signalled the beginning of an enormous network of petrol stations that would eventually fuel Ford cars. Funded by twelve investors, the Ford Motor Company was established in June 1903 in a converted Detroit wagon factory. The first car was sold only a month later and was described as "the most perfect machine on the market.... so simple that a boy of fifteen can run it". Over the following five years, Henry Ford directed a research and development programme that resulted in a plethora of models.

Model T, 1925

↑ *Model T*, 1912

↗ *Model T*, 1913

↑ *Model T*, 1923

↗ *Model T*, 1927

These vehicles, which were coded alphabetically, included two-, four- and six-cylinder cars, some of which were chain driven while others were shaft driven. Not every model made it into production, and of those that did, not all were successful: while the *Model N*, which retailed for a modest $500, sold well, for example, the expensive *Model K* limousine did poorly. Following the failure of the *Model K*, Henry Ford insisted that the company's destiny lay in the manufacture of inexpensive cars. At this time, George Selden held a patent for "road locomotives" powered by internal combustion engines, and most car manufacturers paid him royalties. Ford, however, believed Seldon's claim of exclusivity to automobile technology was invalid and refused to pay any royalties. In 1911, after many years of costly litigation, Ford finally won the suit brought against him by Selden and in so doing freed the entire automobile industry from what was, clearly, an impediment to its progress. Despite this legal distraction, the Ford Motor Company continued to expand and in October 1908 the pivotal *Model T* came into being. Described by Henry Ford as the "universal car", the low-cost and reliable *Model T* became an instant success. The *Model T* was so much in demand that Henry Ford instigated a system of compartmentalized production and came up with the concept of a moving assembly line. This approach to manufacturing, which became known as Fordism, dramatically reduced assembly time and resulted in over 15 million *Model T*s being built between 1908 and 1927. The succeeding *Model A* (named after Ford's first car) placed

Ford *Escort GT*,
1967

Ford *Fiesta 1.6*,
1976

Ford *Focus*,
1999

greater emphasis on safety and comfort than its predecessor and became known as "the baby Lincoln" because of its softer contours. Although sales were impacted by the Great Depression, some 5 million *Model A*s were produced before it was succeeded in 1932 by the *V-8*. In 1948 Ford launched the *F-Series*, a pick-up truck that was "built stronger to last longer". Becoming the best-selling vehicle in North America (a title it still holds) – over 27 million *F-Series* trucks have been manufactured to date. In Europe, Ford became better known for its practical "everyman" cars, such as the *Escort* and *Fiesta*. During the late 1990s, the company evolved a new language of design known as "New Edge Design", which combines "smooth, sculpted surfaces with clear, crisp intersections". This fresh approach led to such innovative vehicles as the *Ka*, *Puma*, *Cougar* and *Focus*, which was voted Car of the Year in Europe in 1999 and in America in 2000. Ford's commitment to design innovation resulted in it commissioning the product designer Marc Newson (b. 1963) to develop a concept car for the 21st century, the Ford *012C*, which features a number of imaginative details such as a slide-out luggage tray and a single LCD headlight. Ford also acknowledges its heritage and its new retro design *Thunderbird* is an intriguing reinterpretation of the legendary 1950s original. Like all motor manufacturers, Ford has become increasingly mindful of environmental issues and is committed to developing a new generation of vehicles that radically reduce CO_2 emissions.

Ford *Ka*, 1996

NORMAN FOSTER

BORN 1935
MANCHESTER, ENGLAND

Lord Foster studied architecture and planning at the University of Manchester, graduating in 1961. He continued his studies at Yale University in New Haven, Connecticut, graduating in 1963. The same Year, Foster established the Team 4 partnership in London with fellow Yale scholarship student Richard Rogers (b. 1933), Georgie Wolton and Wendy Cheesman. In 1967 Foster independently founded Foster Associates in London and from 1968 to 1983 collaborated on several projects with **Richard Buckminster Fuller**. In 1977 Foster received the Royal Institute of British Architects Award for his black glass-clad office building for Willis Faber & Dumas in Ipswich (1973–1975), and has since then become internationally celebrated for his numerous architectural projects, including the new Reichstag (German Parliament)

Enercon E66 Wind
Energy Converter,
c. 1999

in Berlin (1995–1999). Like his buildings, Foster's industrial designs have a High Tech aesthetic, as evidenced by his *Nomos* furniture system (1986–1988) for Tecno and his lighting systems for **Erco** (1986). Other lighting designs by Foster, which have strong architectural qualities, include the *RA* system (1998) for **Artemide** and the *Saturn* down-lighter (1999) for Guzzini. Foster's interest in energy-efficient design led to the development of the elegant *Enercon E66 Wind Energy Converter*, which was selected by the **Design Council** as a Millennium Product in 1999. By combining state-of-the-art materials and technologies within a strictly Modern formal vocabulary, Foster creates inherently logical and beautifully engineered industrial designs.

FRAZER DESIGNERS

FOUNDED 1978
LONDON, ENGLAND

Remote control for
Sky Digital Satellite
Service, 1998

Stephen Frazer (b. 1948) studied industrial design at the Central School of Arts & Crafts and later at the Royal College of Art in London. In 1978 he founded Frazer Designers, which specialized in the design of leading-edge electronic products. From 1983 he began designing products for Psion, including the original Psion organizer. Around 1987 the consultancy began producing designs for the early cell-phones manufactured by NEC – an area that it is still actively involved in. Frazer views cell-phones and other similar products as tools that allow entry to much larger services. The actual production costs of electronic products such as these are today very low – the majority being manufactured in the Far East – and they have become extremely accessible. Over 3.8 million remote controls for Sky's digital satellite service, for example, were distributed within twelve months of its launch in Britain in 1998. Frazer describes the remote control as a "visible and tangible key" to Sky's entertainment service, and believes that design will play an increasingly important role within the electronic products market; as the technology used in one product differs little from that used in another, design will be used more and more as the primary means of product differentiation. Frazer Designers approaches the design process holistically, while considering both the technical and emotional requirements of new products. Its aim is to "stimulate that moment of personal desire by creating visual product identities that will attract and intrigue".

Concept microwave
oven (prototype),
1993

FROGDESIGN

FOUNDED 1969
ALTENSTEIG, GERMANY

Solido headset
for IMAX, 1994

Frogdesign's origins lie in the industrial design consultancy founded by **Hartmut Esslinger** in 1969 in Altensteig, Germany. A leader in the design of consumer products, Esslinger Design, as it was originally called, attracted a clientele which included **Sony, AEG, Zeiss, Olympus** and **Apple Computer**. In 1982 the name Frogdesign (frog = Federal Republic of Germany) was adopted when the consultancy opened an office in Campbell, California, to cater to the design requirements of Silicon Valley. Unlike other design offices, Frogdesign actively promoted itself during the 1970s and 1980s with a strong and eye-catching advertising campaign in specialist design-related journals, which raised its profile enormously. In 1983 it became the first consultancy to win a "million dollar contract" (with Apple Computer) in Silicon Valley. Significantly, the office re-invested in computer-aided machining tools, taking it to the forefront of design technology. Three years later, another international office was opened in Tokyo to meet the demands of the burgeoning Japanese consumer electronics market. Frogdesign developed a house-style that combined Modern Movement functionalism – as pioneered at the **Bauhaus** and later at the Hochschule für Gestaltung in Ulm – with organic forms. This rational vocabulary, however, was occasionally tempered by the use of visually

Z-Lite laptop
computer for Zenith
Data Systems,
1992

expressive and sometimes whimsical references to a product's function – as in the case of the *Personal Communicator* (1992) for AT&T, which has elements that resemble ears. Esslinger sees Frog (as it is now called) as more than just an industrial design firm and believes that design must be both creative and enabling. In Esslinger's words: "The purpose of design is to make our artificial environment more human. My goal is and always was to design mainstream products as art."

FSB

FOUNDED 1881
BRAKEL, GERMANY

Philippe Starck,
PS1 door handle
for FSB, 1991

Founded by Franz Schneider in 1881 to produce devotional objects and brass cabinet-fittings, FSB became renowned for its architectural ironmongery. In the early years of the 20th century, Franz Schneider Brakel GmbH diversified its product line to include door and window fittings. In the 1950s the company introduced a range of "moulded-to-the-hand" products that were created by its in-house designer, Johannes Potente. The first "named" designer to be commissioned by FSB was **Dieter Rams**, who developed a series of twelve fittings that adapted to windows and doors of every size and shape. Reflecting Rams' belief that "unassuming is better than showy, colourless better than gaudy, harmony better than divergence, balanced better than exalted", the design of his aluminium and thermoplastic *rgs 1, 2, 3* Series (1986) was driven entirely by logic. FSB has subsequently produced designs by **Jasper Morrison**, **Philippe Starck**, Franco Clivio, Hartmut Weise, Nicholas Grimshaw (b. 1939) and Erik Magnussen (b. 1940), which range from the sculptural to the functional so as to suit a wide range of tastes. As Otl Aicher (1922–1991), the co-founder of the Hochschule für Gestaltung in Ulm, observed, the uniting feature of all FSB products is the human hand. With his assistance, FSB published the *Good Grip Guide*, which asserted that "any object designed to be held has got to make allowance for the specific needs of the thumb, the forefinger, the ball of the thumb and the palm."

Hartmut Weise, *1025*
door handle for FSB,
c. 1991

NAOTO FUKASAWA

BORN 1956
KOFU, JAPAN

→17" monitor
for NEC, 1998 –
co-designed with
Sam Hecht

Naoto Fukasawa trained as a product designer at Tama Art University and later worked as the chief designer of the R&D Design Group at the Seiko Epson Corporation in Japan. Since 1989 Fukasawa has worked for ID Two (later **IDEO**), leading a design team that has formulated a coherent visual language for NEC products ranging from computer notebooks to LCD projectors, and which have received several G Mark and Hanover iF awards. He has also assisted **Apple Computer** in developing a new vocabulary of design in which the ergonomic form of the product conforms to the user's actions. Fukasawa's *Left Ventricular Heart Assist* (1990), a computer that externally controls an artificial heart for a recovering patient, is not as he describes "a clean, health-goods type of design" but a no-nonsense and robust design that ensures the continuous beating of the heart. Fukasawa, who was design director of IDEO Japan, is fascinated by the interaction between man and technology. He has collaborated with Sam Hecht (b. 1969) on a number of products that attempt to humanize computer technology, and which are born out of his belief that "perfection is admitting to oneself the existence of imperfection."

Left Ventricular Heart Assist for Baxter, 1990 – co-designed with Sam Hecht

RICHARD BUCKMINSTER FULLER

BORN 1895 MILTON, MASSACHUSETTS, USA
DIED 1983 LOS ANGELES, CALIFORNIA, USA

Richard Buckminster Fuller studied mathematics at Harvard University from 1913 to 1915 and in 1917 enrolled at the US Naval Academy in Annapolis. While at the Academy he began his "theoretical conceptioning", which included a proposal for "flying jet-stilts porpoise" transport that was eventually published in 1932. After leaving the military, he founded the Stockade Building Company in 1922, which was a financial failure and led to his personal bankruptcy. He committed himself to finding global solutions to social problems after the death of his four-year-old daughter in 1922, which he believed was due to inadequate housing. In 1933/34 he founded the 4D Company in New York to develop his design concepts, which were driven by his ambition to evolve a "design science" that would bring about the best solutions with the minimum use of energy and materials. He based this concept on the Modern Movement principle of getting the most with the least and named it "Dymaxion", which was derived from "dynamic" and "maximum efficiency". In 1929 he launched the magazine *Shelter* and was its publisher and editor from 1930 to 1932. From 1932 to 1938 he was director and chief engineer of the Dymaxion Corporation, which he set up to develop and produce three streamlined prototype cars that were based on his dymaxion principles and were inspired by aircraft design. It was claimed by Fuller that the Dymaxion car of 1934 could accelerate from 0 to 60 mph in three seconds and provide 30 mpg fuel consumption. The prototype car, however, was not progressed due to several serious design flaws. From 1927, Fuller also developed the Dymaxion House concept and between 1944

Dymaxion car, 1932

Dymaxion Dwelling
Machine (Wichita
House), 1944–1947

and 1947 a prefabricated metal dwelling known as the Wichita House. Although the company he set up to manufacture this industrially produced architecture received an astonishing 38,000 orders after its press launch, he was not prepared to commence manufacture of the Wichita House until its design had been perfected. His backers subsequently grew disheartened with the time it was taking and this product architecture project was shelved. Fuller's most famous invention was the geodesic dome of 1949, which had a wide range of applications from industrial to military to exhibitions. The geodesic dome remains the only large dome that can be set directly on the ground as a complete structure. It is also the most economic space-filling structure and the only practical type of building that has no limiting dimensions. The outcome of Fuller's "more for less" approach, the geodesic dome employed a minimum of materials and could be easily transported and assembled. This remarkable design offered a means of producing ecologically efficient housing for the mass-market, and Geodesics Inc. was set up in 1949 to develop the concept. Fuller was also the first person to coin the expression "spaceship Earth". Prolific communicator, humanist, polymath and futurist, Buckminster Fuller believed that the creative abilities of humankind were unlimited, and that technology and comprehensive, anticipatory design-led solutions could eliminate all barriers to humanity's expansion into a positive future.

ACHILLE GAGGIA

BORN 1895 ITALY
DIED 1961 ITALY

Ambrogio Fumagalli,
*International Two
Group* coffee
machine, 1954

Working as a bartender in Milan, Achille Gaggia grew increasingly dissatisfied with the traditional machines that used steam to brew coffee. He began experimenting in his attic and developed new models that used a piston system to drive boiling water at a high pressure through coffee grounds, so as to make a better-tasting drink. Around 1937 he installed his first machines in several cafés in Milan, and in 1944 he patented his innovative espresso-making process. Three years later, he established a manufacturing company, Brevetti Gaggia, to produce his machines. His first design, known as the *Classico*, was powered by either gas or electricity and was intended only for café use. Around 1948, however, a two-handled espresso maker was developed for domestic use. By pumping the hinged arms, pressure was created in the heated water canister, which

Espresso maker,
c. 1948

in turn forced the water through the grounds. This model was very popular throughout the 1950s, as was the later *Gilda* (1952–1954), named in tribute to a film starring Rita Hayworth. The coffee machine which Ambrogio Fumagalli designed for Gaggia in 1954 was the first to produce two separate quantities of coffee, thus ensuring that a constant supply could be on hand in cafés. By using expressive forms inspired by automotive and aeronautical styling, Gaggia's innovative machines gave the convention of coffee making and coffee drinking a previously unknown glamour.

NORMAN BEL GEDDES

BORN 1893 ADRIAN, MICHIGAN, USA
DIED 1958 NEW YORK, USA

Norman Bel Geddes studied art at the Cleveland Institute of Art and later trained briefly at the Art Institute of Chicago. In 1913 he worked as a draftsman in the advertising industry in Detroit, where he designed posters for Packard and **General Motors**. In 1916 he wrote a stage play and subsequently worked as a theatre-designer for six productions in Los Angeles. In 1918 he became a set designer for the Metropolitan Opera Company in New York, before moving to Hollywood in 1925, where he designed lavish film sets. Influenced by Frank Lloyd Wright (1867–1959) and the German Expressionist architect Erich Mendelsohn (1887–1953), in 1927 Geddes turned to architecture and product styling. While he was a highly successful industrial design consultant – notably working for the Toledo Scale Company in 1929 and the

Soda King Syphon
for the Walter Kidde
Sales Company, 1932

Standard Gas Equipment Corporation in 1931 – he is best remembered as a design propagandist. In 1932 he published a book entitled *Horizons*, in which he outlined his approach to industrial design and his belief in the supremacy of the teardrop shape. Geddes also designed futuristic cars for the Graham-Paige automobile company (1928) and streamlined products such as radios for Philco (1931), radio casings for **RCA**, metal bedroom furnishings for Simmons (1929) and the well-known *Soda King Syphon* for the Walter Kidde Sales Co. (1932). One of his most notable achievements was his standardization of kitchen equipment. The design of his modular and streamlined *Oriole* stove (1931) for the Standard Gas Equipment Company was inspired by the construction of skyscrapers and, as such, had a steel frame onto which white vitreous enamel panels were clipped. Geddes also designed the General Motors' "Futurama" display for the 1939 **New York World's Fair**, which speculated on the future (as it was believed it would be in 1960) and correctly predicted the freeway system. Geddes was one of the greatest exponents of streamlining and was also an important pioneer of industrial design consulting.

GENERAL ELECTRIC

FOUNDED 1892
ALBANY, NEW YORK, USA

Complete line of heating devices, 1907

→ *WD-560S* combination washer-drier, 1954

Model AW1 washing machine, 1933

The origins of this phenomenally successful company can be traced to the inventor **Thomas Alva Edison**, who established the Edison Electric Light Company in 1878 and a year later invented the first practical electric light bulb. This venture, renamed the Edison General Electric Company in 1889, eventually merged with the Thomson-Houston Electric Company in 1892 to form the General Electric Company. Following Edison's earlier idea of establishing "invention factories", the company founded the GE General Engineering Laboratory in 1895 to conduct research into advanced engineering and instrumentation. In 1900 the company established the GE Research Laboratory, one of the first industrial laboratories in America to undertake basic scientific research. These laboratories were crucial to the development of many inventions and discoveries, including the high-frequency alternator (1906) devised by Dr. Ernest Alexanderson (1878–1975) which revolutionized radio broadcasting, the gas-filled incandescent lamp (1913) invented by Dr. Irving Langmuir (1881–1957) that was the forerunner of the light bulbs still used today, and the first practical X-ray tube (1913) invented by Dr. William Coolidge (1873–1975), which was the predecessor of the modern medical X-ray tube. During the late 1920s General Electric also pioneered television broadcasting in America, using technology developed by Alexanderson. In addition, the company was renowned for its labour-saving domestic appliances, from electric heaters and washing machines to vacuum cleaners and refrigerators – all of which literally transformed the

SHES AND DRIES IN ONE COMPLETELY AUTOMATIC OPERATION

Combination Washer-Dryer

UNDERCOUNTER MODEL WD-560S

The first home television reception at the Schenectady residence of Ernest Alexanderson, c. 1928

Arthur BecVar, redesign of cylinder vacuum cleaner for General Electric, c. 1950

American way of life. In the 1940s, GE produced engines that powered the first American jet-propelled aeroplane, the **Bell** *P-59 Airacomet* (1942) and the world's then fastest aeroplane, the **Lockheed** *P-80 Shooting Star* (1945). The company later produced the *J-79*, the first jet engine to propel an aircraft

GE satellite station
in California, c. 1992

at twice the speed of sound. During the 1960s, GE contributed its technical expertise to **NASA**'s Apollo moon-landing mission (1969) and developed weather satellites. In 1975 its research labs designed an essential component for the development of the CAT (computerized axial tomography) scanner and during the 1980s acquired the medical diagnostic imaging company, CGR. Continuing its commitment to television broadcasting, GE purchased **RCA** in 1986, which included the NBC television network, and subsequently launched two new networks, CNBC (1989) and MSNBC (1996). For over a century, General Electric has been driven by technological innovation and its life-changing and life-enhancing inventions have touched us all. Central to its continuing success are the corporation's twelve multi-disciplinary laboratories that reflect the diversity of its operations: from lighting systems to satellite systems, from aircraft engines to washing machines, from state-of-the-art plastics to medical diagnostic equipment. The research and development currently being undertaken in General Electric's laboratories will almost certainly shape the future of industrial design, just as it has done over the past 100 years.

Oldsmobile, 1903

GENERAL MOTORS

FOUNDED 1908
HUDSON, NEW JERSEY, USA

General Motors is the world's largest automotive manufacturer and throughout most of the 20th century was the world's largest industrial corporation. For this reason, it has historically been regarded as a barometer of the American economy. Seven years before its incorporation in 1908, the company manufactured the first car to be produced in quantity in America. The year of its official founding, GM launched the first electric headlamp, and in 1910 introduced "closed bodies' as standard equipment. In 1912 the company pioneered the first all-steel car body, which provided much greater strength and safety than earlier automobiles. In 1924 GM established the first automotive proving-ground test facility in Milford, Michigan. During the early 1920s, the company's chairman predicted the increasing importance of styling in the automotive industry and subsequently commissioned **Harley Earl** to work on the 1927 La Salle, the first mass-produced car to be developed by an automobile stylist. In 1928 GM established its own styling de-

La Salle, 1934

Pontiac, 1935

Cadillac, 1941

Chevrolet, 1955

↖Buick, 1946

partment, which was headed by Earl and was the first of its kind. This influential in-house facility devised annual stylistic changes in an effort to accelerate the aesthetic life cycle of automobile models and, as a result, increase sales. As well as instigating this programme of annual cosmetic changes, GM continued to produce real innovations, such as the first built-in trunk (1933), which revolutionized the overall form of the car. Under Earl's guidance, GM also developed its first concept car, the Buick *Y-Job* (1938). By 1940 GM had produced a massive 25,000,000 vehicles, which sheds some light on the extent of pre-war car ownership in America. After the war GM introduced several innovations to its product line, including the "airfoil' fender (1942–1948), the curved windscreen (1948) and the modern, high-compression overhead valve V-8 engine, the first of which were used on the company's Cadillac and Oldsmobile models. The famous Chevrolet *Corvette* launched in 1953 became the first GM concept car to have evolved unaltered from a Motorama auto show to a full-scale production model. Although GM had set up the first safety-testing laboratory in 1955, the fateful Chevrolet *Corvair* which it produced in 1960 was famously attacked for its lack of safety and tendency to roll over by the young lawyer, Ralph Nader, in his book *Unsafe at any Speed* (1965). Nader's legal action against GM and his subsequent victory led the US Congress to pass 25 pieces of consumer

←Pontiac, 1950

Corvette, 1963

legislation between 1966 and 1973, which opened the floodgates for prod-uct-liability law suits in America. Learning from this difficult experience, GM introduced the first energy-absorbing steering column (1966), the first side-guard door beam (1969) and the first airbags in production vehicles (1974). In 1975 it also began fitting the first-ever catalytic converters on all its cars sold in America. In more recent times, in 1996 GM launched the first modern from-the-ground-up electric car, the *EV1*. This remarkable zero-emission vehicle, with its sophisticated lightweight aluminium body, has the most aerodynamic shape of any production car ever. It is powered by a 137 hp, 3-phase AC induction motor and is offered with two battery tech-nologies: an advanced high-capacity lead acid and an optional Nickel Metal Hydride, which doubles the *EV1*'s range from 55–95 miles to 75–130 miles. Today General Motors boasts many major automobile brands, including Chevrolet, Chevy Trucks, Pontiac, Oldsmobile, Buick, Cadillac, GMC and Saturn, and employs over 397,000 people with operations in 73 countries. As a truly global corporate giant and with the will to innovate such revolu-tionary vehicles as the *EV1*, General Motors seems likely to continue to dominate the automotive industry as it has done for nearly 100 years.

EV1, 1996

Giacinto Ghia

→De Tomaso
Mangusta, 1966

↘Maserati *Ghibli*,
1968

VW *Karman Ghia*,
1961

GHIA

FOUNDED 1915
TURIN, ITALY

An Italian legend, Ghia SpA is one of the most famous *carrozzeria* or coachbuilders in Italy. Founded in Turin in 1915 by Giacinto Ghia (1887–1944), who had previously learned about car production in Diatto's workshop, the company initially manufactured designs for **Bertone**. It later became acknowledged for its own designs for luxury cars and racing cars. Among its competition car designs, the aerodynamic "torpedo" form came to characterize Ghia's work. After the death of Giacinto Ghia, young designers, most notably Mario Boana, revived the company's outlook. During the 1950s, the Ghia works created specially designed bodies for Bentley, Delahaye and Talbot. The prototype of the Ghia *Gilda* (1955) was one of the first to undergo aerodynamic wind-tunnel testing, and the data collected resulted in the addition of fins for increased stability – a feature that was later "borrowed' by American manufacturers for stylistic purposes. During this same period Ghia also styled many American cars, and in so doing influenced the car designs of Detroit-based manufacturers in general. Throughout the 1960s, the company continued styling cars for mainstream volume-manufacturers such as **Fiat**, **Volkswagen**, **Alfa Romeo**, **Renault**, **Chrysler** and

Ford. While Ghia designs such as the VW *Karman Ghia* (1961), the De Tomaso *Mangusta* (1966) and the Maserati *Ghibli* (1968) had a recognizable character, they did not match the elegance of cars styled by some other Italian *carrozzerie*. In 1972 Ghia became a wholly-owned subsiduary of the Ford Motor Company and subsequently styled the first Ford *Fiesta* (1973). Today, as part of Ford's Advanced Design group, Ghia is a proponent of "New Edge Design", which blends "the smooth, sculpted surfaces of traditional aero design with clean, crisp intersections to produce a sharper, more defined image."

DANTE GIACOSA

BORN 1905 ROME, ITALY
DIED 1996 TURIN, ITALY

Drawing of Fiat *500A*

Dante Giacosa began working for **Fiat** in 1926 and around 1932 was asked to design a small yet stylish car. His work resulted in the *500A*, which was launched in 1936. Also known as the "Topolino" (Little Mouse), this compact and streamlined vehicle marked one of the greatest advances in small car design. The *500A*'s rounded form broke with the convention of the box shape, while its lightweight metal body was highly suited to mass production. Between 1938 and 1939, it was Italy's biggest-selling vehicle export to America. The *500A* was the world's smallest mass-produced car, with 519,000 being built between 1936 and 1955. Giacosa went on to design modified versions of the Topolino – the *500B* saloon and station wagon (1948) and the *500C* (1949), which incorporated a number of new features such as interior heating. These models were succeeded by Giacosa's diminutive *Nuova 500* (1957), of which 3.6 million were built. He also designed the Fiat *124*, the Fiat *128* and the Fiat *130*. While the *500A* helped democratize car ownership in Italy, Giacosa's *Nuova 500* became a potent symbol of Italy's post-war economic miracle.

Fiat *500A*, 1936

GILLETTE

FOUNDED 1901
BOSTON, MASSACHUSETTS, USA

King Camp Gillette

Reared in Chicago, King Camp Gillette (1855–1932) became the star salesman at the Crown Cork & Seal Company, which mainly manufactured cork
bottle caps. His mentor at the company, who noticed his predilection for
mechanical tinkering, suggested that he "try to invent something like the
Crown Cork product which, when used once, is thrown away" so that the
customer would keep coming back. Gillette became obsessed with this idea
and one morning in 1895, while struggling to hone his straight razor, he had
a vision of a razor with a separate handle that clamped a thin double-edged
disposable steel blade between two plates. After six difficult years, in 1901
Gillette eventually founded the American Safety Razor Company, which in
1902 became the Gillette Safety Razor Company. While not the first "safety
razor", Gillette's innovative refillable razor and blade system offered much
greater convenience. It was first manufactured in 1903 and patented a year
later. During the First World War, the company supplied 3.5 million razors
and 36 million blades to the US armed forces. In 1932 the company introduced the famous double-edged *Blue Blade*, which dominated the blade
market for many years. Gillette diversified its interests with its acquisition of
the Paper Mate Pen Company in 1955 and **Braun** AG in 1967. As a leading

⌐Early Gillette safety
razor, c. 1904

Patent drawing of
King Camp Gillette's
first safety razor,
1901 – the final
production model
varied slightly

No. 775,134.
PATENTED NOV. 15, 1904.
K. C. GILLETTE.
RAZOR.
APPLICATION FILED DEC. 3, 1901.
NO MODEL

manufacturer of electric razors, Braun was renowned for its product design team, which was headed by **Dieter Rams**, and which now began developing products for Gillette. In 1971 the company introduced the world's first twin-bladed cartridge system, the *Trac II*, which became an immediate best-seller. Later, after ten years' development, Gillette launched the *Sensor* system (designed by the Braun design team) in 1990. This revolutionary product featured thin spring-mounted blades set in a pivoting plastic cartridge. In the year of its introduction, 24 million razors and an astonishing 350 million cartridges were sold. As the most successful product launch in Gillette's history, the lessons learned from *Sensor* were clear: it reaffirmed the company's commitment to "spend whatever it takes to gain technological supremacy in a category, and then produce innovative products that will capture consumers, even at a premium price". This philosophy led directly to Gillette's latest flagship product, the remarkable triple-bladed *Mach 3*. By far the most expensive wet razor systems product today, the ultra-thin *Mach 3* blades were the first ever to be coated with a diamond-like film. Throughout the 1980s and 1990s, Gillette continued diversifying by purchasing Oral-B, **Waterman**, **Parker Pen** and Duracell, with the result that, by the year 2000, it had become the world's second most profitable brand after Coca-Cola.

GIORGETTO GIUGIARO

BORN 1938
GARESSIO, ITALY

→→ *Macchina*
Sportiva sports
watch for Seiko, 1995

After studying technical drawing and graphic design at the Accademia di Belle Arti in Turin, Giorgetto Giugiaro initially worked in **Fiat**'s design department. While at **Fiat**, he also trained at an engineering technical college in Turin. From 1959 he headed the styling department at **Bertone** in Turin, where he collaborated with Nuccio Bertone (b. 1914) and designed several cars, including the **BMW** *3200 CS* (1961), the **Alfa Romeo** *Giulia GT* (1963) and the Fiat *850 Spider* (1965). In 1965 Giugiaro became director of the **Ghia** automotive-styling studio, where he designed the Fiat *Dino Coupé* (1967). Three years later, he founded Ital Styling (later re-named ItalDesign) with Aldo Mantovani (b. 1927). This enormous design consultancy, which now employs over 750 people, offers car manufacturers a complete design service from pre-production studies to the construction of working prototypes. As well as designing sleek sports cars, such as the Maserati *3200 GT Coupé* (1998), Giugiaro also designed more utilitarian vehicles such as the **Volkswagen** *Golf* (1974), the Fiat *Panda* (1980) and the Fiat *Uno* (1983). In 1981, ItalDesign's industrial design section was established as an independent office, Giugiaro Design. It has since designed numerous products and vehicles, including the *Logica* sewing machine for Necchi (1982), the **Nikon** *F4* and *F5* cameras (1988 & 1996), the *High Speed Ferry* catamaran for SEC

Logica sewing
machine for Necchi,
1982

(1992), the *ETR460 Pendolino* locomotive for the Italian Railway Company (1995), the *Macchina Sportiva* watch for Seiko (1996), and *Grand Prix* ski

boots for Nordica (1997). Giugiaro Design also works in the areas of signage and packaging. Driven by the "pursuit of function", Giugiaro skilfully combines physical requirements with those of production methods, while being acutely aware of marketing considerations. He does not attempt to impose a formal design language on his products, but instead works out "the aesthetic and functional answers on the basis of need".

GK DESIGN GROUP

FOUNDED 1957
TOKYO, JAPAN

YFM350FWX
all-terrain vehicle
for the Yahama
Motor Co., 1995

→→Signage for Expo
'70, Osaka, 1970

→→Street furniture
and signage for Expo
'85, Tsukuba, 1985

The GK Design Group is one of Japan's foremost multi-disciplinary design consultancies. Founded in 1957 by **Kenji Ekuan**, the GK (Groupe Koike) Design Group initially specialized in industrial design. Over the succeeding years, the office widened its remit and became involved in communication, graphic and environmental design. Believing in an integrated approach to design, sometimes referred to as "total design", the studio has built up an impressive network of subsidiaries so as to offer a comprehensive range of services from research (market surveying and product planning) to promotion (corporate identity and event planning). Over the last 40 years, GK Design has designed motorcycles, audio equipment, a snowmobile, a personal watercraft and an all-terrain vehicle for **Yamaha**; locomotives for the East Japan Railway Company, packaging for Kikkoman, Konica and the Kirin Brewery; and public toilets for Restatio. It has also designed a diversity of products for other manufacturers, including a parking meter, a sewing machine, saucepans, watches, dental equipment and cosmetic packaging. GK Design is additionally noted for its design of signage and urban planning, especially its work for Expo '70 in Osaka and Expo '85 in Tsukuba. The work of the GK Design Group is distinguished by exquisite engineering and a functionalist formal vocabulary.

Komachi train for the
East Japan Railway
Company, 1997

KENNETH GRANGE

BORN 1929
LONDON, ENGLAND

Kenneth Grange trained at the Willesden School of Arts & Crafts in London between 1944 and 1947. Later, while completing his national service, he was trained as a technical illustrator by the Royal Engineers. He subsequently worked as an assistant for various architecture and design practices before founding his own London-based design office in 1958. Two years later, he redesigned a food mixer for Kenwood that had been originally launched in 1950. His resulting *Chef* food mixer (1960), which was restyled in just four days, was inspired by earlier kitchen appliances designed by Gerd Alfred Müller (b. 1932) for **Braun**. Grange worked as a design consultant to Kenwood for over 40 years and regularly updated this classic British design. In 1972 Grange, together with Theo Crosby, Alan Fletcher, Colin Forbes and Mervyn Kurlansky formed the design partnership **Pentagram**. Among Grange's many notable designs as part of Pentagram are the **Kodak** *Pocket Instamatic* camera (1975), the **Parker** *25* range of pens (1979), the *Royale* and *Protector* wet razors for Wilkinson Sword (1979 & 1992) and the *ST50* travel steam iron for Kenwood (1995). Grange also designed the exterior body of British Rail's *125 Intercity* high-speed train (1971–1973), Adshel bus shelters for **London Transport** (1990) and the restyled classic London black cab for London Taxis International (1997). Grange has received numerous awards for his esign work, including ten **Design Council** awards and the Duke of Edinburgh's Prize for Elegant Design in 1963. From 1985 to 1987 he served as

⌄*Royale* razors for
Wilkinson Sword,
1979

Kenwood *Chef A701*
food mixer, 1960

ST50 travel steam iron for Kenwood, 1995

Master of the Faculty of Royal Designers for Industry and was appointed president of the Chartered Society of Designers in 1987. His work was also the subject of a one-man show at the Boilerhouse at the Victoria and Albert Museum, London, in 1983. In 1997 he retired from Pentagram and established his own independent design office. Grange regards design as a means of innovation and believes that design should be an integral part of the manufacturing process. Grange combines the traditional attributes of British design – honesty, integrity and appropriateness – to create high-quality, meticulously-detailed products that meet the criteria for "good design". Throughout his long career Grange has shown a remarkable consistency – his gently elegant designs have all had an underlying robustness – which reflects his skill in balancing functional requirements with those of aesthetics. In many ways, Grange's work eloquently defined British design during the last half of the 20th century.

"The Exhibition as a Lesson in Taste"

THE GREAT EXHIBITION

1851 LONDON, ENGLAND

The *Report on Arts and Manufactures* drawn up by a Select Committee in 1836 concluded that, so far as industrial design was concerned, Great Britain was being surpassed by its European rivals. During the 1840s Queen Victoria's husband, Prince Albert of Saxe-Coburg Gotha, and **Henry Cole**, an official in the Public Records Office who had founded the short-lived but highly influential Summerley's Art-Manufactures, attempted to address this situation. They campaigned tirelessly for design reform that would align art more closely with industrial production, and endeavoured to revitalize the Society of Arts with a series of exhibitions dedicated to industrial design. When these efforts proved unsuccessful, they decided to organize an exhibition to promote their cause themselves. In 1848 they began planning an international fair that had a three-fold purpose: "exhibition, competition and encouragement". Drawing enthusiastic support both nationally and internationally, the organizers arranged an open competition for the design of the great hall that would house all the exhibits. In 1850, after a cumbersome design had been officially selected, an inspired plan by Joseph Paxton (1801–1865) was substituted. Paxton, who was the superintendent of the gardens at Chats-

Southern Entrance to the Transept

worth, the Derbyshire home of the Duke of Devonshire, ingeniously met all the rigorous specifications, from material cost to construction time, with an enormous glass house, which was based on an earlier design of his for a conservatory. This elegant structure, which became known as the Crystal Palace, comprised a remarkable construction of industrially prefabricated iron components and provided 33 million cubic feet of space. Sited in Hyde Park, the building enclosed an area of 19 acres and easily incor-

THE INDUSTRY OF ALL NATIONS

The silver works of Mr. HIGGINS, of London, are such as come within the province rather of a spoon and fork manufacturer than of a maker of the articles which Mr. Higgins exhibits, as specimens of his best ordinary production, are worthy of more public attention than such as have been prepared expressly for the present occasion. Our illustrations commence with an APOSTLE SPOON, surmounted by a figure of St. Peter, being one of a series of twelve, which are elevated on a rotatory pedestal. With this is a simple but graceful DESSERT FORK, of which the stem and prongs are of silver in imitation of twisted silver plate. From his very numerous contributions we select several, chiefly commending those designs which are taken exclusively from natural objects; and we may remark that SPOON and a DESSERT-FORK taken from different sets, one composed of vine branches and the other of conventional ornament. In the centre column we engrave a light and elegant CREAM-LADLE, the design of which appropriately consists of stems, leaves, and flowers of the common buttercup. The plant is fashioned for its purpose branches, and the handle composed of agate. On the opposite side of the page are a DESSERT with the best possible taste, and the effect of the work is greatly enhanced by the gilding, which

Silverware exhibited by Mr Higgins of London – typical of the highly decorated designs shown at the exhibition

porated the park's mature trees. It could be argued that Paxton's exceptionally innovative building was the highlight of the event, which was named the "Great Exhibition of the Industry of All Nations" and was held in 1851. Over 14,000 exhibitors participated in the exhibition, nearly half of whom were non-British. While some of the most popular exhibits were industrial machines, such as hydraulic presses, steam engines, pumps and automated

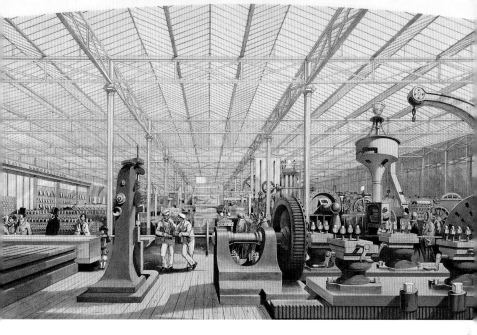

Whitworth's stand of machine tools at the Great Exhibition

cotton mules (spinning machines), many household items, such as furniture, glass, ceramics and metalware, also attracted much attention. The domestic exhibits, however, were for the most part notable for their profusion of superfluous decoration, which was drawn from an eclectic range of styles – Greek, Egyptian, Renaissance, Rococo, Baroque, Celtic – and the "vegetable world". Everything from scissors to carriages was encrusted with ornament, and what was meant to be a "lesson in taste" had become, much to the horror of the organizers, a celebration of decorative excess. Although the exhibition was extremely popular, attracting over 6 million visitors, and was a political and economic success, it was to a large extent seen by its organizers as a vindication of their concerns. Afterwards, some of the exhibits were used to form a study collection for the teaching of design that eventually became the basis of the Victoria & Albert Museum. Between 1852 and 1854 the Crystal Palace was moved to Sydenham Hill in Upper Norwood, London, where it was re-erected in a somewhat different form. It remained there until it was destroyed by fire in 1936.

WALTER GROPIUS

BORN 1883 BERLIN, GERMANY
DIED 1969 BOSTON, MASSACHUSETTS, USA

Walter Gropius studied architecture at the Technische Hochschule, Munich, and the Technische Hochschule, Berlin. Between 1908 and 1910 he worked in **Peter Behrens**' office, where he designed offices and furniture for the Lehmann department store in Cologne. In 1910 Gropius established an architectural partnership with Adolf Meyer (1881–1929) and became a member of the **Deutscher Werkbund**. He was an active member of the Werkbund and initially opposed Hermann Muthesius' (1861–1927) advocacy of standardization and design "types". In 1911 Gropius designed his well-known Fagus factory, which incorporated an innovative curtain wall suspended from the building's vertical elements. He also designed a steel and glass model factory for the 1914 Deutscher Werkbund exhibition in Cologne. After the appalling devastation of the First World War and influenced by the success of Henry **Ford**, Gropius came to realise the need for standardization and an industrial approach to design if society was to be modernized. He subsequently merged the two art schools in Weimar to form the Staatliches

Walter Gropius &
Adolf Meyer, Nickel-
plated brass door
furniture, 1928

Bauhaus in an effort to unify the arts and to "provide artistic advisory services to industry, trade and craft". As director of the design school from 1919 to 1928, Gropius instigated a system of workshops headed by "Masters". During this period he created several pieces of white-painted furniture, co-designed metal door furniture with Adolf Meyer, and developed a prefabricated house for the Weißenhofsiedlung estate built as part of the Werkbund exhibition "Die Wohnung" (The Dwelling) in Stuttgart in 1927. The purpose-built premises which Gropius designed for the Bauhaus following its move to Dessau in 1925 dramatically embodied the school's shift towards industrial modernity.

In 1934 Gropius moved to London, where he worked in partnership with the architect E. Maxwell Fry (1899–1987) until 1937. While in London, Gropius also worked for Jack Pritchard's company, **Isokon**, where he was appointed controller of design in 1936. In 1937 he emigrated to America and became professor of architecture at Harvard University. One of the greatest proponents of the Modern Movement, Gropius was also one of the most influential industrial design educationalists of the 20th century.

HANS GUGELOT

BORN 1920 CELEBES, INDONESIA
DIED 1965 ULM, GERMANY

After studying architecture, Hans Gugelot worked with Max Bill until 1954.
From 1954 to 1965 he worked in the design department at **Braun**, where he
helped develop a house style with a strong visual identity based on functionalism and essentialism. Between 1954 and 1965 Gugelot also directed the
product design department 2 at the Hochschule für Gestaltung in Ulm,
where he promoted the "form follows function" approach to design. Among
his best-known products, the *Phonosuper SK 4* radio-phonograph (1956) was
co-designed with **Dieter Rams** and nicknamed "Snow White's Coffin" because of its clear acrylic lid and hard-edged geometric formalism. Other notable designs for Braun included the *Sixtant* electric shaver (1962). Gugelot
also designed the *S-AV 1000* carousel slide projector for **Kodak** (1962) and
built-in furniture for Bofinger (1954). While Gugelot's career was short, his
influence – particularly upon the development of German product design –
was immense.

Sixtant electric
shaver for Braun,
1962 (co-designed
with Gerd Alfred
Müller)

JAMES HARGREAVES

BORN C. 1720 NEAR BLACKBURN, LANCASHIRE, ENGLAND
DIED 1788 NOTTINGHAMSHIRE, ENGLAND

Hargreaves' *Spinning Jenny*, 1764

Around 1760, James Hargreaves was employed by Robert Peel to co-design an improved carding machine for combing wool. Later, while working as a weaver, Hargreaves developed the first practical spinning machine. Like many weavers, he had his own loom and spinning wheel at home. When his daughter Jenny accidentally knocked over the wheel one day, so the story goes, Hargreaves observed how the spindle continued revolving. This gave him the idea of how a whole line of spindles could be worked off a single wheel. He built the first *Spinning Jenny* around 1764, which used eight spindles to spin thread and thus increased output eightfold. Hargreaves' invention, which could be easily operated by children, was initially used only at his home. When he later began selling the machine in 1768, Lancashire hand-spinners, who were fearful of its impact on their livelihoods, marched on his house and destroyed the machinery. He eventually moved to Nottingham and established a small cotton mill. As Hargreaves did not apply for a patent until 1770, the *Spinning Jenny* was copied and improved by others, which resulted in its spinning capacity being increased from eight to 80 threads. At the time of his death, there were over 20,000 *Spinning Jennies* in operation in Great Britain, making it one of the great inventions of the Industrial Revolution.

Woman working at the *Spinning Jenny*, c. 1870

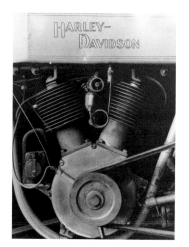

HARLEY-DAVIDSON

FOUNDED 1902
MILWAUKEE, WISCONSIN, USA

45-degree V-Twin
engine, 1909

In 1901 William Harley (1880–1943) and Arthur Davidson (1881–1950) began experimenting with ways to "take the work out of bicycling". A year later, they founded the Harley-Davidson Motor Company in a shed that belonged to Davidson's father. By 1903 the friends had produced their first motorcycle, which was powered by a single 167cc engine, and later the same year they produced another two motorcycles. In 1906 the company was incorporated and the following year its production had risen to 150 motorcycles per annum. The famous 45-degree V-Twin engine, which could produce a top speed of 60 mph, was introduced in 1909. By 1913 Harley-Davidson was dominating motorcycle racing in America and its production leapt to 12,904 motorcycles for that year. Having previously been proven effective by the US military in border skirmishes with Pancho Villa, over 20,000 Harley-Davidson motorcycles were deployed during the First World War. Following the war, the company grew rapidly and by 1920 had become the largest manufacturer of motorcycles in the world. The repu-

Early model, c. 1913

tation of Harley-Davidson's motorcycles for speed was enhanced when a works bike became the first to win a race with an average speed of 100 mph in 1921. During the 1920s, the Teardrop gas tank and front brake were introduced and in 1936 the overhead valve Knucklehead engine came into being. During the Second World War, 90,000 motorcycles were produced for Al-

Motorcycle with Panhead engine, produced 1948–1965

lied use, and in 1948 the famous Panhead engine was launched. 1957 saw the introduction of the *Sportster*, described as the "father of superbikes", and a year later the smooth-riding *Duo Glide* with its hydraulic suspension was launched. In 1965 an improved model known as the *Electra Glide* was ushered in, and a year later the Shovelhead engine succeeded the V-Twin. The 1970s saw a motorcycle boom in America and many Harley-Davidson enthusiasts began to customize their bikes, often in the chopper style. By 1975 the company was producing over 75,000 motorcycles a year. While facing increasing competition from Japanese manufacturers, the brand identity of the company continued to strengthen and today it enjoys a 56% share of the US market, with growing demand in Europe, Japan and Australia. For many years Harley-Davidson has powerfully symbolized American individualism, and this has no doubt contributed to its success and its recognition the world over as the "King of the Road".

Road King
motorcycle, 2000

JEAN HEIBERG

BORN 1884 OSLO, NORWAY
DIED 1976 OSLO, NORWAY

Plaster prototype
of the *DBH 1001*
telephone for
Ericsson, c. 1930

Jean Heiberg studied painting in Munich and then trained under Henri
Matisse (1869–1954) in Paris from 1908 to 1910. He remained in Paris until
1929, becoming a follower of the Fauves and receiving widespread recognition for his own paintings and sculptures. After his return to Oslo, he was
commissioned to design the casing of a newly developed Bakelite telephone,
which was to replace the existing metal types. The internal layout of this
early plastic telephone had been devised by the Norwegian engineer, Johan
Christian Bjerknes, for a joint venture between the Norsk Elektrisk Bureau
and the Swedish telephone manufacturing company, **L. M. Ericsson**. Heiberg's sculptural housing for the *DHB 1001* telephone (1930), which was
developed using plaster prototypes, was less angular than earlier models.
The coherent relationship between the handset and the main instrument
body established the standard for telephone design until the 1950s. The
DHB 1001 was produced from 1932 and was widely distributed throughout
Scandinavia, Britain, Italy, Greece and Turkey. The telephone was also manufactured under licence by **Siemens** in England as well as in France and
America. Heiberg's design later inspired **Henry Dreyfuss**' *Model 300* tele-
phone (1937) for **Bell** Laboratories.

DBH 1001 telephone
for Ericsson, put into
production in 1932

D. J. De Pree

HERMAN MILLER

FOUNDED 1923
ZEELAND, MICHIGAN, USA

Robert Propst (b. 1921) wrote of Herman Miller: "Our aspiration has always been to design things that are fashion-proof; that have more quiet, enduring qualities; that try to retreat to more elementary qualities". The origins of this remarkable company can be traced back to 1923, when D. J. De Pree and his father-in-law, Herman Miller, acquired the majority holding in the Michigan Star Furniture Company and subsequently renamed it the Herman Miller Furniture Company. Like many furniture manufacturers at the time, Herman Miller produced reproduction furniture that was popular within the mainstream market. In 1930, however, when De Pree met the designer Gilbert Rohde (1894–1944), the company's destiny changed. Rohde proposed a high quality range of Modern furniture that was of the utmost simplicity, so that all the value went into materials and construction rather than decorative surface treatment. The subsequent success of Rohde's Modern furniture led Herman Miller to abandon the manufacture of reproduction furniture altogether in 1936, and by 1941 the company had opened a New York showroom to display its new Modern designs. In 1946 **George Nelson** succeeded Rohde as design director and subsequently brought in other talented designers, including **Charles Eames**, Isamu Noguchi (1904–1988) and Alexander

Robert Propst,
Action Office II
system, 1968

Don Chadwick & Bill Stumpf, *Aeron* office chair, 1992

Girard (1907–1993), to create high-quality Modern furnishings for the company. Over the following years, Herman Miller produced designs that met De Pree's criteria for good design – "Durability, Unity, Integrity, Inevitability". These included Charles and Ray Eames' moulded plywood chairs (1945–1946) and plastic shell series of chairs (1948–1950) and George Nelson's *Comprehensive Storage System* (1959) and *Action Office I* (1964–1965). 1968 saw the launch of Robert Propst's landmark *Action Office II*, which literally transformed the office landscape. Since then, the company has been a leader in the contract furniture market and has produced a series of pioneering office chairs by Bill Stumpf (b. 1936) and Don Chadwick (b. 1936) – the *Ergon* (1976), the *Equa* (1984) and the revolutionary *Aeron* (1992). As one of America's most admired corporations, Herman Miller continues to flourish through its quest for design and manufacturing excellence.

GEOFF HOLLINGTON

BORN 1949
ESSEX, ENGLAND

Proficia internet telephone for Camelot Inc., 1997

Geoff Hollington trained as an industrial designer at Central School of Arts & Crafts and later studied environmental design at the Royal College of Art, London, graduating in 1974. Between 1976 and 1978 he was a member of the design team for the new town of Milton Keynes, and as such was responsible for landscaping and street furniture. In 1980 he founded Hollington Associates with the aim of concentrating on industrial design – from consumer products and medical equipment to personal electronics and office furniture systems. Since then he has worked on numerous projects for such companies as **Herman Miller**, **Parker Pen**, **Gillette**, **Matsushita**, Cable & Wireless, NEC, Filofax, Lloyd Loom, **Ericsson**, Hille, Artifort and **Kodak**. In 1992 his groundbreaking *Relay* furniture group (1990) for Herman Miller received a Gold Award from the Industrial Designers Society of America. Hollington has readily embraced the introduction of computer-aided design and high-tech engineering tools, yet still believes in the primary importance of "the front end, innovative, thoughtful part" of the design process. More recently, the office has added "interaction design" to its remit and has designed around 100 interactive exhibits for the Wellcome Wing at the Science Museum in London.

Digital TV remote control for Cable & Wireless, 1999

KNUD HOLSCHER

BORN 1930
RØDBY, DENMARK

Knud Holscher studied architecture at the Royal Danish Academy of Art, Copenhagen, and from 1960 to 1964 worked as an assistant to **Arne Jacobsen**. Together with Alan Tye, he subsequently designed the *Modric* range of sanitary fittings, which received a British Design Award and the British Aluminium Industry's prize in 1966. After returning to Denmark from Great Britain, he became a partner in the Krohn & Hartvig Rasmussen architectural office in 1967. In 1970 he won another British Design Award for an office signage system, and two years later designed his highly successful *d-line* range of stainless-steel door furniture, which has won numerous awards and has grown to become the world's most extensive range of architectural ironmongery and bathroom fittings. In 1975 Holscher designed stainless-steel salad servers and tongs for Georg Jensen that were remarkable for their simplicity. Today, Knud Holscher Industrial Design is one of Scandinavia's leading design consultancies and numbers **J C Decaux**, **Erco**, **Philips**, Fritz Hansen and Facit among its clients. The office has also won numerous competitions for street furniture and signage systems.

d-line staircase
system, 1999

HONDA

FOUNDED 1948
HAMAMATSU, JAPAN

*Dream 246cc
motorbike, 1960*

↘*CUV-ES Canopy
(Clean Urban
Vehicle), early
1990s*

"You meet the nicest
people on a Honda"
– advertising cam-
paign for the *Super
Cub*, 1962

From an early age, Soichiro Honda (1906–1991) was fascinated with anything mechanical. As a young man, he built and drove racing cars and won several races before retiring from competition after seriously injuring himself. In 1937 he founded a piston ring manufacturing company, which was later acquired by **Toyota**, and in 1946 established the Honda Technical Research Institute, which produced the first Honda product, the *A-type* bicycle motor, a year later. This venture was re-named the Honda Motor Company in 1948 and in 1949 launched its first motorcycle, the *Dream D-type*. During the 1950s Honda introduced its first *H-type* power engine (1953) and its lightweight 100cc *Super Cub* motorcycle (1958). It subsequently established sister companies in America (1959) and Europe (1960), and in 1962 finished the construction of the famous Suzuka circuit, near its factory. The same year, the company began marketing the *Super Cub* motorcycle with its well-known "You meet the nicest people on a Honda" advertising campaign in America, which helped to alter attitudes towards motorcycle riding. The result was that, by 1967, some 5 million *Super Cubs* had been sold, making it the best-selling motorcycle in the world (some 25 million have been sold to date). In 1963 Honda introduced its first sports car, the *S500*, and the *T360* light truck. As part of its research and develop-

You meet the nicest people on a Honda. And the remarkable thing is the low cost of it all. Prices start about $215? Insurance is painless. Upkeep negligible. Honda's four-stroke engine demands 200 miles from a gallon of gas. And gets it. Plenty of drive. That's how you stay at the top of the class. World's biggest seller. HONDA

ment programme, Honda entered Formula 1 racing in 1964, winning its first
victory in 1966 in Mexico. Since then, F1 racing has remained an important
aspect of the company's business. Honda also went on to become the first
company to win the Constructor's Championship in all motorcycle classes
at the 1966 World Grand Prix. Apart from its racing triumphs, Honda's mini-
compact *N360* (1966) became the best-selling car in Japan in 1968. Another
highly successful compact car was launched in 1972, the omnipresent fuel-
efficient Honda *Civic*, which sold 1 million units within its first two years of
production and 10 million units by 1995. The larger Honda *Accord* was introduced in 1976 and achieved a similar sales record. During the 1980s, Honda
continued to expand abroad and introduced several new car models, including the mini-compact *Today* (1985) and the luxury-end *Legend* (1985). In 1989
the *Accord* became the best-selling car in America. In 1992 the company established a huge high-tech manufacturing plant in Swindon, England. Ever
mindful of environmental concerns, in 1993 Honda became the first manufacturer to produce engines that met the new Californian emissions regula-
tions. Honda's quest for technological innovation continues to produce
challenging designs in the areas of motorcycles, cars and power products,
and since 1986 the company's engineers have been developing a walking
two-legged "humanoid robot" that has a degree of artificial intelligence.

HOOVER

FOUNDED 1908
NORTH CANTON, OHIO, USA

Vacuum cleaner for
the Electric Suction
Sweeper Company,
1908

⬎Henry Dreyfuss,
Hoover vacuum
cleaner *Model 150*,
1946

Advertisement for
the Hoover vacuum
cleaner *Model 150*,
1936

The first Hoover vacuum cleaner was based on a design
by a department store janitor, James Murray Spangler
(1848–1915). He realised that if the sweeping action of
a carpet cleaner was combined with the relatively new
idea of using suction to remove dirt and dust from carpets, performance would be dramatically improved.
His wood and tin prototype used a pillowcase for the bag and although it
was a relatively clumsy device, it worked fairly well. Lacking the funds to
commercialize this new invention, Spangler managed to interest his friend,
W. H. Hoover (1849–1932), who was a manufacturer of leather goods. In
1908 the Electric Suction Sweeper Company (later re-named the Hoover
Suction Sweeper Company in 1910 and then The Hoover Company in 1922)
was established and Spangler became its production manager. Frank Mills
Case subsequently re-designed the cleaner, giving it an aluminium casing.
These first electric vacuum cleaners were initially produced at the rate of six
to eight machines a day and weighed nearly 40 pounds, but the development in 1909 of a small high-speed universal motor by Hamilton & Beach,

A Totally New Conception
of the
World's Greatest Cleaner

MADAM · HERE'S THE MOST HELPFUL CLEANER EVER BUILT

Hoover *Constellation*
vacuum cleaner,
1955

Racine, allowed their weight to be decreased to around five or six pounds.
The same year, Hoover began a research and development programme,
which resulted in the introduction of the "Agitator" to all its upright models in 1926. This innovative feature comprised a rotating bar with spiralled
bristles that helped loosen dirt and dust caught in carpets through a beating action. The famous Hoover slogan of 1919, "It beats, as it sweeps, as
it cleans" was used very effectively to market this improvement in vacuum
cleaner performance. Also in 1926, another feature appeared for the first
time – the power switch was moved to the handle on the *Model 700* cleaners. From the 1930s to the early 1950s, the company commissioned **Henry
Dreyfuss** to design a range of streamlined cleaners, including one that featured an illuminated lighting strip. In 1955 the futuristic looking *Constellation* cleaner was introduced which, when operating, floated on a cushion
of air just like a hovercraft. Extremely popular in America, the design of this
model reflected contemporary interest in future space technology. In testimony to the brand's continuing influence, the Hoover name is still recognized around the world, although the Hoover Company is no longer an international company. Maytag Corporation, which acquired Hoover in 1989, di-
vested its European operations (including UK operations) in the mid-1990s.

ELIAS HOWE

BORN 1819 SPENCER, MASSACHUSETTS, USA
DIED 1867 BROOKLYN, NEW YORK, USA

Elias Howe's sewing machine, c. 1845

Elias Howe was interested in mechanics from childhood, and later worked as a machinist in a factory producing machinery for the cotton industry. While there, it was suggested to him that a fortune could be made by developing a machine that could sew. For the next five years, he undertook experiments that eventually led in 1845 to the construction of the first practical sewing machine. The following year, he patented his handwheel-operated lock-stitch sewing machine, which carried the cloth on a spiked feed-plate past a vertically positioned needle, so that the eye-pointed needle could then pierce the cloth in the same direction as the handwheel was turned. His invention received little attention in America at first, so his brother Amasa took it to England, where the patent rights were sold to William Thomas for

Detail of Elias Howe's sewing machine

£250 (then equivalent to US$1,250). Around this time, Howe also moved to England, where he worked for a pittance while continuing to perfect his invention so that it could be used for the sewing of leather and corsets. In the meantime, sewing machines were being widely manufactured in America that infringed his patents – including those produced by Isaac **Singer**. After three years in Britain, Howe returned to the United States and began litigation. His rights were eventually recognized in 1854 and he subsequently received royalties from all sewing machines manufactured in the United States. Howe's innovative design heralded the birth of the sewing machine industry and quite literally revolutionized clothing manufacture in both homes and factories around the world.

IBM

FOUNDED 1911
NEW YORK, USA

Eliot Noyes, IBM
Model A typewriter,
1948

→Eliot Noyes, Main-
frame computer
system *360*, 1964

→↘IBM personal
computer, c. 1984

Eliot Noyes, *705
Electronic Data
Processing Machine*,
1954

In 1911 three firms – the Tabulating Machine Company (producing electrical machines that processed data using a punched card system), the Computing Scale Company (holding the patent for Julius Pitrat's computing scale) and the International Time Recording Company (manufacturing a mechanical time recorder) – merged and later re-named International Business Machines (IBM) in 1924. Four years later, the data capacity of punched cards almost doubled from 45 to 80 columns of information, which heralded the development in the early 1930s of a new series of machines that could not only perform addition and subtraction, but could also undertake full-scale accounting calculations. In 1935 IBM launched its first electric typewriters and, a year later, provided accounting machines for America's Social Security Program, which was referred to as "the biggest accounting operation of all time". Developed in collaboration with Harvard University, the ASCC (Automatic Sequence Controlled Calculator) was introduced in 1944 and was the first machine capable of automatically executing lengthy computations. Four

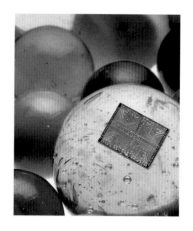

years later, IBM launched the first computer that combined electronic computation with stored instructions. During the 1940s, **Norman Bel Geddes** advised IBM on the design of its office products. 1952 saw the introduction of the first computer capable of scientific calculations, the IBM *701*, which used magnetic tapes to store data equivalent to 12,500 punched cards. **Eliot Fette Noyes**, who had previously worked in Geddes' office, designed the *705 Electronic Data Processing Machine* in 1954 and two years later was appointed corporate design director of IBM. Noyes was responsible for several revolutionary products, including the *Selectric I* typewriter (1961) with its innovative golf ball typing head. He also developed a strong corporate identity for IBM, both through the design integrity of his own products and by commissioning graphics from Paul Rand and buildings from leading architects such as **Marcel Breuer**. During the 1950s and 1960s, IBM continued to develop state-of-the-art computing systems for a range of uses, from accounting and ticketing operations to helping **NASA** with its space programme. Computer processors and peripheral units, such as those of IBM's landmark *System/360* (1964), were still huge in size but were becoming increasingly powerful. In 1968 IBM introduced its *3986* Braille typewriter – an important design for disability – and two years later entered the photocopying market with the IBM *Copier*. During the 1970s, IBM developed the *3614 Automatic Banking Machine* and the *3660 Supermarket System* that could read universal product codes. The late 1970s saw the development of computers that used the new semiconductor technology and electronic typewriters with microprocessors. IBM introduced its first personal computer in 1981, but by the mid-1980s was finding it difficult to compete with **Apple Computer's** more user-friendly models. As one of the largest corporations in America, IBM was slow to react to technological change and as a result began losing market share. As a pioneer in the field of information processing, IBM has nevertheless affected all areas of life – from banking and shopping to energy generation and weather forecasting. The character of the company is hallmarked by technological innovation and a strong design ethos that reflects its core values – respect for the individual, customer service and excellence.

128,00-BIT CMOS SRAM microchip, early 1990s

IDEO

FOUNDED 1969
LONDON, ENGLAND

Tracy Currer & Nick
Dormon, Digital
Radio for BBC, 1997

IDEO's founder, Bill Moggridge (b. 1943), studied industrial design at the Central School of Arts & Crafts, Lon-
don, graduating in 1965. He later undertook a research fellowship in typography and electronic communications at the Hornsey School of Art. In 1969
he established the design office, Moggridge Associates, and ten years later
opened a second office in San Francisco, called I. D. Two, to cater to Silicon
Valley's fledging computer industry. One of his first products was the GRiD
Compass laptop computer (1980). Over the succeeding years, believing that
the design of an interface must be an integral part of a product's development, Moggridge christened and pioneered a new discipline known as
"Interaction Design". In 1991 Moggridge's office merged with David Kelley

Christopher Loew,
SyncMaster 400 TFT
flat screen monitor
for Samsung, 1996

Design and was renamed IDEO. By offering a full product development
service, the consultancy's interdisciplinary teams translate ideas into ready-
to-manufacture products. IDEO's non-linear design methodology, which
includes brainstorming sessions, can at times appear
chaotic; as the business guru Tom Peters noted: "IDEO
is a zoo.... Experts of all flavours commingle in 'offices'
that look more like cacophonous kindergarten class-
rooms than home of one of the world's most successful
design firms". This way of working, however, generates
a powerful synergy that harnesses both creativity and
cutting-edge technology to produce design solutions
that are both novel and imaginative. While at the forefront of technological developments, IDEO also realises
that emphasis must be placed on human needs, and
has therefore adopted a "user-focused' approach to design so as to create user-friendly products. IDEO has
designed over 2,000 medical, computer, telecommunications, industrial, furniture and consumer products
for a wide-ranging clientele, including **Apple Computers**,
Black & Decker, **Nike**, British Airways, Baxter Healthcare, Amtrak, Deutsche Telecom, Microsoft, NEC,
Nokia, Samsung, **Siemens** and **Steelcase**.

Rick Lewis, *Visor*
communicator for
Handsprung Inc.,
1999

TAKENOBU IGARASHI

BORN 1944
HOKKAIDO, JAPAN

Hyvälysti travel kit for Kai International, 1992

Takenobu Igarashi trained at Tama Art University, graduating in 1968, and went on to further studies at the University of California, Los Angeles in 1969. He later established his own design office specializing in graphic design, corporate identity and product design. Igarashi's early poster designs were remarkable for the way in which he was able to translate simple numbers and letters from the Western alphabet into striking eye-catching compositions. He also created sculptures that were almost architectural in their complexity. In 1985 he founded the Takenobu Igarashi Studio and began producing sculptural yet restrained designs, such as his *Zao* stool (1992) and series of iron platters (1990). Interestingly, Igarashi is able to imbue three-dimensional objects with a strong graphic quality, as illustrated by his *YMD* paperknife (1992) for ClassiCon, which has a handle that casts a wave-like shadow on the blade. Igarashi's sensitivity to both volume and line is, however, best expressed in his design of compact personal objects such as the *Hyvälysti* gardening and grooming kits. Igarashi is a professor at the Tama Art University and a visiting professor at the University of California, as well as vice-president of the Alliance Graphique International.

Hyvälysti gardening kit for Kai International, 1990–1991

IITTALA

FOUNDED 1881
IITTALA, FINLAND

Göran Hongell,
Aarne glassware,
1954

Kaj Franck, *Kartio*
glassware, 1958

The Swedish glassblower Per Magnus Abrahamsson founded the Finnish glassworks, Iittala, in 1881. The company's early glassware revealed influences from Europe and mainly consisted of simple undecorated designs for household use. In 1917 Iittala was acquired by the Karhula Glassworks, as part of the Ahlström Group, and numerous improvements were subsequently instigated at the factory. By the 1930s the glassworks was concentrating primarily upon the production of blown-glass rather than pressed glass. Aino Aalto's range of functional glassware (1932) and Alvar Aalto's famous *Savoy* vase (1936) were initially manufactured by Karhula and were only later produced by Iittala. In an effort to develop a range of art glass, in 1946 Iittala organiz-

ed a competition which was won jointly by Tapio Wirkkala (1915–1985) and Kaj Franck (1911–1989). The unconventional work of both designers, such as Wirkkala's *Kantarelli* vases (1946), heralded a new direction for Iittala and a new period in Finnish glass design. Iittala was later joined by Timo Sarpaneva (b. 1926), who produced sculptural art glass and in the mid-1950s also designed the company's first modern utility glassware, the *i-Collection*, which was suited to mass production. "Frosted" designs by Wirkkala and Sarpaneva subsequently exerted much influence on mainstream glassware in the 1960s and 1970s. In 1990 Iittala was purchased by the Hackman Group and has since commissioned designs from Konstanin Grcic (b. 1965), Marc Newson (b. 1963) and Harri Koskinen (b. 1970).

IKEA

FOUNDED 1943
ÄLMHULT, SWEDEN

Morten Kjelstrup &
Allan Østgaard,
Mammut child's
chair, 1993

Knut & Marianne
Hagberg, *Puzzle*
children's furniture,
1988

With a loan from his father of £150, the 17-year-old Ingvar Kamprad (b. 1926) founded IKEA in the town of Älmhult, a few miles from his home village, in 1943. Situated in the Småland region, historically one of the poorest parts of Sweden, the company originally sold a variety of products through mail order, from cattle soap to stockings. In the early 1950s, IKEA began retailing furniture at factory prices through mail-order catalogues. This move was met with such resistance by the established furniture trade that IKEA was not allowed to exhibit at the large furniture trade fair held in Stockholm, and suppliers threatened to boycott the company. Observing that "most nicely designed products were very, very expensive", Kamprad wanted to manufacture well-designed practical products that the majority of people could afford. With this "social mission" to democratize design, Kamprad realised that there were three essential priorities – aesthetics, function and suitability for mass production. During the late 1950s he recruited his first designers, Gillis Lundgren, Bengt Ruda and Erik Wørts, who, while working within the Modern idiom, designed furniture for self-assembly – a revolutionary development in furniture design. Lundgren's *Regal* bookshelf (1959) was one of the first IKEA products sold in flat-pack form. Another concept which IKEA's designers helped to pioneer was "modular thinking" – the development of modular components and interlinking systems rather than individual one-off products. This led to the company's co-ordination of its product range and promotion of an integrated home-furnishing look – an early and important example of lifestyle retailing. The advent of

Noburo Nakamura,
Poem chair, 1977

particleboard in the 1960s heralded the arrival of really low-cost furniture and influenced IKEA's product-line immensely. An inexpensive yet relatively hard-wearing material, particleboard continues to be used extensively by IKEA for all kinds of self-assembly furniture. While the 1970s were a period of general insecurity, IKEA prospered and expanded. During this time the company retailed several products in plastics, including the *Telegono* lamp (1970) by the avant-garde Italian designer, Vico Magistretti (b. 1920). Throughout its history, however, IKEA has mainly concentrated on producing interpretations of well-known classic designs – from the ubiquitous **Thonet** café chair to the *Polyprop* by **Robin Day**. The 1970s were also the decade of stripped pine and brightly coloured textiles, and IKEA developed this informal look within its product range specifically for younger consumers. The following decade, IKEA began producing stylish yet practical furniture designed by Niels Gammelgaard (b. 1944) – his *Moment* table (1987) won the Excellent Swedish Design prize. Other award-winning designs included the *Puzzle* range of children's furniture (1988) designed by Knut and Marianne Hagberg, which combined primary colours and simple forms and became a best seller. Throughout the 1990s, IKEA continued to expand its international operations and popularize its version of Scandinavian Modernism. Today the company boasts annual sales of £3.8 billion. Kamprad is not content with just producing home furnishings, however, and plans to diversify his product line into small pre-fabricated flats. As the design theorist Victor Papanek wrote: "One thing is certain: IKEA will continue in the forefront – ecologically, socially and culturally – of making things that work, possess beauty and are affordable."

INDIAN

FOUNDED 1901
SPRINGFIELD, MASSACHUSETTS, USA

Detail of Indian *Chief*
motorbike, 1947

Having built several racing bicycles around 1900, Oscar Hedstrom constructed an engine-powered "pacer" bike, which was used for the training of cyclists. This practical two-wheeler caught the attention of the entrepreneur George Hendee. In 1901 Hendee and Hedstrom founded the Hendee Manufacturing Company (later renamed Indian Motocycle Manufacturing Company), and the following year the first Indian motorcycles were manufactured. These single-cylinder bikes were the first commercially marketed gasoline-powered motorcycles in the United States. In 1903 the company launched a V-Twin engine-powered motorcycle, which was updated over the succeeding years to incorporate two- and three-speed gearboxes and swingarm rear suspensions. The revolutionary *Hendee Special* introduced in 1913 was the first motorbike to have an electric starter and a full electrical system. Prior to the First World War, Indian was the largest motorcycle manufacturer in the world. Later it created such well-known designs as the *101 Scout* (1919), the *Chief* (1920) and the *Sport Scout* (1935). Before the Great Depression there were over 20 American motorcycle manufacturers, but only Indian and **Harley-Davidson** survived. Indian later supplied large numbers of motorcycles to the US military, but in 1953 it went out of business, leaving a design legend that has refused to die.

Indian *Chief*
motorbike, 1946

INTERDESIGN

FOUNDED 1982
STOCKHOLM, SWEDEN

Tom Ahlström &
Hans Ehrich,
toothbrush for
Colgate, 1986

In 1968 Tom Ahlström (b. 1943) and Hans Ehrich (b. 1942) founded **A&E Design** in Stockholm, a design consultancy that became celebrated for its design for disability and won numerous awards for its life-enhancing and ergonomic products. In 1982 Ahlström and Ehrich established InterDesign as a subsidiary of A&E Design, and in 1997 the venture became an independent industrial design consulting company. Although mainly specializing in plastics, InterDesign has worked in a wide range of fields. It has designed sanitaryware for Gustavsberg (1984), a locomotive for ASEA-Traction (1985), a range of toothbrushes for Colgate (1986), automatic injection devices for SHL Medical (1997) as well as dental floss and tooth pick dispensers for Jordan, electric circuit breakers for Norwesco and lighting fixture components for Belysia. In 1992 the consultancy received the Swedish Environment Design Award for its returnable milk bottle (1992) designed for Arla. In the development of its product designs, InterDesign uses computer-aided design (CAD) systems to rapidly determine product structure, and stereolithography to validate the ergonomics, aesthetics and integration of components – all before tooling is engaged. According to InterDesign: "Designing the first product for a company is not without consequences. Designing the second and subsequent products means implementing a design strategy that includes product image in perfect coherence with user ergonomics. The whole must satisfy and attract the direct client (the distributor) as well as the client of the client (the end-user). This is the way to build fidelity, notoriety and competitive differentiation."

Tom Ahlström, Hans
Ehrich & Jochen
Ratjen, returnable
milk bottle for Arla,
1992

ISOKON

FOUNDED 1931
LONDON, ENGLAND

Marcel Breuer,
chaise longue for
Isokon, 1935–1936

Having studied engineering and economics at Cambridge University, Jack Pritchard (1899–1992) began working as a designer in 1925 for Venesta Plywood, a leader in the development of this new and exciting material. While there, he commissioned leading architects and designers such as Le Corbusier (1887–1965), László Moholy-Nagy (1895–1946) and **Wells Coates** to design exhibition stands for Venesta. In 1931 Pritchard, together with his wife and Coates, founded Isokon with the aim of bringing "modern functional design to houses, flats, furnishings and fittings". For a while, **Marcel Breuer** acted as the design controller of the firm and produced five plywood furniture designs between 1935 and 1937. **Walter Gropius** also worked for Isokon and was its design director from 1936 to 1937. One of Isokon's best-known products was the small *Penguin Donkey* (1939) bookcase designed by Egon Riss specifically for the storage of paperbacks published by **Penguin Books**. Benefiting from the exodus of **Bauhaus**-associated designers during the mid-1930s, Isokon became a leading exponent of Modernism in Britain.

Cover of Isokon
catalogue, c. 1936

THE NEW ISOKON CHAIR

ALEC ISSIGONIS

BORN 1906 SMYRNA (IZMIR), TURKEY
DIED 1988 BIRMINGHAM, ENGLAND

→Austin *Mini Seven*,
1959

Born in Turkey as the son of a Greek merchant, Alec Issigonis emigrated to England in 1922. Although he apparently received no formal training as an engineer, he did study for a time at Battersea College, London. He was subsequently employed as a draughtsman by Rootes Motors in Coventry and later moved to Morris Motors in Oxford, where he worked as a suspension designer. He later became Morris's chief engineer. 1948 saw the launch of his revolutionary Morris *Minor*, which was influenced by American automotive styling from the 1930s and 1940s. This diminutive yet curvaceous vehicle had a unitary body construction which made it suitable for large-scale mass production. As the first Modern British car, the Morris *Minor* was immensely popular and became the first all-British car to exceed sales of one million. It remained in production until 1971 and to this day has a devoted following of collectors and enthusiasts.

In the 1950s Issigonis briefly worked for other companies before returning to Morris, which had by now been renamed the British Motor Corporation (BMC). In response to petrol shortages resulting from the Suez Crisis and

Advertisement
featuring the 1949
model of the Morris
Minor

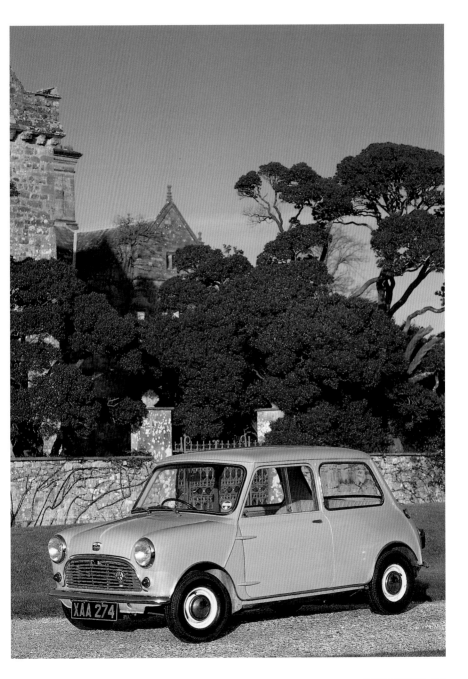

the resurgence of interest in the **Volkswagen** *Beetle*, Issigonis designed the small, fuel-efficient and inexpensive Morris *Mini*. Launched in 1959, this front-wheel drive, box-like vehicle was only three metres long. It employed a radical layout that included a space-saving transverse engine, which allowed the compact cab to accommodate four adult passengers in relative comfort. The *Mini* was influential not only in terms of its design but also culturally – it was the first truly "classless" car, as likely to be driven by celebrities as by ordinary citizens. No car came as close to epitomizing the Swinging Sixties as the Mini, although in later years its safety record was seriously questioned. The *Mini* was sold in several versions, including the wooden-framed, van-like *Traveller*. In 1962 Issigonis designed the slightly larger *1100* which was similarly manufactured under Morris, Austin and badges. Like his earlier models, this design was also influential upon later compact cars such as the **Renault** *5* and the Volkswagen *Golf*. In 1967 Issigonis was made a fellow of the Royal Society and two years later was knighted for his immense contribution to the British car industry. He eventually retired in 1971. By 1988 – the year of his death – over five million *Minis* had been sold.

Drawing of the
Morris *Mini Minor*.
c. 1959

JONATHAN IVE

BORN 1967
LONDON, ENGLAND

Jonathan Ive studied design at Newcastle Polytechnic and later worked for the London-based industrial design consultancy **Tangerine**, where he designed a wide range of products, from televisions and VCRs to sanitaryware and hair combs. At Tangerine he also assisted in the development of the *PowerBook* for **Apple Computer** in 1991. While working on this project, Ive noted that the bland product identity of computers was the result of their arbitrary configurations, and concluded that there was a huge opportunity in creating exciting new products that did not conform to the conventional grey, beige or black boxes. At this stage, the computer industry was mainly concerned with the internal aspects of its machines – processing speed and memory capacity – and little if any thought was being given to their external form. As a consequence, the industry as a whole was suffering from what Ive describes as "creative bankruptcy". Frustrated that, as an external consultant, he could only have a small impact on the future development

Advertisement for
iMac, 1999

of computers, in 1992 Ive joined the design team at Apple, where he became director of design. It was not until Steve Jobs' return to Apple, however, that the design team was given the freedom to concentrate on the "pursuit of nothing other than good design". Jobs realised that Apple needed to recapture its once strong identity, which had been diluted by a "design by focus group" mentality. Significantly, his first day back at the helm marked the beginning of the *iMac* project. The translucent turquoise *iMac* (1998) with its unified curvaceous organic form, broke all conventions. At long last, here was a computer that looked cool and had a strong identifiable character. At its launch, it became obvious that the rather beleaguered Apple (whose market share had shrunk to a miserable 3 % in 1997) had backed a winner that could restore its fortunes – an astonishing 150,000 *iMacs* were sold over the weekend following its introduction. Helped by a high-profile advertising campaign, which included the in-sightful slogan "Chic, not Geek", *iMac* became the best-selling computer in America – impressively, because of its design rather than its technology. Ive's designs have managed to powerfully differentiate Macs from PCs, and it may well be that the people who buy into Apple Computer products on the strength of their looks will become hooked on the company's excellent user-friendly operating system.

ARNE JACOBSEN

BORN 1902 COPENHAGEN, DENMARK
DIED 1971 COPENHAGEN, DENMARK

→→ *Series 7* chairs for
Fritz Hansen, 1955

Arne Jacobsen initially trained as a mason before studying architecture at
the Kongelige Danske Kunstakademi, Copenhagen, from where he graduated in 1927. He later worked in the architectural practice of Paul Holsøe
from 1927 to 1929. Around 1930 he established his own design office in
Hellerup and began practising independently as an architect and interior
designer. His early work was influenced by Modernists such as Le Corbusier
(1887–1965), Gunnar Asplund (1885–1940) and Ludwig Mies van der Rohe
(1886–1969). Jacobsen became one of the first designers to introduce Modernism to Denmark through such projects as the "House of the Future",
which he co-designed with Fleming Lassen in 1929. His first important architectural commissions were for the Bellavista housing project in Copenhagen (1930–1934) and the functionalist Rothenborg House, Ordrup (1930),
which was conceived as a *Gesamtkunstwerk*. For his most fully integrated
project, the SAS Air Terminal and Royal Hotel in Copenhagen (1956–1960),
Jacobsen designed every detail from textiles to light fittings. Prior to this,
in 1945 Jacobsen had also begun designing for industrial production and
this led most notably to his *Ant* chairs (1951–1952) and his *Series 7* chairs
(1955) for Fritz Hansen, which remain among the most commercially successful seating programmes ever produced. Jacobsen also designed the *AJ*
range of lighting (1955–1960) for **Louis Poulsen**, the *Cylinda-Line* metalware

Cylinda-Line teapot
for Stelton, 1967

collection (1967) for **Stelton**, cutlery for Georg Jensen, bathroom fittings for
I. P. Lunds and textiles for August Millech, Graucob Textilen and C. Olesen.

From 1956 until 1965, he was a professor at the Skolen for Brugskunst
in Copenhagen. In the 1960s, Jacobsen's most important architectural
scheme was St Catherine's College,
Oxford – a wholly unified project
for which he designed site-specific
furniture. Jacobsen's elegant essentialist industrial designs possess a
timeless appeal, and because of
this they have enjoyed remarkable
success.

JAGUAR

FOUNDED 1922
BLACKPOOL, ENGLAND

C-type on its way to victory at Le Mans, 1953

As a young motorcycle enthusiast, William Lyons (1901–1985) met William Walmsley, who had constructed a stylish motorcycle sidecar. Realising the sales opportunity of such a design, Lyons founded the Swallow Sidecar Company in Blackpool in 1922. The company's torpedo-shaped aluminium sidecars became an instant success and the firm grew rapidly. Lyons spotted another business opportunity in 1927, when Herbert Austin launched his revolutionary Austin *Seven*, the first small and inexpensive British car intended for the masses. Lyons subsequently constructed a chic two-seater body that was mounted on an Austin *Seven* chassis. Known as the Austin *Seven Swallow*, the car provided inexpensive luxury and proved so popular that a saloon version was introduced in 1928. Having moved the Swallow factory to Coventry around 1929, Lyons and Walmsley decided to design their own chassis, which were built by the Standard Motor Company. The resulting *S1* and *SS2* Coupés caused a sensation when launched in 1931. With their long bonnets and low bodies, they looked luxurious and expensive but were actually modestly priced. After Walmsley left the company in 1934, Lyons recruited the design engineer, Harry Weslake, and established an engineering department, which was headed by the youthful William Heynes. Working together, Weslake and Heynes developed a completely new saloon and sports car range,

Jaguar *XK-E* roadster, 1965

JAGUAR

A special kind of motoring which no other car in the world can offer

THE MARK TEN SALOON

Powered by the XK'S' Type
3.8 litre engine with 3
carburettors. Independent
suspension front and
rear. Disc Brakes on
all 4 wheels

THE MARK 2 SALOON

Available as 2.4, 3.4
or 3.8 litre models powered
by the world famous
XK engine. Disc Brakes
on all 4 wheels

THE 'E' TYPE G.T. MODELS

Available as Fixed Head Coupe
or Open Two Seater (with detachable hard
top as optional extra). 3.8 litre XK'S'
Type engine with 3 carburettors.
Independent suspension front and rear.
Disc Brakes on all 4 wheels.

Jaguar advertisement
from *The Motor*, 1963
– showing the *Mark-Ten* saloon, the *Mark 2* and the *E-type GT* model

which was launched at the 1935 London Motor Show. In tribute to their feline grace, agility and power, these cars were named "SS Jaguar". At the outbreak of the Second World War, car production ceased and the company manufactured sidecars for military use and also built aircraft. After the war, car production resumed, the SS name was dropped and the company became known as Jaguar Cars. In 1948 the first post-war model was launched, the rather conservative *Mark V*. Meanwhile, Jaguar had developed a new XK engine that had an impressive output of 160 bhp. For the Motor Show that same year, Lyons designed a suitable body for this new powerful engine and the result was one of the greatest sports cars of all time, the classic *XK120*. Claimed to have a top speed of 120 mph, the *XK120* was marketed as the fastest production car in the world. To prove the claim to the press, Jaguar speed-tested the car in Belgium, where it clocked 126 mph; when the windscreen was removed, it went even faster, achieving 133 mph. Orders soon flooded in, and it was not long before a new five-seater saloon with the XK engine, the *Mark VII*, was introduced at the 1950 Motor Show with an eye to the American market. Following the racing successes of the *XK120*, Jaguar decided to produce a real racing car, the *XK120C*, or *C-type* as it became known. To reduce weight on the car, a triangulated frame was adopted and a sleek aerodynamic body was designed by Malcolm Sayer. Although only 53 *C-types* were built, it was an influential design which took first, second and fourth place at the 1953 Le Mans. The similarly styled *D-type* followed, but from 1955 Sayer began developing a new car which would be both a superlative road car and racing car. The legendary *E-type* was subsequently unveiled at the Geneva Motor Show in 1961. Arguably the most beautiful car ever designed, with its seductive **styling** and top speed of nearly 150 mph, the *E-type* set new standards in car refinement. Costing £1,830, the *E-type* was half the price of an Aston Martin and a third of the

XK8 coupé and convertible, 1996

price of a **Ferrari** and stunningly embodied the value-for-money ethos that Lyons had always tried to perpetuate at Jaguar. In 1963 the stately *S-type* saloon was launched, followed five years later by the unmistakable *XJ6* saloon. Around this time Jaguar developed a new high-performance yet "silky smooth" engine, the famous V12, which was used to power later *E-types* and the *XJ12* launched in 1972, which was the world's fastest four-seater production car. As the luxury car market became increasingly competitive during the 1980s, Jaguar realised the absolute necessity of collaborating with one of the automotive giants. In 1990 it was acquired by the **Ford** Motor Company, since when it has undergone a renaissance. In 1991 the beautiful *XJ220* supercar was put into limited production and in 1996 the exquisite *XK8* was launched at the Geneva Motor Show. Both these cars, as well as the *XJ-Series* saloon (1994) and the *F-type* concept roadster (2000), were designed by **Keith Helfet**, who has skilfully evolved Jaguar's "signature" styling while embracing state-of-the-art technology. Jaguar's commitment to design has given the marque a strong visual identity and enabled the company to build rather than rest upon its long and illustrious heritage.

F-type concept roadster, 2000

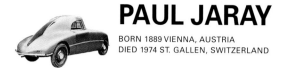

PAUL JARAY

BORN 1889 VIENNA, AUSTRIA
DIED 1974 ST. GALLEN, SWITZERLAND

Audi front chassis
with streamlined
body designed by
Paul Jaray, 1930s

Engineer Paul Jaray studied in Vienna and Prague before moving to Germany, where he worked as an aircraft designer for Flugzeugbau Friedrichshafen from 1913 to 1914. Between 1914 and 1923 he worked for Luftschiffbau **Zeppelin**, developing airships. His hydrogen *LZ38* airship was over 163 metres long and had a gas volume of 33,780 m³. His later helium *LZ126* airship was even larger, measuring over 200 m in length. This model became known as the *Los Angeles* and was the first airship to cross the Atlantic during the winter months. As an important pioneer of aerodynamics, Jaray designed a wind tunnel for Zeppelin in 1919 and was responsible for the adoption of the teardrop shape in airship design. Jaray also established scientific principles and mathematical models for streamlining, and in 1922 filed a patent for streamlined automobile bodies. His revolutionary enclosed car design, with its wrap-around windscreen, integrated headlamps and flush door handles, was at least ten years ahead of its time and was extremely influential upon later designs such as the 1936 **Volkswagen Beetle** and the 1938 Tatra *V-8*. **Audi** also produced several experimental designs based on Jaray's patents. In 1921 Jaray patented an aerodynamic bicycle known as the *J-Rad*, of which 2,000 were produced. The rider sat in a semi-recumbent position and ropes rather than chains drove the bicycle. Jaray's *J-Rad* design was an early antecedent of later fully recumbent high performance vehicles, such as those designed by **Mike Burrows**.

Patent drawing of an
aerodynamic car,
1922

JCB

FOUNDED 1945
UTTOXETER, STAFFORDSHIRE, ENGLAND

Joseph Cyril Bamford

JCB was founded in 1945 by Joseph Cyril Bamford. His first design, a trailer, was produced in an Uttoxeter lock-up garage out of war surplus and sold for £90. Three years later, having seen the potential of hydraulics, Bamford introduced what was probably the world's first two-wheeled hydraulically-operated tipping trailer. In 1951 JCB launched the *Master-Loader* for use on farms and two years later developed the *Mark I* – the world's first backhoe loader. 1954 saw the introduction of the powerful *Hydra-Digger*, heralding the true commercial beginnings of JCB. It was the launch of the *3C* backhoe loader in 1963, however, that reinforced the company's position as market leader. Two years later, JCB entered a new market with the development of its first 360° crawler excavator. By 1969, 50% of its production was being exported and JCB continued to grow through its commitment to design and engineering excellence. In 1976 Anthony Bamford took over the chairmanship of the company from his father and in 1980 oversaw the successful launch of the

JCB *Mark I* excavator and major loader, 1953

JCB *3CX* – a new backhoe loader to replace the earlier 19-year-old model, of which the company had sold an astonishing 72,000 units. The *3CX* took ten years to develop and was an immediate commercial success. By 1984, JCB had a turnover in excess of £150 million and captured 17% of the total world backhoe market, with exports accounting for 65% of production. Throughout the 1990s JCB continued producing significant innovations, including the world's first 4 x 4 x 4 backhoe loader. In 1996 the company won a **Design Council** Award for Innovation for the JCB *Landpower*, while its JCB *Teletruk* (telescopic lifting truck) was named "Innovation of the Year" by the Institute of Transport Management in 1997. In total, JCB has won over 50 major awards for engineering excellence, design, exports, marketing, management and environmental concerns, including 14 Queen's Awards for Technology and Export achievement. As a consequence of the company's innovative research and design, extensive testing and stringent quality control, JCB machines have become renowned all over the world for their performance, value and reliability.

JCB *3CX* excavator, late 1990s

JEEP (WILLYS-OVERLAND)

FOUNDED 1909
TOLEDO, OHIO, USA

Willys-Overland
Jeep, 1940 in action

Described by General George C. Marshall as "America's greatest contribution to modern warfare", the Jeep is an iconic design with a name that has become generic for small four-wheel drive vehicles. Following the First World War, the US military realised the need for an all-terrain reconnaissance vehicle that was both lightweight and fast, to replace motorcycles. To this end, in 1940 the US Army issued specifications for such a vehicle and invited 130 automobile manufacturers to present working prototypes within just 49 days. Only three manufacturers responded to this challenge, and it was Willys-Overland, which had been founded in 1909 by John North Willys (1873–1935), whose *MB* Jeep with its powerful "Go Devil" engine ultimately won the Army contract. Over 700,000 Jeeps were subsequently manufactured under licence by **Ford** and 368,000 were produced by Willys-Overland. This versatile four-wheel drive vehicle, which was capable of carrying a 660-pound payload, was used for ferrying troops and carrying the wounded. It was also used as an all-purpose pick-up truck and "limousine" for top-ranking officers. One of its greatest assets was undoubtedly its robust design, allowing it to be easily disassembled for transportation and quickly re-assembled for rapid deployment. There are several theories as to how the name Jeep came about, but whatever its origin, Willys-Overland managed to obtain the Jeep trademark in 1950. Civilian Jeeps (known as CJs) have also been manufactured in various models since 1945, including the famous *CJ-5*, which was introduced in 1955 and remained popular well into the 1980s. Other land-

Willys-Overland
Jeep, 1940

mark Jeep sport utility vehicles include the *J*-series *Wagoneer* (1962), the *J*-series *Cherokee* (1974) and more recently the *Grand Cherokee* (1993). Willys-Overland was purchased by Kaiser Manufacturing in 1953 and became known as Willys Motors and then later as the Kaiser Jeep Corporation. This company was purchased by the American Motors Company (AMC) in 1970, which was in turn acquired by the **Chrysler** Corporation in 1987, which still owns the rights to the Jeep name.

JAKOB JENSEN

BORN 1926
COPENHAGEN, DENMARK

Jakob Jensen studied industrial design at the Kunsthandvaerkerskolen, Copenhagen, graduating in 1952. Between 1952 and 1959 he was the chief designer for Denmark's first industrial design consultancy, founded in 1949 by **Sigvard Bernadotte** and **Acton Bjørn**. From 1959 to 1961 Jensen lived in the USA, where he established a design office with Richard Latham (b. 1920) and others. During this period he also taught at the University of Illinois, Chicago. On returning to Copenhagen in 1961, Jensen founded his own industrial design office and three years later began designing audio equipment for **Bang & Olufsen**. His sleek, high-performance designs set the aesthetic and technological standards for audio systems. His *Beogram 1200* hi-fi (1969) was awarded a Danish ID prize for its harmonious balance between "apparatus" and "furniture", while his *Beogram 4000* record player (1972) was notable for incorporating a tangential pick-up arm. Although best known for his pioneering designs for Bang & Olufsen, Jensen has also designed products for other Danish manufacturers, including the *E76* pushbutton telephone (1972) for Alcatel-Kirk, an office chair (1979) for Labofa, ultrasound scanning equipment (1982) for Bruel & Kjaer and a wristwatch (1983) for Max René. One of Scandinavia's most famous industrial designers, Jensen has received numerous awards for his innovative work.

Beogram 4000
record player, 1972

WILHELM KÅGE

BORN 1889 STOCKHOLM, SWEDEN
DIED 1960 STOCKHOLM, SWEDEN

Liljeblå (Blue Lily)
tureen, 1917

Wilhelm Kåge studied painting and graphic design before being recruited by **Gustavsberg** in 1917, following a campaign by the Svenska Slöjdföreningen (Svenska Form) to encourage manufacturers to employ the services of artists, and thereby improve the design of their products. Kåge's subsequent *Liljeblå* (Blue Lily) dinner service (1917), with its folk-inspired decoration, was so affordable that it became known as the "Workers' Service". For the 1930 Stockholm exhibition, which heralded the new "funktionalmen" movement in Sweden, Kåge designed the *Pyro* range that was the first oven-to-table ware produced by Gustavsberg. His later *Praktika* earthenware service (1933), however, was a far more revolutionary design. But while it was extremely practical with stackable, different sized elements, *Praktika* was not a commercial success – the result, no doubt, of its severe utilitarian aesthetic. Design critics nevertheless deemed *Praktika* "the" functionalist service of the era. In 1937 Kåge designed the less austere *Soft Shapes* service. With its subtle organic forms and simple grey-striped pattern, *Soft Shapes* sold extremely well for more than 20 years. While Kåge was also celebrated for his studio ceramics, his designs for industrial production were of far greater significance as they predicted the basic functional shapes that have come to dominate large-scale ceramics manufacture.

Praktika ware for
Gustavsberg, 1933

Gino Colombini,
KS 1171 plastic dish
rack with plastic
dishes, 1954

KARTELL

FOUNDED 1949
MILAN, ITALY

Kartell, a plastics consumer products manufacturing
company, was founded by Giulio Castelli (b. 1920) in
1949. Castelli started from the view that: "The public is willing to accept new
forms for machines performing new functions, but when century-old objects
like a spoon, a chair, etc. are involved, it's not easy to make new aspects accepted. If men fear novelty, let's give them something even more novel."
The first product launched by Kartell was an innovative ski-rack designed by
the architect and industrial designer, Roberto Menghi (b. 1920). This was
followed by a range of household articles, many of which were designed by
Gino Colombini. From lemon squeezers and dustpans to washing-up bowls
and baby's baths, Kartell transformed every-day objects into sleekly designed
polyethylene products that were startlingly forward-looking. The company's
design and materials innovations were widely celebrated and its products
won numerous Compasso d'Oro awards at Milan Triennale exhibitions. Kartell also produced furniture, including a metal and plastic sectional cupboard system (1956) designed by Gino Colombini and Leonardo Fiori, but it
was not until the 1960s that it became widely known in this field, in particu-

Anna Castelli Ferreri,
4870 stacking chair,
1987

lar through its manufacturing of **Marco Zanuso** and **Richard Sapper**'s *Model No. 4999/5* stacking child's chair (1961–1964), **Joe Colombo**'s revolutionary *Model No. 4860 Universale* (1965–1967) which was the first adult-sized fully injection-moulded plastic chair, and the injection-moulded ABS *4953–54–55–56* stacking storage cylinders (1970) designed by Anna Castelli Ferrieri (b. 1920). Kartell also produced lamps by Sergio Asti (b. 1926), Marco Zanuso and **Achille** and **Pier Giacomo Castiglioni**. The company maintained a high profile in the 1980s through its production of furniture designed by **Philippe Starck**, such as the characterful *Dr. Glob* chair (1988). In the 1990s, Kartell's translucent *Mobil* drawer unit (1995), designed by Antonio Citterio (b. 1950) and Glen Oliver Löw and manufactured in injection-moulded PMMA, was awarded a Compasso d'Oro. In 1997 the company began producing **Ron Arad**'s highly successful *Book Worm* shelving in extruded and injection-moulded technopolymers. Kartell's in-house design studio, Centrokappa, which is directed by Giulio Castelli's wife, Anna Castelli Ferrieri, has also produced some notable designs, including the *5300, 5312, 5320* system of children's furniture. When he founded Kartell in 1949, Castelli stated that he "intended to arrive at the difficult synthesis between technology and design, between economy and the response to social demand". Having successfully struck this balance over its 50-year history, Kartell today can boast that one of its products is sold somewhere in the world every 30 seconds.

Antonio Citterio,
Mobil drawer
system, 1995

KAWASAKI

FOUNDED 1896
HYOGO, JAPAN

ZX600-G2
motorcycle, 1999

In 1878 Shozo Kawasaki (1837–1912) founded a shipyard, which was incorporated as the Kawasaki Dockyard in 1896. In 1906 the yard diversified and began manufacturing locomotives, freight and passenger carriages. The following year, marine steam turbines were also introduced. In 1918 the company expanded into aircraft, building the first metal Japanese aeroplane, and it was the Kawasaki Aircraft Co. that eventually produced the first Kawasaki motorcycle in 1961. By the time the company was restructured in 1969 to form Kawasaki Inc., it had gained a firm foothold in the American motorcycle market, and by the early 1970s was a major force within the industry. 1973 saw the launch of the legendary 900 cc *Z1*, which was voted "Machine of the Year" by *Motorcycle News* for four successive years. During the 1970s and 1980s, Kawasaki introduced several off-road models and enjoyed numerous racing triumphs. The company also developed other sports vehicles, most notably the *Jet Ski* launched in 1973, which although not the first personal watercraft, was the most successful. Kawasaki is also known for its ATVs (All Terrain Vehicles), off-road utility vehicles and loaders. The **GK Design Group** has been responsible for many of Kawasaki's product designs.

Kawasaki *Jet Ski*,
late 1990s

WALTER MARIA KERSTING

BORN 1889 MÜNSTER, GERMANY
DIED 1970 WAGING, GERMANY

*Volksempfänger
(People's Radio)
VE301 receiver, 1936*

*Kleinempfänger,
1938 – a smaller
version of the
Volksempfänger*

Walter Maria Kersting studied at the Technische Hochschule in Hanover from 1912 to 1914. He worked as a graphic designer and subsequently as art director for a printing works in Hanover. From 1922 to 1932 he worked as an architect, first in Weimar and then in Cologne, and also designed lighting and furniture. Recognizing the exciting potential of Bakelite as a material highly suited to the mass production of radio casings, in 1928 Kersting designed the first version of his well-known plastic-housed radio. His choice of material was inspired, as was his design for the layout of the controls, which he believed should be easy to operate and understand. Inexpensive and reliable, Kersting's radio came to be known as the *Volksempfänger* (People's Radio) – and with sales estimated to have topped 12.5 million by 1939, there was indeed a Kersting radio in virtually every German home. Manufactured by Hagenuh, the radio was produced in several versions, including the smaller *Kleinempfänger* and the model *VE 301*, which commemorated the date (30 January 1933) that Adolf Hilter became Chancellor of Germany. Kersting's concept of an affordable, standardized radio was subsequently co-opted by the National Socialist government as a means of delivering state propaganda to German households; the fact that the somewhat utilitarian radio was only powerful enough for home broadcasts and could not receive Allied transmissions was simply an added advantage. In the 1950s Kersting went on to design a wide range of products, from telephones and sewing machines to lighting and industrial machinery.

KODAK

FOUNDED 1881
ROCHESTER, NEW YORK, USA

George Eastman

At the age of 24, George Eastman (1854–1932) decided to visit Santo Domingo and bought some photographic equipment with which to record his travels. In those days the paraphernalia needed for photography was extensive – an enormous camera and heavy tripod, a dark-room tent and a large assortment of chemicals. In the end Eastman never made the trip, but he became captivated by photography and determined to simplify the process. Having read that British photographers were making and using gelatine emulsions that, when dry, could be exposed at leisure, Eastman made his own using a formula outlined in a magazine. After three years' experimentation, in 1878 he perfected a dry-plate gelatine emulsion. The following year he developed an emulsion-coating machine, which allowed him to begin the

Brownie camera,
c. 1902

Springtime is Kodak time

Advertisement for Kodak cameras, c. 1910

commercial mass production of his dry plates. In 1881 he founded the Eastman Dry Plate Company together with Henry A. Strong (b. 1919). In 1884 the firm introduced Eastman Negative Paper and developed a roll holder for the papers, and in 1885 pioneered the first transparent photographic film. Three years later, the first Kodak camera was launched; costing just $25, it was marketed under the slogan: "You push the button – we do the rest." This revolutionary and easy-to-use camera heralded the advent of amateur photography around the world. In 1889 Eastman marketed the first commercial transparent and flexible film roll, paving the way for the subsequent development of **Thomas Alva Edison**'s motion picture camera (1891). The company was renamed the Eastman Kodak Company in 1892, and in 1895 introduced the *Pocket* Kodak camera, which used roll film. 1900 saw the launch of the famous *Brownie* camera, which was constructed of a pressed cardboard box with a wooden end. Inexpensive and cheap to use, the *Brownie* took perfectly acceptable photographs and made photography accessible to virtually everyone. In 1908 Kodak pioneered the first commercial safety film using a cellulose acetate rather than a flammable cellulose nitrate base. Four years later, it established one of the first industrial research laboratories in America, dedicated to the evolution of new cameras and films. 1923 saw the introduction of 16mm-reversal film, the 16mm *Cine-Kodak* motion picture camera and the *Kodascope* projector, all of which made amateur cinematography possible. 16mm Kodacolour film was launched five years later, together with the first microfilm system for the storage of bank records. Through the 1930s and 1940s, Kodak continued to develop new products, including the first commercially successful colour slide film (1935) and the highly successful *Bantam* camera range, which included the streamlined Kodak *Bantam Special* camera designed by **Walter**

Bantam Colorsnap
camera, 1955–1959

Kodak *Instamatic*
50 camera, 1972

Dorwin Teague. Its cameras were often developed by outside design consult-ants in conjunction with the Kodak Styling Division and the engineering department, and grew increasingly compact and easier to use. Thus the Ban-tam *Colorsnap* camera of 1954, with its "dial-the-weather" exposure setting, simplified colour photography to such as extent that it was nicknamed "Auntie's Camera". Advances in colour film meant that lightweight snap-shot cameras were being manufactured by the early 1960s. Kodak *Instamatic* cameras, which first appeared in 1963, required only very simple exposure controls – just sunny or cloudy flash settings – and featured easy-to-load cartridge film. By 1970, more than 50 million *Instamatics* cameras had been sold, revolutionizing popular photography. By using narrower cartridge film, Kodak further reduced the size of the *Instamatic* so that it could fit inside a pocket. Designed by **Kenneth Grange**, the Kodak *Pocket Instamatic* was ex-tremely popular and over 50 million cameras were sold within seven years. In 1987 Kodak introduced its first one-time-use disposable camera, the Kodak *Fun Saver*, and by 1995 over 50 million of these, too, had been sold. 1995 also saw the launch of the Kodak Digital Camera *40*, the first full-feature digital camera to retail for less than $1000. Kodak's historic success has been the result of its "everyman" approach to photography. With film-based photog-raphy increasingly becoming a thing of the past, it seems certain that Kodak will continue to be at the forefront of affordable digital camera technology.

Kenneth Grange,
Kodak *Instamatic*
130, 230 and *330*
cameras, 1975

ERWIN KOMENDA

BORN 1904 JAUERN, AUSTRIA
DIED 1966 GERLINGEN, GERMANY

Porsche *356*
cabriolet, 1952

One of the most influential automotive stylists ever, Erwin Komenda initially worked for Daimler-Benz as chief designer in the body development section. He later joined the automotive design office of Ferdinand **Porsche** I in Stuttgart, where his first brief was to design a body for Auto Union's world record-breaking *Type A* racing car (1934). Obviously influenced by earlier aerodynamic studies undertaken by **Paul Jaray** and Wunibald Kamm (1893–1966), Komenda later designed a streamlined body for the original **Volkswagen** *Beetle* (1934), which was first produced in 1938. Komenda also undertook his own aerodynamics research in the wind tunnels at the **Zeppelin** works in Friedrichshafen and at Kamm's in Stuttgart. His interest in this field was translated into streamlined forms that were designed for maximum efficiency rather than just styled to look beautiful. For the styling of Porsche's first car, the *356* (1949), Komenda was asked by Ferdinand "Ferry" Porsche II to ensure that the car's nose was kept as low to the ground as possible. The resulting compact and unified form was extremely influential upon later car design. Komenda designed a number of other models for Porsche, concluding with the great design classic, the *911* (1963–1964).

Porsche *911*, 1964

M572 Water Tapper,
1988

M581 Sandworks,
1988

KOMPAN

FOUNDED 1970
RINGE, DENMARK

Kompan is one of the world's leading manufacturers
of playground equipment and was founded by Tom
Lindhardt Wils and Hans Mogens Frederiksen in 1970.
One of its earliest products was a seat in the shape of
a stylized animal mounted on a rocking spring, called
the *M100/Classic* (1970). Another early product was the *M401 Sprout* wave-
like ladder (1972) – a relatively simple yet ingenious design that required
only a small area for its installation. An improved version of this design was
later introduced with two platforms and a small crawl space at its bottom.
Among Kompan's most successful designs is the *Sandworks M581* (1988).
This multi-activity play centre was designed for children between the ages of
three to eight years old and has numerous features, such as two crane-like
elements which can be manipulated to scoop, pour and lift sand. Its sculptural form and cheerful combination of primary colours encourages interaction and provides play that is both fun and educational. For older children

aged between six and twelve years old, Kompan launched the *Sirius GXY 8011* "jungle-gym"
in 1998. Constructed of galvanized steel and
polyurethane, it has a more sombre aesthetic
than the equipment for younger children, yet
still invites interactive play. Guided by a design philosophy that recognizes play as an important part of life, Kompan enters the world
of play from the child's perspective in order to
work with issues of great importance to the
child: "Play Value, Play Rights, Play Safety and
Play Together". In recognition of its innovative
products, Kompan was awarded a European
Community Design Prize in 1994 and an Industrial Designers Society of America award
in 1999.

KRUPS

FOUNDED 1846
SOLINGEN, GERMANY

Jörg Glasenapp, *3 Mix* hand-blender, 1960

3 Mix 3 7007 Pro hand-blender, 1996

The German domestic appliance manufacturer Krups was founded in 1846 by Robert Krups, initially to produce rotating scale and spring balance weighing machines. Over the decades, the company remained under family control and continued specializing in these types of products. It was not until 1956 that it began to diversify its product line with the introduction of an electric coffee grinder, which it launched under the Onko brand in collaboration with the well-known Bremen-based coffee company, Hag. Within a year, over one million of these coffee mills had been sold. In 1958 Krups began the production of electric motors, and a year later began manufacturing electric hand-mixers. 1960 saw the lauch of the *3 Mix* mixer designed by Jörg Glasenapp, which became an icon of 1960s German design. A year later, the company introduced an electric coffee-maker that was a forerunner of the hugely successful *T8* machine of 1970. During the 1960s, Krups began producing a wide range of electrical appliances, including hair dryers, meat slicers, clocks, toasters, egg boilers, juicers, yoghurt makers and razors. In 1983 it launched its first electric espresso makers. In 1991 Krups was acquired by the Moulinex Group, which sells 45 million products a year (or one product every second). In 1996 new versions of the famous *3 Mix* mixer and *Nespresso Programatic* coffee machine were launched. Krups is renowned for household products which exemplify the attributes of German design: functionality, logic and superb build quality. Its commitment "to make good products even better" has led Krups to become one of the leading brands not only in Germany, but also in other European countries and in the USA.

Tea Time tea-maker, late 1990s

LAMBORGHINI

FOUNDED 1960
RENAZZO DI CENTO, ITALY

Bertone, Lamborghini
Miura P400, 1966

After the Second World War, Ferruccio Lamborghini (b. 1916) began converting surplus military vehicles for civilian use. In 1949 he founded Lamborghini Tractors to produce high-powered agricultural vehicles, and in 1960 established Lamborghini Bruciatori to manufacture boilers and air-conditioners. By now a leading Italian industrialist, he next decided to construct his own car. The engineer Giotto Bizzarrini subsequently designed a powerful V-12 engine, while a body was styled by Franco Scaglione, resulting in the *350 GTV* (1963). In 1965 Lamborghini began building a new supercar styled by **Bertone**, the *Miura*, named after a legendary Spanish fighting bull. It had aggressive modern lines and was the fastest series production car in the world. In 1971 the Bertone-styled *Countach* was launched at the Geneva Motor Show and caused a sensation. Futuristic, masculine and angular, the *Countach* was a missile-like testosterone-driven dream car. Its successor, the *Diablo* (1987), designed by Marcello Gandini, was more curvaceous yet retained the strong and menacing character synonymous with the marque. Lamborghini is now a subsidiary of **Chrysler**.

Bertone, Lamborghini
Countach LP400, 1973

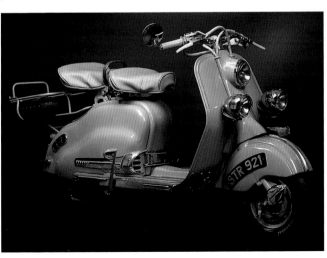

AUGUST 1961

LAMBRETTA

FOUNDED 1926
ROME, ITALY

Cover of *Scooter* magazine, August 1961

Lambretta's origins can be traced back to a steel pipe manufacturing business founded in Rome by Ferdinado Innocenti (1891–1966). In 1933 the company (then known as Innocenti Brothers) began manufacturing a revolutionary scaffolding system designed by its founder. The same year it moved to the Lambratte district of Milan and eventually began manufacturing automotive components. After the Second World War, the company decided to develop an inexpensive means of transport, and so Innocenti, together with the company's director Giuseppe Lauro and the engineer Pierluigi Torre, set about designing a scooter. Taking its name from the local district, the resulting Lambretta *Model A* was launched at the 1947 Paris Motor Show and boasted a maximum speed of around 70 kph. It was followed in the 1950s and 1960s by numerous Lambretta scooters in a wide range of colours, amongst them the *Grand Prix* model launched in 1969, which was styled by **Bertone**. Despite their having become synonymous with youth culture, production of Lambretta scooters ceased in 1971.

Lambretta *LD150* scooter, 1957

LEAR

FOUNDED 1964
WICHITA, KANSAS, USA

Learjet *Model 23*
on its first flight,
7 October 1963.
The jet was prod-
uced from 1964
to 1966.

In the late 1950s, William Lear (1902–1978) founded the Swiss American Aviation Corporation (SAAC) in Switzerland to develop a small corporate jet. In 1963 the operation was moved to Wichita, Kansas, and in 1964 was renamed the Lear Jet Corporation. The first compact jet, the Learjet *Model 23*, made its maiden flight that same year and in 1965 became the first business jet to be put into production. 1966 saw the launch of the Learjet *Model 24*, which circled the globe in an impressive flying time of 50 hours 20 minutes. The Learjet *Model 55* launched in 1983 broke six altitude and "time-to-climb" records in its weight class. In 1985 Learjet's aerospace division built components for the main booster rocket on **NASA**'s Space Shuttle. The Learjet Corporation was purchased by **Bombardier** in 1990, and four years later the Learjet *60* (1990) became the best-selling mid-sized corporate jet in the world. In 1998 the company's super-sleek eight to nine-passenger Learjet *45* (1992) became the first business aircraft to receive certification from the European Joint Aviation Authorities (JAA).

Learjet *Model 45*,
1992

LEGO

FOUNDED 1932
BILLUND, DENMARK

Duplo figure,
introduced in 1983

In 1932 Ole Kirk Christiansen (d. 1958) founded a workshop to manufacture stepladders, ironing boards and wooden toys. Two years later he came up with a name for both his toys and workshop – Lego, a contraction of the Danish words "leg godt" meaning "play well". During the 1930s and 1940s, the company produced well-built wooden toys which reflected the company motto that "Only the best is good enough". The Second World War accelerated research into new materials and production-methods, including the injection moulding of plastics. In 1947 Lego became the first Danish company to invest in this new technology for high-volume plastic manufacture. Two years later it brought out its *Automatic Binding Bricks*, which were forerunners of today's Lego bricks but which were only sold in Denmark. In 1953 these were renamed "Lego Bricks" and a year later the "Lego" name was officially registered. The "stud-and-tube" brick-coupling system was invented and patented in 1958. This new means of connecting the bricks offered greater stability for models and opened up unlimited building possibilities. Over the succeeding years, the range was expanded to include sloping roof bricks (1958), wheels (1961), the Lego family (1974) and mini figures (1978).

Lego bricks, first
launched in 1947

For pre-school children, Lego developed the larger and easier to use brick systems *Duplo* (1967) and *Primo* (1995). Lego is a unique and hugely successful toy – between 1949 and 1998, approximately 203 billion Lego elements were moulded. While appealing to both genders and to a wide range of ages, Lego provides stimulating play that teaches children the fundamental principles of construction and, therefore, design.

LEICA

FOUNDED 1849
WETZLAR, GERMANY

Advertisement for
Oskar Barnack's
revolutionary *Leica I*
camera, 1930

Leica's origins can be traced back to 1849, when a young mathematician, Carl Kellner, founded an "Optical Institute" to develop and market lenses and microscopes. In 1865 the mechanic Ernst Leitz became a partner in the firm. He later took over the company and expanded it under his own name. Around 1914 **Oskar Barnack** perfected the world's first 35mm camera, the *Ur-Leica*, but its production was hampered by the outbreak of the First World War. After his father's death in 1920, Ernst Leitz II decided to begin mass-producing Barnack's design in 1924 (although it was not marketed until the following year). Known as the *Leica I* – its title being derived from LEItz CAmera – it was the first commercially successful 35mm camera and proceeded to revolutionize the art of still photography. The Leitz works earned widespread regard for its precision-engineered cameras and the exceptional quality of its lenses. In 1930 the company was the first to pioneer a system of interchangeable accessories and

Leica camera, 1914

Leica camera, 1955

screw-fit lenses that were compatible between camera models. The *Leica II*, launched in 1932, incorporated a built-in rangefinder that enabled faster and more accurate focusing. The *Leica M3* (1954) introduced a bayonet fitting that allowed for even quicker lens changes. In response to Japanese competition, Leica began producing single lens reflex (SLR) cameras from 1964. Throughout its history, Leica has consistently stood for design and manufacturing excellence. Today, the company continues to produce precision-engineered cameras and high optical performance lenses which rank among the very best.

LOCKHEED MARTIN

Kelly Johnson
standing in front
of a *U-2*, 1955

LOCKHEED	MARTIN
FOUNDED 1916	FOUNDED 1912
SANTA BARBARA,	LOS ANGELES,
CALIFORNIA, USA	CALIFORNIA, USA

Glenn Martin, an early aviation pioneer, flew his first aeroplane in 1909 and three years later performed the world's first major over-ocean flight in his own hydroplane. That same year he established his own aircraft manufacturing company in Los Angeles. Among the firm's earliest contributors were the renowned design engineers James **McDonnell**, Donald **Douglas** and Chance Vought. A similar venture was set up in Santa Barbara in 1916 by the brothers Allan and Malcolm Loughead, who had developed and flown their own seaplane in 1913. In 1918 their Loughead Aircraft Manufacturing Company (renamed the Lockheed Aircraft Corporation in 1926) produced the *F-1* Flying Boat, while Glenn Martin constructed the first twin-engine bomber, the *MB-2*, that same year. During the 1920s and 1930s, both companies enjoyed high profiles – Amelia Earhart became the first woman to fly the Atlantic solo in 1932, piloting a

Lockheed *F-104*
Starfighter, 1958

SR-71 Blackbird, 1964

F-117A Stealth Fighter, c. 1988

Lockheed *Vega* designed by John Northrop (1895–1981), while Pan-Am used the Martin *M-130* (nicknamed the "China Clipper") to make the first scheduled trans-Pacific flight in 1935. During the war, Martin's *B-26 Marauder* (1940) became renowned for its survivability rate, while Lockheed's *XP-80 Shooting Star* (1944) became the first American jet fighter. The *XP-80* was developed at Lockheed's experimental development facility, the famous "Skunk Works" established in 1943 by the company's legendary chief engineer, Clarence "Kelly" Johnson (1910–1990). The Skunk Works (now officially known as the Lockheed Martin Advanced Development Programme) was, according to Johnson, "a concentration of a few good people solving

problems far in advance – and at a fraction of the cost – by applying the
simplest, most straightforward methods possible to develop and produce
new products". It was set up to develop a fighter that could compete with
the **Messerschmitt** *Me 262* jet aircraft (1943). The resulting *XP-80* was the
first American aircraft to exceed 500 mph. Located in Palmdale, California,
Johnson's Skunk Works was responsible for some of the greatest advances
in military aviation, including the *F-104 Starfighter* (1958), the first operational aircraft to sustain a speed of Mach 2; the *U-2* Spy Plane (1955), which
is still in operation; the *SR-71 Blackbird* (1964), which is still the world's
fastest and highest flying aeroplane with a maximum speed of Mach 3.5
and an operational ceiling of over 100,000 feet; the *F-117* Stealth Fighter
(c. 1988), the first operational aircraft with a cross-section invisible to radar;
and the *F-22 Raptor* (1991) designed to ensure American air-dominance in
the 21st century with its thrust vectoring system. Apart developing aircraft,
Lockheed also introduced the first FAA-approved "Black Box" flight data
recorder in 1958. The Martin Company also specialized in military development programmes – notably the *Pershing* Missile project begun in 1958 –
and worked extensively on **NASA** space programmes. In 1995 a "merger
of equals" took place between Lockheed and Martin, creating one of the
world's largest aerospace and defence companies.

F-22 Raptor
Advanced Tactical
Fighter, 1991

RAYMOND LOEWY

BORN 1893 PARIS, FRANCE
DIED 1986 MONACO

Portrait of Raymond
Loewy on the cover
of *Time* magazine,
1949

→Design for *L300*
refrigerator for
Electrolux, 1939

Raymond Loewy was the greatest pioneer of industrial design consulting
and is still remembered for the humorous adage that summed up his approach to design – "Never leave well enough alone". At the age of 15, Loewy
designed, built and flew a toy model aeroplane that won the then-famous
James Gordon Bennett Cup. Around this time, he also designed and patented a model plane powered by rubber bands. Having subsequently sold the
rights to this toy, called the *Monoplan Ayrel*, to a company that marketed it
across France, Loewy learnt that "design could be fun and profitable". With
the money he earned from this venture, Loewy was able to study at the Uni-
versité de Paris and later at the École de Lanneau, from where he received
an engineering degree in 1918. During the First World War he served in the
French army as a second lieutenant. After his demobilization, he travelled

First redesign of the
Gestetner duplicator
machine, 1929

Coldspot refrigerator
for Sears, 1934

to America in 1919. Arriving in New York, he was utterly amazed by what he later described as "the chasm between the excellent quality of American production and its gross appearance, clumsiness, bulk and noise". He initially found work as a window dresser for Macy's, Saks Fifth Avenue and Bonvit Teller. From around 1923 until 1928, he enjoyed a "pleasant but superficial career" as a fashion illustrator for *Vogue*, *Harper's Bazaar* and *Vanity Fair* among others. In 1923 he also designed the trademark for the Neiman Marcus department store. Loewy eventually left the fashion world to set up his own industrial design office in New York in 1929. Always a self-promoter, Loewy had a card printed that read: "Between two products equal in price, function and quality, the better looking will outsell the other", and sent it to everyone he knew. Soon afterwards he received his first brief, namely to redesign the casing of Sigmund Gestetner's duplicator machine. For this project, he used modelling clay to create a sleek form – a technique he later employed to great effect for his automotive designs. The resulting machine not only looked better but, thanks to its simplified form, was also easier to manufacture and maintain than earlier models. In 1932 Loewy designed his first car, the *Hupmobile*, which was less boxy than existing automobiles. His improved, tapering model of 1934 featured innovative integrated headlamps and predicted the streamlined forms for which he later became so famous. That same year, Loewy also designed the *Coldspot* refrigerator for **Sears Roebuck**, which was the first domestic appliance to be marketed on the

←Design for floor-polisher for Electrolux, 1939

Design for vacuum cleaner for Electrolux, 1939 – this model was never put into production

↑Studebaker, 1939

↗Studebaker, 1950

↑Studebaker
Starliner, 1953

↗Studebaker *Avanti*,
1961

strength of its aesthetic appeal – contemporary advertisements invited consumers to "Study its Beauty". The *Coldspot* was also alluring to customers because its improved design had reduced its cost of manufacture, and this was reflected in its competitive retail price.

Also in 1934, the Metropolitan Museum of Art in New York displayed a mock-up of Loewy's glamorous design office, thus illuminating his rise in status from product designer to celebrity industrial design consultant. From 1935, Loewy was commissioned to re-plan several large department stores, including Saks Fifth Avenue. During this period he was also designing aerodynamic locomotives, such as the *K4S* (1934), the *GG-1* (1934) and the *T-1* (1937), and in 1937 published a book entitled *The New Vision Locomotive*. In 1946 he famously remodelled coaches for Greyhound and in 1947 designed his innovative *Champion* car for Studebaker – a precursor of his later European-styled *Avanti* car, also for Studebaker. Not only could streamlining make transportation look more appealing, it also often improved performance and efficiency because of the effect of aerodynamics. Loewy regarded the design of a casing or sheath as an opportunity to allow a machine or a product to express itself. Unlike so many Modernists who allowed form to be completely dictated by function, Loewy balanced engineering criteria with aesthetic concerns in order to achieve what he believed to be the optimal solution. Loewy also became celebrated for his corporate identity work, and in particular for his repackaging of Lucky Strike cigarettes (1942). His presti-

Greyhound
Scenicruiser bus,
1954

Interior of a
Greyhound bus,
c. 1946

Logo for Exxon, 1966

→ Logo for Shell, 1967

Redesign of Lucky Strike packaging for the American Tobacco Company, 1942

→ Redesign of Coca-Cola dispenser *Dole Deluxe*, 1947

gious clientele included Coca-Cola, Pepsodent, the National Biscuit Company, British Petroleum, Exxon and Shell. By 1939 his design office had branches in Chicago, São Paulo, South Bend and London. In 1944 he founded Raymond Loewy Associates with four other designers. In 1949 Loewy became the first designer to be featured on the front cover of *Time* magazine; his picture accompanied by the memorable caption: "He streamlines the sales curve." That same year, Loewy expanded his operations and founded the Raymond Loewy Corporation to undertake architectural projects. During the 1960s and 1970s, Loewy worked as a design consultant to the United States Government, most famously redesigning Air Force One for John F. Kennedy and designing interiors for **NASA**'s *Skylab* (1969–1972). Always putting the consumer first, Loewy's MAYA (Most Advanced, Yet Acceptable) design credo was crucial to the success of his products. Undoubtedly the greatest pioneer of streamlining in the 20th century, Raymond Loewy clearly demonstrated that the success of a product depends as much on aesthetics as it does on function. Few design consultants have been as influential or prolific as Loewy, nor as misunderstood; for although he was a styling genius, he also skilfully improved the design of many products and pioneered numerous design innovations. Signficantly, Raymond Loewy glamorized the practice of design and in so doing raised the status of industrial design as a whole.

→ Stowable hammock for Habitablity Study for Saturn Five Space Station, c. 1972

Exploded view of
Skylab for NASA,
c. 1970

Another arrangement for a crew memb
a desk and credenza are mounted to a s
desk accommodates personal equipment
credenza. Above the desk a hinged han
use as shown in the right margin of the
provides stowage for a pressure suit.

STOWED
HAMMOCK

SATURN FIVE SPACE STATION
HABITABILITY STUDY
RAYMOND LOEWY/WILLIAM SNAITH

Edward Johnston, Standard Underground Bullseye design, version shown post-1933

LONDON TRANSPORT

FOUNDED 1933
LONDON, ENGLAND

Renowned worldwide for its distinctive corporate identity, London Transport's integrated design programme was pioneered by Frank Pick (1878–1941), who joined the Underground Group in 1906 and became its Traffic Manager in 1909. Pick believed that public transport should be both functional and aesthetically pleasing, and so began instituting a unified and standardized corporate identity that was based on a "fitness for purpose" ethos. In 1913 he commissioned the typographer Edward Johnston (1872–1944) to design a new typeface for the Underground Group, which resulted in an easy-to-read sans serif font. Johnston also redesigned the Underground's eye-catching "roundel" symbol, which had first appeared in 1908. Pick also commissioned posters from well-known artists and engaged the architect Charles Holden (1875–1960) to modernize existing stations and design new ones. In total, Holden designed over 50 London Underground stations, which were influenced by Modern Movement architecture in Scandinavia and the Netherlands and

Enid Marx, *Shield* upholstery textile designed for London Transport, 1948

represented a high point in English rational design. He also designed shelters, signs, lamp-standards and platforms for LT. In 1932 Pick became president of the Design and Industries Association and in 1933 Vice Chairman and Chief Executive Officer of the newly formed London Transport company, which controlled London's bus, tram and underground train networks. Pick continued to pursue his "house-style" design mission for London's transport systems, and the same year the London Underground map, redesigned by the engineering draftsman, Henry Beck (1903–1974), was issued. Ingeniously, Beck used a diagrammatic approach for his new, highly legible map, which showed the spatial relationship rather than the geographical distance between the stations. In 1937 Enid Marx (1902–1993) designed a range of hard-wearing upholstery textiles with Modern patterns for London Transport, which further enhanced the visual coherence of its trains, trams and buses. 1938 saw the design of new Underground rolling stock which

ARRANGEMENT OF 56 SEATER DOUBLE DECK BUS (Central)
Approved Colour Scheme.

also fitted Beck's "form and function" criteria. The best-loved and most famous LT design, however, must be the red *Routemaster* bus designed by
Douglas Scott in 1954. While red double-decker buses had been used exclusively in London from 1933, it was the *Routemaster* that came to symbolize
the capital. This was LT's last in-house bus design and while today it no
longer designs its own vehicles, LT remains committed to integrated design.

LOUIS POULSEN

FOUNDED 1892
NYHAVN, DENMARK

Poster showing a
PH lamp designed
by Poul Henningsen,
late 1920s

Poul Henningsen,
PH4–3 table lamp,
1966

The well-known Danish lighting company Louis Poulsen & Co evolved from a wholesale cork-trading and general ironmongery business established by Ludwig R. Poulsen, which from 1892 had an "electrical department" that supplied carbon arc lamps. Louis Poulsen, the founder's nephew, headed the business from 1906 and two years later took over premises in Nyhavn. The turning-point for the company occurred in 1911, when Sosphus Kaastrup-Olsen acquired a 50% stake in Poulsen's electrical division. Olsen was a true believer in electric power as well as a political radical. In 1917 Louis Poulsen retired and Olsen became the sole proprietor of the company. Almost immediately, he began benefiting from the increase in demand brought about by the end of the First World War and the return of the long over-looked Southern Jutland region to Denmark. By 1918, Louis Poulsen & Co. had a considerable turnover of 5 million kronen, of which 600,000 kronen was clear profit. Although the 1920s were a financially turbulent period for the company, it gained valuable market share after Olsen came into contact with the lighting designer Poul Henningsen (1894–1967). Henningsen produced lighting designs for the company from 1924 and in 1925 some of his designs won a gold medal at the Paris "Exposition Internationale des Arts Décoratifs et Industriels Modernes". A year later, he designed the famous *PH* series of lamps, which remain in production today. The *PH* lamps were startlingly modern for their day and appealed in particular to the avant-garde – Alvar Aalto (1898–1976), for example, incorporated them into his design for the Auditorium at Turku in Finland. These "shadow-free" lamps featured three highly effective light-diffusing shades, and were used in large public buildings such as Cologne railway station (1929) as well as in smaller domestic and commercial premises. By 1931, 30,000 *PH* lamps had been sold around the world. Louis Poulsen & Co. managed to survive the Second World War and continued to expand during the post-war years, establishing international subsidiaries and commissioning other

Louis Poulsen, *Enamel* pendant lamp, 1930

well-known designers, such as **Arne Jacobsen**, King – Miranda and Alfred Homann (b. 1942), to design domestic and contract lighting products for the company. By 1997 Louis Poulsen & Co. had a workforce of 1,090 and wholly owned subsidiaries in Germany, Sweden, Norway, Finland, Britain, Holland, France, Switzerland, Australia, Japan and the USA.

Alfred Homann, *Nyhavn* pendant lamp (maxi), 1933

ROSS LOVEGROVE

BORN 1958
CARDIFF, WALES

→ *Pod* lights for
Luceplan, 1996–1997

Ross Lovegrove studied industrial design at Manchester Polytechnic and later trained at the Royal College of Art, London, receiving an MA in 1983. Between 1983 and 1984 he worked for the industrial design consultancy **Frogdesign** in Altensteig, where he was assigned to projects that included the design of Walkmans for **Sony** and computers for **Apple**. From 1984 to 1987 he worked as an in-house designer for Knoll International in Paris, for whom he created the successful *Alessandri* office system. Having returned to Britain in 1986, Lovegrove established a design partnership with Julian Brown (b. 1955) and designed the highly successful *Basic* thermos flask for Alfi Zitzmann (1990). After the partnership was dissolved in 1990, Lovegrove established his own London-based design office, Studio X. In addition to furniture for such clients as **Kartell**, Ceccotti, Cappellini, Fasem, Moroso, Driade and Frighetto, Lovegrove's designs include an innovative range of lighting for Luceplan, featuring the solar-powered *Solar Bud* (1996–1997). He has also designed luxury leather goods for Connolly Leather (1994), a kettle for Hackman (1999), *Handy* kitchen tools and knives for Fratelli Guzzini (1999), the *Fluidium* lava lamp for Mathmos (1999) and accessories for TAG Heuer (2000). Lovegrove has also worked on a proposal for product architecture involving a solar-powered "nomadic" structure known as a *Solar Seed*. By "seeking perfection in everyday tools for living", Lovegrove combines sculptural and ergonomic forms with state-of-the-art materials and industrial techniques, to create designs that are at once emotionally seductive and technologically persuasive. Inspired by the natural world and driven by ecological concerns, Lovegrove strives to achieve organic unity of design wherever possible while working within an essentialist formal vocabulary. Very much more a leader and risk-taker than a follower in industrial design, Lovegrove pushes materials, technology, function and even aesthetics to the limits – often breaking new ground – in his pursuit of new and better performing forms that touch the souls of the people who use them. Lovegrove's holistically derived and highly considered products reveal the direction of design for the 21st century.

Ammonite communicator (study project) for Apple Computer, 1995–1996

DIETRICH LUBS

BORN 1938
BERLIN, GERMANY

Dietrich Lubs studied shipbuilding and later trained as a designer at **Braun**. In 1962 he began working for the company's design department, headed by **Dieter Rams**. From 1971 he has been responsible for product graphics and has also been active as a product designer. He has designed numerous pocket calculators, alarm clocks and watches for Braun, all of which are distinguished by geometric forms and functional layouts. The majority of his clock designs have matt black plastic bodies, which allow the white and yellow numerals and hands to stand out so that they can be easily read. Since 1995 Lubs has been deputy director of the Braun Design Department. His many product designs are typical of the rational language of design pioneered by Rams at Braun, while also being characteristic of the attributes generally associated with German industrial design – functionalism, logic and high quality engineering.

Pocket calculator, variant of *ET 44* co-designed with Dieter Rams, 1978

Quartz clock,
c. 1990

ABR 314 df time
control digital radio,
1997

MAG INSTRUMENTS

FOUNDED 1974
LOS ANGELES, CALIFORNIA, USA

Anthony Maglica

Born in New York during the Depression, Anthony Maglica was brought up in Europe, where he received machinist training before returning to the United States in 1950. Five years later, he invested $125 of his savings in setting up a small "job shop" in Los Angeles. With his well-honed skills as an experimental machinist, coupled with his unyielding perseverance, Maglica soon gained an excellent reputation for the industrial, military and aerospace components that he produced. During the 1970s he turned his attention to flashlights, which at that stage were generally considered poor-quality items, frequently thrown away because of their unreliability. Recognizing a market opportunity for a more durable and better-performing flashlight, in 1974 Maglica founded Mag Instruments and went on to produce a machined aluminium flashlight called the *Mag-Lite* (1978). Designed to be an essential tool for police officers and fire fighters, professions such as these were soon insisting upon the distinctive *Mag-Lite* flashlight as standard issue. Consumers also began buying into the design's superior reliability. In 1984 a smaller flashlight was introduced, the *Mini Maglite*, and in 1987 an even smaller version was launched, followed a year later by a micro-flashlight that was ideal for use on key chains. Like the **Airstream** trailer and the Zippo lighter, the water and shock-resistant *Mag-Lite*, with its cam-driven beam that can be transformed from spot to flood, is considered a classic example of American design.

2-cell AA Mini Maglite flashlight, 1984

3-cell D Black Mag-Lite flashlight, 1978

Salad bowls and
salad servers for
Stelton, 1986

ERIK MAGNUSSEN

BORN 1940
COPENHAGEN, DENMARK

Erik Magnussen trained as a ceramicist at the Kunsthandvaerkerskolen in Copenhagen until 1960. From
1962 he worked as a designer for the porcelain manufacturer Bing & Grøndahl, designing simple geometric
tableware such as his *Form 25* service of 1965. The eleven pieces that made
up this service could be stacked and employed a self-insulating double-wall
construction. In 1967 Magnussen was awarded the prestigious Lunning
Prize for his ceramic designs, which combined artistic sensibility with an
understanding of industrial production techniques. The following year, he
designed the **tubular metal** and canvas folding *Z-chair* for Torben Ørskov &
Co., and subsequently created a range of school furniture in collaboration
with Per Kragh-Müller. In 1969 he designed an innovative teapot, featuring
a handle that was incorporated into the main body of the vessel. From 1976
Magnussen worked for **Stelton**, for whom he produced a number of well-
known designs, including the classic *Thermal Carafe* (1976). From 1978 he
also designed hollow-ware for Georg Jensen. Magnussen's designs from
the 1980s and 1990s include the *301* kerosene lamp (1988) for Harlang &
Dreyer, a clock and barometer for Georg Christensen, and a table and chair
(1989) for Paustian, the frame of which comprised one continuous piece of
bent tubular metal.

Thermal Carafe set,
1976

GERHARD MARCKS

BORN 1889 BERLIN, GERMANY
DIED 1981 BURGBROHL, GERMANY

Gerhard Marcks started teaching himself to sculpt in 1906, and from 1908 to 1912 trained in the Berlin workshop of sculptor Richard Scheibe (1879–1964). In 1914 his sculpture was featured in the entrance of **Walter Gropius'** installation at the **Deutscher Werkbund** exhibition in Cologne. Marcks subsequently joined the Novembergruppe, a radical art association founded in Berlin in 1918. After his military service, Marcks taught at the Berlin School of Applied Arts from 1918 to 1919. In 1919 he became a member of the Arbeitsrat für Kunst (Working Council for Art) in Berlin and also began teaching at the Weimar **Bauhaus**. In 1920 Gropius appointed him "Formmeister" (artistic director) of the Bauhaus' newly established ceramics workshop at Dornburg, some 30 kilometres from Weimar. His own style nevertheless remained relatively ornamental throughout his tenure. It was only in 1928,

Sintrax coffee-maker and stand for Jenaer Glaswerke, Schott & Gen., c. 1928

when he became director of the **Burg Giebichenstein** School of Applied Arts in Halle, that he shifted towards an increasingly rational style, one better suited to industrial manufacture. That year, he designed a prototype of his *Sintrax* coffee-maker, which was subsequently manufactured by the Schott & Gen. glassworks in Jena. This quintessentially Modern utilitarian design, which was constructed of heat-resistant glass, plastic, wood and metal and intended for large-scale mass production, was utterly in keeping with the "Form ohne Ornament" (Form without Ornament) ethos that had been promoted earlier by the Deutscher Werkbund. In 1930 **Wilhelm Wagenfeld** designed a new straight handle for Marck's *Sintrax* coffee-maker, giving it an

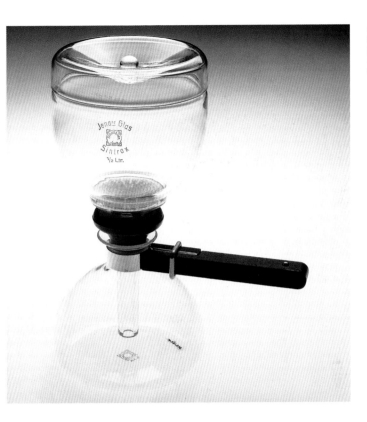

Sintrax coffee-maker
for Jenaer Glaswerke,
Schott & Gen., 1930
– handle designed by
Wilhelm Wagenfeld

even more rational appearance. Marcks went on to design the *Tiergarten* tea
set (1932) for the Staatliche Porzellanmanufaktur (formerly the Königliche
Porzellanmanufaktur and still known as KPM) in Berlin. While rationally conceived with a number of interesting ergonomic features, the *Tiergarten* tea
set was never put into production. In 1933 Marcks was dismissed from his
post in Halle by the National Socialists and subsequently moved to Nieha-
gen, near Wustrow in the Mecklenburg region. From 1936 he attempted to
work as a freelance artist in Berlin but was banned from exhibiting. Marcks
suffered further persecution when his studio in the Nikolassee suburb of
Berlin was destroyed in 1943. After the Second World War, he taught sculpture at the Hamburg Landeskunstschule from 1946 to 1950 and then moved
to Cologne in 1950, where he worked as an independent artist. The Gerhard
Marcks Foundation was established in Bremen in 1971 and in 1987 a retro-
spective of his work was held in Bremen, Cologne and Berlin.

GUGLIELMO MARCONI

BORN 1874 BOLOGNA, ITALY
DIED 1937 ROME, ITALY

Guglielmo Marconi was educated in Bedford, England and Florence and later spent two years at a technical school, the Leghorn Lyceum, in Livorno. While there, he studied physics and undoubtedly investigated the work of scientific pioneers such as James Clerk Maxwell (1831–1879), Heinrich Hertz (1857–1894) and Sir Oliver Lodge (1851–1940) – all of whom had conducted research into either electro-magnetism or radio receivers. He later attended lectures at the University of Bologna by special arrangement and from 1894 began conducting his own experiments. Using relatively primitive apparatus, Marconi was able to demonstrate that, by using an aerial, the range of a wireless signalling system could be extended by up to 2.4 km. He also discovered that if reflectors were attached to the aerial, the radiated electrical energy could be concentrated into a beam. Convinced that his discoveries would lead to a new system of wireless communications, but unable to generate any commercial interest in them in Italy, he travelled to London where he filed his first patent in 1896. Between 1896 and 1897, he conducted a series of successful demonstrations, some of which involved balloons and kites to increase the height of his aerials and thereby their signalling range. To exploit his invention, he established his own company (renamed Marconi's Wireless Telegraph Company in 1900) and on Christmas Day 1898 transmitted ship-to-shore telegraph signals over a 12-mile stretch between the moored East Goodwin Lightship and the South Foreland Lighthouse. In 1899 the lightship was involved in a collision with a steamer and the use of Marconi's wireless telegraph helped avert the

Marconi's first transmitter, 1897

Marconi ribbon microphone, used by the BBC from 1934 to 1959

loss of life. This incident did much to publicize the life-saving potential of his invention. The same year, the first wireless telegraph messages were sent across the English Channel from South Foreland to Wimereux near Boulogne. In 1900 Marconi went on to file his famous patent for "Improvements in Apparatus for Wireless Telegraphy". Although several eminent mathematicians believed that the curvature of the earth would limit the signalling range of wireless telegraphs to between 161 and 322 km, in 1901 Marconi succeeded in transmitting the first trans-Atlantic signal from St. John's in Newfoundland to Poldhu in Cornwall, thus heralding the beginning of global communications. Over the following years, Marconi continued increasing both the range of his wireless transmissions and the quality of their reception and received numerous accolades for his work, including the Nobel Prize for Physics in 1909. In 1920, the first advertised public broadcast – a song recital by Dame Nellie Melba – was transmitted from the Marconi works in Chelmsford. Two years later, Marconi and five other companies together established the British Broadcasting Company (BBC) and began broadcasting from London. In 1937 the company received an order from the British Government for "Chain Home" stations – the first air defence radar network established in Britain. Marconi subsequently became a world leader in defence radar systems, amalgamating with the English General Electric Company in 1968 and later merging with British Aerospace to form BAE Systems, the second largest defence contractor in the world. Marconi was the most important pioneer of radio broadcasting and his revolutionary communications designs quite literally changed the world.

ENZO MARI

BORN 1932
NOVARA, ITALY

Between 1952 and 1956, Enzo Mari trained at the Accademia di Belle Arte di Brera, Milan. While there, he published his first article on art and after his studies, became involved in the design of books and games. In 1957, he began working for **Danese** and one of his first projects was a simple yet ingenious toy – a child's puzzle made up of sixteen different stackable wooden animals (1957). In 1959, he began experimenting with plastics and these researches resulted in a number of products that were manufactured by Danese, including a PVC cylindrical umbrella stand (1962) and the *Pago-Pago* vase (1969) made of ABS. He joined the radical Nuove Tendenze movement in 1963 and taught at the Scuola Umanitaria, Milan. In 1970, his radical theories were published in *Funzione della ricerca estetica* and in 1974, he exhibited a number of "tecnica povera" (poor technology) furniture pieces. Mari has also designed several notable chairs including the wire framed *Sof Sof* (1971) for Driade, the *Box* (1975–1976), a knock-down design

EM02 colander for Alessi, 1997

that was sold in a box, for Anonima Castelli and the elegant *Tonietta* (1980–1985) for Zanotta, which was awarded a Compasso d'Oro in 1987. As a leading design theorist, Mari was president of the ADI (Associazione per il Disegno) from 1976 to 1979 and has taught in Milan, Rome, Parma and Carrara. Mari's most recent designs for **Alessi** such as the *EM02* colander (1997) and the *Titanic* lemon squeezer (2000) reveal his continuing search for new forms. Alberto Alessi has stated, "I am truly glad that Enzo Mari continues to exist and to work, a courageous and irreducible paranoic in a world … liable to become too meekly commercial."

MATSUSHITA

FOUNDED 1918
OSAKA, JAPAN

Yoichi Takahashi,
MS4/M40 video
camera for Panasonic,
1995

Konosuke Matsushita (1894–1989) founded the Matsushita Electric Housewares Manufacturing Works to produce an adapter socket that he had designed. He later developed an improved two-way socket that was better than any other on the market, and was soon flooded with orders. In 1922 he built a factory in Osaka and a year later developed a revolutionary bullet-shaped, battery-powered bicycle lamp that was marketed under the "National" name. In 1931 his factory also started manufacturing radio and dry cell batteries. Konosuke Matsushita was an industrialist with a social mission who believed that mass production would contribute to "the growth of human civilization". In 1935 the company began researching television technology and was incorporated as the Matsushita Electric Industrial Co. During the war, MEI constructed ships and aeroplanes for the Japanese military and was subsequently branded a *zaibatsu* company by the Allied powers. By 1950, however, such restrictions had been lifted and in 1951 Konosuke Matsushita travelled to America for the first time. MEI produced its first washing machine that same year, followed by its first black-and-white television (1952), refrigerator (1953), tape recorder (1958), colour television (1960) and microwave oven (1963). Today, Mat-

Model R-72S wrist-
radios, 1969

sushita is the world's largest manufacturer of consumer electrical products, which it markets under the brand names of Panasonic, Quasar, Technics, Victor and JVC. The Matsushita concept of product design "incorporates both function and beauty, maximizing the interface between user and machine to enhance the comfort and convenience of people's lives". The company's basic principles of product design are: fun and ease of use, innovation, universality, environmental responsibility and facilitative of new lifestyles.

CYRUS HALL MCCORMICK

BORN 1809 ROCKBRIDGE COUNTY, VIRGINIA, USA
DIED 1884 CHICAGO, ILLINOIS, USA

Cyrus Hall McCormick was the eldest son of Robert McCormick, a farmer, blacksmith and inventor. While the elder McCormick had developed several agricultural implements, he failed in his attempts to construct a viable reaping machine. At the age of 22, however, Cyrus succeeded in designing and building the first practical mechanical reaper, which he patented in 1834. This pivotal invention could cut a swath of wheat over 70 yards long in just over a minute. By providing an inexpensive means of gathering crops, the reaper helped turn the West into a "bountiful wheatland" and transformed the North's economy during the American Civil War, which contributed to the Union's victory over the Confederacy. It was not until the mid-1840s, however, that the reaper was perfected and fully commercialized. By the 1850s, McCormick's reaper was being mass-produced and sold throughout the United States. In 1851 it was exhibited at the **Great Exhibition** in London, where it was described by the *Times* newspaper as a "cross between an Astley chariot, a wheelbarrow and a flying machine". By 1856, McCormick was selling over 4,000 reapers per annum and in the process, revolutionized agricultural production.

Engraving showing a horse-drawn harvester, originally developed by Cyrus Hall McCormick in c. 1831

MCDONNELL DOUGLAS

FOUNDED 1967
ST. LOUIS, MISSOURI, USA

Douglas *DC-3*, 1935

The McDonnell Douglas Corporation was formed in 1967 as a result of a merger between the McDonnell Aircraft Corporation, founded by James S. McDonnell (1899–1980) in St. Louis in 1939, and the Douglas Aircraft Company, founded by Donald W. Douglas (1892–1981) in Santa Monica in 1920. As a civil engineering assistant at the Massachusetts Institute of Technology's department of aerodynamics, Douglas helped develop one of the first wind tunnels for the testing of aircraft (1914–1915). In 1920 he founded his company and the same year designed the *Cloudster*, the first aerodynamically streamlined aeroplane and the first aeroplane able to carry a load that exceeded its own weight. Recognizing the passenger and cargo-carrying potential of aeroplanes over long distances, in 1932 Douglas began developing his famous *DC* (Douglas-Commercial) series of aircraft. This included the immortal *DC-3* (1935), which was the world's first successful commercial airliner. Powered by twin Pratt & Whitney engines, it could carry 21 passengers at a speed of 195 mph.

Douglas *DC-3*
operated by KLM
airlines, late 1930s

Assembly line
at the Douglas
Aircraft Company
in California,
early 1940s

FH-1 Phantom, 1948

During the Second World War, the *DC-3* was converted for military use and re-designated the *C-47*. In total, the Douglas Aircraft Company supplied 29,000 warplanes to the US military (including over 10,000 *C-47*s) – one-sixth of the airborne fleet. After the war, the company continued developing commercial passenger aircraft, including the advanced piston-engined *DC-7*, whose range made non-stop coast-to-coast services possible for the first time. With the advent of passenger jet aircraft, Douglas developed the *DC-8* (1958) and *DC-9* (1965), but nevertheless found itself lagging behind **Boeing**. As a result of this, the merger with McDonnell was sought.

Having been founded on the eve of the Second World War, the McDonnell Aircraft Corporation grew quickly and became a major US defence supplier. During the war it began research into jet propulsion, which led to the development of *FH-1 Phantom* (1946), the world's first operational carrier-based jet fighter. As the first US Navy aircraft able to attain a speed of 500 mph, the *FH-1* was the progenitor of a long and distinguished line of McDonnell fighter aircraft, including the *A-4 Skyhawk* (1954), the super-versatile Mach 2 *F-4 Phantom* (1958), the *F-15 Eagle* (1972) – still the US Air Force's premier fighter – and the *F/A-18 Hornet* (1978). Both McDonnell and Douglas were at the forefront of Space Age technology and made important contributions to the Mercury, Gemini and Sat-

urn/Apollo space programmes. Today, McDonnell Douglas' commercial transports, such as the impressive *MD-11* wide-body tri-jet (1990), combat aircraft and space vehicles add new lustre to both companies' long history of aerospace accomplishments. In 1997 the McDonnell Douglas Corporation merged with Boeing.

F-4A Phantom, 1961

F-15C Eagle, 1978

F-15E Eagle, 1989

ALBERTO MEDA

BORN 1945
LENNO TREMEZZINA/COMO, ITALY

← Meda chair for
Vitra, 1996

Alberto Meda studied mechanical engineering at the Politecnico di Milano, graduating with a MA in 1969. In 1973 he became technical manager for **Kartell**, where he was made responsible for furniture projects and also instigated a research programme into laboratory equipment made of plastics. In 1979 he began working as a freelance industrial designer, acting as a consultant to **Alfa Romeo**, Alias, Cinelli, Colombo Design, **Gaggia**, Ideal Standard, Mandarina Duck, Luceplan, **Philips**, Vitra and others. Between 1983 and 1987, Meda lectured on industrial technology at the Domus Academy and from 1995 taught industrial design at the Politecnico di Milano. He has received two Compasso d'Oro awards for his innovative lighting, co-designed with Paolo Rizzato (b. 1941) – for the *Lola* lamp in 1989 and the *Metropoli* lighting system in 1994, both manufactured by Luceplan. Meda has also produced several notable furniture designs, including his remarkably light yet strong *Light Light* chair (1987), which is constructed of a carbon-fibre matrix with a Nomex-honeycomb core. During the 1990s, Meda designed a range of elegant cast-aluminium seating for Alias, including the *Armframe* chair and *Longframe* lounger (1996). In 1997 his *Meda* office chair for Vitra was assigned "Best of Category" in *I. D.* magazine's design review. The structural rightness and visual coherence of Meda's work testifies to his engineering background. This and his talent for exquisite detailing and proportion make him one of the most interesting industrial designers working today.

Madi handle for
Colombo Design,
1990s

MERCEDES-BENZ

FOUNDED 1926
STUTTGART, GERMANY

AVOIR SA BENZ!!

Benz poster, 1910

Mercedes-Benz is the luxury car brand of the Daimler-Benz company, which was formed in 1926 through the merger of two pioneering German automobile companies, Benz & Co founded by Karl Benz (1844–1929) in 1883, and the Daimler-Motoren-Gesellschaft founded by Gottlieb Daimler (1834–1900) in 1890. Karl Benz initially established his company to manufacture stationary internal-combustion engines. In 1886, however, he launched the three-wheeled *Motorwagen*, which was the world's first practical car powered by an internal-combustion engine, and patented it the following year. Meanwhile, Gottlieb Daimler and Wilhelm Maybach (1846–1929) had also been developing internal-combustion engines from 1883 onwards, and in 1885 they patented a four-stroke version. That same year, they mounted one of their engines onto a wooden-framed bicycle and thus created the world's first motorcycle. In 1886 they motorized a previously horse-drawn carriage; this was a revolutionary design in that it was propelled by an "invisible" power source. In 1890 Daimler finally founded the Daimler-Motoren-Gesellschaft. In 1894

Mercedes car, 1904

Mercedes-Benz *260 Stuttgart* roadster, 1928

Mercedes-Benz *150 H*, 1934

Mercedes-Benz *190 SL*, 1955

Mercedes-Benz
500 SL, 1991

the company launched a belt-driven car, featuring a four-speed belt transmission connecting the engine to the back wheels, which allowed smoother
gear changing. Benz, meanwhile, had its own new car to launch that year,
the *Benz Velo*, which was the world's first series production car. By 1899 it
had developed a racing car, one of the finest and most reliable of its time.
Daimler followed this in 1906 with its first Mercedes racing car. In 1926 the
two companies merged to create the Daimler-Benz company, which continued to produce cars under the Mercedes-Benz name. These included the
famous *SSK* sports car (1928) and the *260 Stuttgart* (1928), which was the
company's mainstay during the late 1920s. During the 1930s, as well as producing elegantly styled saloons and sport cars such as the *159 H* roadster
(1935), the company also designed and built aerodynamic racing cars such
as the *W 25* (1934) and the *W 125* (1937), the latter scoring 27 victories. During the Second World War, Daimler-Benz's factories were turned over to military production. Car production resumed in 1946, but it was not until 1950
that the company launched a new model, the *170 S* cabriolet, which became
a powerful symbol of Germany's economic recovery. 1955 saw the introduction of the *180* – the first Mercedes with a modern three-box body – and
the elegant *190 SL*, which was described as "the ideal vehicle for sporting
ladies". Over the succeeding decades, Daimler-Benz continued to produce
cars, vans, trucks and buses and by 1965 it was the largest manufacturer
of commercial vehicles in Europe. In 1988 Mercedes-Benz produced its
ten millionth car. Having merged with the **Chrysler** Corporation in 1997 to
become DaimlerChrysler, the company continues its long tradition of producing high-quality innovative vehicles – from the top-of-the-range *S-Class*
to the compact *A-Class* – that exemplify German design at its best.

MESSERSCHMITT

FOUNDED 1938
AUGSBURG, GERMANY

Wilhelm "Willy" Messerschmitt (1898–1978) studied engineering at the Munich Institute of Technology, graduating in 1923. Three years later, he became chief designer and engineer at the Bayerische Flugzeugwerke in Augsburg. While there, he designed the low-wing, single-engine, single-seat *Bf 109* monoplane, which was re-designated the *Me 109* when the company became the Messerschmitt-Aktien-Gesellschaft in 1938. The famous *Me 109* fighter was powered by a 1150-hp Daimler-Benz engine and had a top speed of 350 mph. During the Second World War, around 35,000 *Me 109*s were manufactured for the German Luftwaffe. Other wartime designs included the *Me 110* bomber/night fighter, the *Me 163* (1943), the first operational rocket-propelled aeroplane, and the *Me 262* (1943), which was the first operational jet-propelled plane. After the war, Messerschmitt diversified its production to include a sewing machine and pre-fabricated housing, and in the mid-1950s became a pioneer of bubble cars, such as the three-wheeled *KR 200* (1956) designed by Fritz Fend and the four-wheeled rocket-like *TG500* (1957).

MINALE
TATTERSFIELD

FOUNDED 1964
LONDON, ENGLAND

Agip petrol station,
1996

The concept of multi-disciplinary design consultancy originated in the United States during the 1930s, with designers such as **Raymond Loewy**, **Norman Bel Geddes** and **Walter Dorwin Teague** producing designs across a wide range of fields – from heavy industrial to graphic design. Today, however, multi-disciplinary design consultancies are more common in Europe, while American design practices tend to specialize in specific areas such as communications design or corporate identity. One of the most interesting and successful multi-disciplinary design consultancies is Minale Tattersfield & Partners, which was founded by Marcello Minale and Brian Tattersfield in 1964. From the beginning, the office placed an emphasis on innovation and attracted high-profile clients such as Johnson & Johnson, Harrods and Thames Television. Since then, it has grown into a huge international consultancy. As well as working in the realms of corporate identity, packaging and multi-media, the firm also designs furniture, products and interiors. As a "design strategy group", the success of Minale Tattersfield & Partners lies in its ability to solve problems in a stylish and often witty way, while "expressing the maximum possible with the simplest means".

Packaging design for
Shark energy drink,
1996

MINTON

FOUNDED 1793
STOKE-ON-TRENT, ENGLAND

Christopher Dresser, artwork border for transfer-printed ceramics for Minton & Co., 1871

Having previously worked for Josiah Spode, Thomas Minton (1765–1836) founded his own factory in Stoke-on-Trent in 1793. Three years later, he began mass-producing transfer-printed earthenware and was responsible for the popularization of the well-known blue and white "Willow Pattern" decoration. The factory also became renowned for its fine bone china, introduced in 1821, and for its elaborate wares of the 1850s, which were influenced by a variety of historical styles. Herbert Minton took over from his father in 1836 and began instituting innovative methods of production. He also widened Minton's product range to include Parian, porcelain and majolica wares. The company also produced encaustic tiles that were exhibited at the **Great Exhibition** of 1851 and used in the interiors designed by A. W. N. Pugin (1812–1852) for the new Houses of Parliament. After Herbert Minton's death in 1858, the company was divided into two new ventures: Minton, Hollins & Co. (which produced encaustic floor tiles and mosaics) and Minton & Co. (which manufactured decorative tiles and chinaware). The latter was a pioneer of "art-manufactures" and commissioned leading designers such as **Christopher Dresser** to create

C. Bevan (attributed), tile for Minton & Co., c. 1860

wares suitable for serial production. Around 1900, Minton manufactured designs influenced by the Art Nouveau style. It was eventually acquired by the Doulton company in 1968.

REGINALD MITCHELL

BORN 1895 TALKE, NEAR STOKE-ON-TRENT, ENGLAND
DIED 1937 SOUTHAMPTON, ENGLAND

Spitfire LFVc, 1936

Reginald Mitchell trained as an engineer prior to joining the Supermarine Aviation Works in Southampton in 1916. Three years later, he was appointed chief designer and engineer for the works, a position he held until his death. During his time there, Mitchell designed high-speed seaplanes for the Scheidner Trophy races (1922–1931) that influenced his later aircraft designs. He subsequently designed the legendary *Spitfire* (1936), a single-engine, single-seat, low-wing monoplane, according to Air Ministry Specifications. The *Spitfire* was both a day and night fighter and the model that first entered service in 1938 had a top speed of 360 mph. As one of the fastest fighters deployed in the Second World War, it was highly effective as a defence interceptor during the Battle of Britain. The *Spitfire* was continually modified throughout the war and served a variety of roles, including photo-reconnaissance. The *Spitfire XIV*, one of the last models of the war, had a remarkable top speed of 440 mph and an operational ceiling of 40,000 feet (12,200 m). Renowned for its aerodynamic sleekness and high manoeuvrability, the *Spitfire* remained in active service with the Royal Air Force until 1954.

Supermarine *Spitfire MK16e,* c. 1940

JASPER MORRISON

BORN 1959
LONDON, ENGLAND

Jasper Morrison studied at the Kingston Polytechnic Design School from 1979 to 1982 and at the Royal College of Art in London from 1982 to 1985. He also studied at the Academy of Arts in Berlin on a scholarship in 1984. Morrison became well known in the early 1980s for his experimental furniture and in 1986 established his own London-based industrial design practice, Office for Design. His celebrated indoor/outdoor *Thinking Man's Chair* (1987) for Cappellini was followed by his *Ply-Chair* (1989) for Vitra, which predicted the de-materialistic approach to design that is characteristic of his more recent work. In 1992, in collaboration with James Irvine (b. 1958), Morrison organized the *Progetto Oggetto* for Cappellini – a collection of household items created by a group of young designers. The same year, Morrison published a book entitled *A World Without Words* and received a Bundespreis für Produktdesign award for his anodized aluminium door-handle range designed for **FSB** (1990). One of his most important commissions to date has been the design of the new Hanover tram *TW 2000* (1995–2000) for Expo 2000, which was awarded the IF Transportation Design Prize and the Ecology Award. Other recent projects include his stainless steel *Tin Family* kitchenware for **Alessi** (1998), the *Glo-ball* suspension light for **Flos** (1998), a flat-screen television and hi-fi system for **Sony** (both 1998) co-designed with John Tree, and furniture for the Tate Modern at Bankside (1999). Morrison's aesthetically pure and highly functional work exemplifies what has become known as the new essentialism in design.

Hanover tram
TW 2000 for
the Expo' 2000,
1995–2000

ALEX MOULTON

BORN 1920
BRADFORD-ON-AVON, ENGLAND

Raleigh *Moulton MK3*
bicycle, 1970

Richard Grigsby of
the Bath Cycling Club
riding a Moulton
bike, 1970s

Alex Moulton studied mechanical engineering at Cambridge University and subsequently worked as technical director of his family's rubber manufacturing company until 1956. The same year, he founded Moulton Developments, which worked in close collaboration with Dunlop developing rubber automotive components, including the innovative Hydrolastic fluid suspension system used in **Alec Issigonis'** *Mini* and Morris *1100*. Between 1956 and 1960, Moulton endeavoured to design and manufacture "a better bicycle system". His first series of Moulton bicycles was launched in 1962, and that same year he founded Moulton Bicycles Ltd. to produce his small-wheeled bicycle design, which featured high-pressure tyres and a full suspension.

The Moulton bicycle was claimed to be easier to control because of its low centre of gravity and increased stability. Over 100,000 of these high-performance mini-bikes were built and exported to widespread acclaim. In 1965 Raleigh began producing an imitation of the Moulton bike and two years later acquired the licence to produce it (an agreement that was terminated in 1975). Undeterred, Moulton launched a range of deluxe folding bicycles in 1983 and in 1998 introduced the *New Series* Moulton as "more pleasurable to have and effective to use than any other". With a price tag of over £2,700, however, it is intended only for the seriously committed bicycle *aficionado*.

GERD ALFRED MÜLLER

BORN 1932 FRANKFURT/MAIN, GERMANY
DIED 1991 ESCHBORN, GERMANY

SM 3 electric shaver, 1960 (co-designed with Hans Gugelot)

Gerd Alfred Müller apprenticed as a joiner prior to studying interior design at the Werkkunstschule in Wiesbaden. Between 1955 and 1960 he worked in the prestigious design department at **Braun** headed by **Dieter Rams**. While there, he designed many of the company's best-known household appliances and electric shavers. Typical of his work for Braun are his *Blender MX 3/MX 32* liquidizer (1958/1962) and *Multimix KM 3/KM 32* food processor (1957/1964), whose designs are distinguished by a remarkable clarity and underlying logic that made them easy to use and at the same time aesthetically pleasing. The neutral and rational styling of these products exemplified Braun's essentialist approach to design, which was founded on Rams' belief that it is the responsibility of the designer to create order in a world of visual noise and that products should therefore be as unobtrusive as possible. In 1960 Müller established his own office in Eschborn and began working as a freelance industrial and graphic designer. The solidity and purity of Müller's products were highly influential upon the design of succeeding generations of household appliances.

Multimix KM 3 food processor for Braun, 1957

Blender MX 32
liquidizer for Braun,
1962

PETER MÜLLER-MUNK

BORN 1904 BERLIN, GERMANY
DIED 1967 PITTSBURGH, PENNSYLVANIA, USA

Plastic table radio for Sylvania Electric Products Inc., c. 1950

Normandie pitcher for Revere Copper & Brass Inc., 1935

Peter Müller-Munk trained as a silversmith under Waldemar Rämisch at the Kunstgewerbeschule, Berlin. In 1926 he emigrated to the United States and worked briefly as a silversmith for Tiffany & Co. His silver designs were shown at the 1929 "Modern American Design in Metal" exhibition held at the Newark Museum. That same year, he published an article entitled *Machine-Hand* in which he asked: "Why does nobody ever try to define the process of the machine in action and deduct the forms and decorations most closely related to it?" In the same article, he also urged for "a new art resulting from the harmony of technique and object". Around this time, he established his own design practice in New York, which specialized in the design of products for industrial manufacture. Most notably, he designed metalware for the Revere Copper & Brass Company, including the well-known *Normandie* pitcher (1935), named after the luxurious French ocean-liner. The simple streamlined form of this design was dictated both by functional requirements and by its suitability to industrial production. The clean lines and strong geometric forms that are typical of Müller-Munk's work were strongly influenced by the European Modern Movement. Although best known for his metalware designs, such as his copper-bottomed saucepans for **Ekco** (c. 1950), he also designed various appliances, including a plastic table radio for Sylvania Electric Products Inc. (c. 1950), a retractable ball-point pen and a retractable lip-brush for the B-B Pen Company (c. 1950), a portable sander for the Porter-Cable Machine Company (c. 1950) and the well-known Waring blender. He established another office in Chicago in the 1930s and his work was included in an exhibition at Metropolitan Museum of Art, New York, in 1939. Müller-Munk also designed interiors, exhibitions and vehicles, but is best remembered as an early pioneer of American product design who made the transition from skilled craftsman to professional industrial design consultant.

NASA

(NATIONAL AERONAUTICS AND SPACE ADMINISTRATION)

FOUNDED 1958
WASHINGTON, DC, USA

Apollo 1 spacecraft,
1961

The National Aeronautics and Space Administration (NASA) was founded in 1958 in direct response to the Soviet Union's launch of the *Sputnik 1* satellite the previous year. Having evolved out of the National Advisory Committee for Aeronautics, which had been established in 1915, NASA was (and still is) an independent US government agency committed to the research and development of space flight, aeronautics, space science and space applications. The US space programme was boosted in 1961 by President John F. Kennedy's pledge to put a man on the moon before the end of the decade. The resulting *Apollo* programme, after several setbacks and fatalities, eventually managed to achieve this goal with the *Apollo 11* mission on July 20, 1969. For this sensational historic event to be made possible, NASA drew technical expertise from the cream of American industry. In 1975 NASA launched the unmanned *Viking I* and *Viking II* spacecraft to investigate Mars, while its *Voyager* probes launched in 1977 provided further insights into Jupiter, Saturn, Uranus and Neptune. More recently, the *Galileo* unmanned mission launched in 1989 has brought even greater understanding of Venus, Jupiter and the outer solar system. Apart from its exploratory

Apollo 11 – Edwin (Buzz) Aldrin descending the Lunar Excursion Module's ladder to become the second man on the moon, 20 July 1969

→Apollo 9, launched 3 March 1968 – this image shows the lunar module docking with the command module for the first time

missions, NASA has also undertaken cutting-edge aeronautics research into aerodynamics and wind shear, as well as research into space habitability and earth-orbital space stations, such as the *Sky-Lab* and the *Saturn Five* Space Station (both of which **Raymond Loewy** helped design in the 1970s). 1981 saw the launch of the first Space Shuttle, heralding the advent of partly re-usable space vehicles and opening up the future possibility of commercial space travel. NASA has also developed and launched numerous communication and weather satellite systems, such as *Echo, Telstar* and *Syncom* in the 1960s and *Landsat* in the 1970s. With nearly 40 years "blue sky" research behind it, NASA's programmes have not only fundamentally altered man's perspective of the universe but also changed his material environment on earth. NASA continues to push the frontiers of space and technology in its quest for knowledge. Significantly for the realm of industrial design, the advanced technologies pioneered by NASA continue to result in "spin-offs" across wide-ranging scientific, technical and commercial fields of endeavour.

GEORGE NELSON

BORN 1907 HARTFORD, CONNECTICUT, USA
DIED 1986 NEW YORK, USA

Prolon line of mela-
mine dinnerware
for the Pro-Phy-Lac-
Tic Brush Co.,
1952–1955

George Nelson was one of the foremost design practitioners, theorists and communicators of the 20th century. After studying architecture, he became an associate editor at *Architectural Forum* and wrote profiles of leading Modernist architects for *Pencil Points*. In 1941 Nelson joined the architecture faculty at Yale University, where he developed numerous innovative concepts, including the pedestrianized shopping mall in his "Grass on Main Street" proposal (1942). He also pioneered the concept of built-in storage with his *Storagewall* (1944). In 1946 Nelson became design director of **Herman Miller** – a position he held until 1972. During his time there, Nelson designed Modern furnishings for the company, including the *Comprehensive Storage System* (1957) and the revolutionary *Action Office I* system (1964–1965). As a prolific industrial designer, Nelson also created the *Prolon* melamine line of dinnerware for the Pro-Phy-Lac-Tic Brush Co. (1952–1955), numerous wall and table clocks for the Howard Miller Clock Company (late 1940s and early 1950s), the *Bubble* lamps made of self-webbing plastic (1947–1952) and the *Omni* extruded aluminium pole system for Dunlap. Nelson was also interested in product architecture and designed a modular plastic-domed *Experiment House* (1957). Nelson's concepts were extremely influential and forward-looking – in 1978 he correctly predicted that the advances in computer technology would result in greater "miniaturization, ephemeralization, dematerialization". In 1979 he stated: "Human needs are variable and frequently unpredictable; they certainly are not quantifiable; they are complex, subtle and mysterious."

NEW YORK ★ WORLD'S FAIR

1939

TRYLON AND PERISPHERE

W 101

NEW YORK WORLD'S FAIR

1939
NEW YORK CITY

Elektro, the mechanical man with his pal *Sparko*, the world's first robotic dog, which could perform 27 different tricks, at the New York World's Fair

←Postcard showing the *Trylon & Perisphere* at the New York World's Fair, 1939

With the effects of the Great Depression still fresh in people's minds, the New York World's Fair of 1939 offered an opportunity to look towards a better future. At this time there was also an overwhelming desire to assert America's cultural identity before a global audience. Many of the exhibitors were among the country's leading corporate giants; they included AT&T, **Chrysler**, Dupont, Eastman **Kodak**, Firestone, **Ford**, **General Electric**, **General Motors**, **RCA** and **Westinghouse** — all of whom had major exhibits that showed the "wonders of modern research" and used the machine as a potent symbol of the future. General Motors' towering Automobile Salon had glamorous Hollywood connotations, while the "No.1 hit of the Fair" was the Futurama, which provided "a magic flight through time and space into the world of 1960". **Norman Bel Geddes'** *City of the Future* and **Henry Dreyfuss'** *Democracity* offered utopian visions of futuristic lifestyles that included robots, space travel, televisions and freeway systems. Among the most memorable symbols of the fair, however, were the enormous *Perisphere* and soaring *Trylon* structures, which reflected America's optimism and faith in the future.

Giant *National Cash Register* (designed by Walter Dorwin Teague) at the New York World's Fair, showing daily attendance figures

NIKE

FOUNDED 1971
PORTLAND, OREGON, USA

Tailwind trainer, 1979
– Nike's first air-
cushioned shoe

In 1957 Bill Bowerman, a coach, and Phil Knight, an athlete, met at the University of Oregon, Eugene. Knight subsequently attended Stanford Business School, while runners at the University of Oregon broke records wearing shoes that had been handcrafted by Bowerman. In 1962, under the name of BRS (Blue Ribbon Sports), Knight started importing athletic shoes made to Bowerman's specifications by the Onitsuka Tiger Company in Japan. Bowerman calculated that, over a distance of a mile, every ounce that could be reduced in a runner's shoe would reduce the total weight lifted by 200 pounds. To this end, in 1971 he began developing prototypes of a new and highly innovative running shoe using rubber moulded in his wife's waffle iron. The following year, after a dispute with Tiger, BRS launched the new Nike brand (named after the Greek goddess of victory) and by 1974 the now fully developed *Waffle Trainer* became the best-selling training shoe in America. In 1978 the company signed up the star tennis player, John McEnroe, in the first of many high-profile sports sponsorship deals. A year later, the landmark *Tailwind* shoe was launched and took over as the best-selling running shoe in America. The *Tailwind* was the first trainer to incorporate Nike-Air cushioning – a system of sponge rubber wedges (or mid-soles) sandwiched between the upper and outer sole that act as shock absorbers. The stress put on the body by sport can be immense. A basketball jump shot, for example, will exert a force ten times that of the player's weight, while a runner's foot will endure over 25,000 heel strikes when running a 26.2 mile marathon. 1987 saw the introduction of Nike-Air shoes,

Tuned Air Max
trainer, 1999

which had a revolutionary cushioning system that comprised flexible urethane bags filled with a pressurized gas that reduced the force of impact. After absorbing the force, the air-filled bags quickly returned to their original form and volume in readiness for the next impact. Also in 1987, Nike sold its five millionth pair of *Air Pegasus* shoes (introduced in 1983) and launched

Air Warp Flex in-line
skates, 1999

Triax watch, c. 1998

the high-profile "Just Do It" advertising campaign. By 1989 Nike's annual
net income had risen to $167 million, largely as a result of its product en-
dorsements by top sports personalities. In 1995 it launched a new high-
performance cushioning system, called *Zoom Air*, featuring an 8mm gas
cushion which provided the necessary impact protection yet which kept
the foot close to the ground for increased feel and response. Today, Nike
is the world's number one sports brand and its product range includes
footwear, equipment and apparel.

NIKON

FOUNDED 1917
TOKYO, JAPAN

Giorgetto Giugiaro,
Nikon *F3* SLR
camera, 1980

In 1917 Japan's three leading optical manufacturers merged to form the Nippon Kogaku company and subsequently began researching and producing optical glass, resulting in the issue of its miniature *Mikron* binoculars (1921) and its *Joico* microscope (1925). In 1932 the company adopted the brand name Nikkor for its camera lenses and around 1945 began manufacturing cameras. From 1946 it marketed its 35mm cameras under the Nikon name and two years later introduced the *Nikon I* camera. Gaining a reputation for its superior cameras and lenses, the company founded the Nikkor Club to promote photography and in 1953 established Nikon Inc. in America. 1957 saw the launch of the *Nikon SP* rangefinder model, which was an important precursor of the landmark *Nikon F* which appeared in 1959. The *Nikon F* quite literally revolutionized photojournalism and marked a shift from rangefinders (RF) to single lens reflex (SLR) cameras as the professionals' preferred choice. With its reflecting mirror, the SLR system offered much more precise framing than the rangefinder system, which only gave an approximate idea of what the image would look like. Although not the first SLR camera on the market – the Contax *S* had been available since 1948 – the *Nikon F* was far lighter than earlier Contax models and had the added advantage of being much easier to load

Giorgetto Giugiaro,
Nikon *F5* SLR
camera, 1996

with film. The *Nikon F* – or *The Nikon*, as it became known – not only ushered in widespread consumer acceptance of "Made in Japan" cameras, but also heralded the adoption of 35mm film among professionals and a move towards complete camera systems. *The Nikon* subsequently became the ultimate professional camera and in 1971 the improved *F2* was launched. This was followed in 1980 by another landmark model, the *F3*, designed by **Giorgetto Giugiaro**. Featuring a number of automatic rather than manual functions, the *F3* led automation to become accepted among professional photographers. The design, however, retained Nikon's concept of modularity and could therefore be "stripped" and customized with a wide array of accessories to suit photographers' needs. Three years later, Nikon introduced its first autofocus SLR camera based on the *F3*. Described as "a non-offensive tool to a photographer on duty because it serves more than commands", the *F3* became a modern classic celebrated for its ease of control and handling, extraordinary reliability and system compatibility. Giugiaro also designed the *F4* (1988), the *Nikonos RS* (1992), the first underwater SLR with autofocus, and the now top-of-the-range su-per-fast *F5* (1996), which has an eight frames per second capability and an automatic focus tracking system. Giugiaro declared of the *F5*: "this is a top-class camera designed with thorough emphasis on ease of use. I was con-vinced that originality could be produced from within a condensed form." Over 2,300 products currently carry the Nikon name, including a range of digital still cameras known as *Coolpix* (1998) and the state-of-the-art profes-sional *D1* digital SLR camera (1999) which, like its illustrious predecessors, offers high-performance to "pro-sumers".

↖Nikon *F60* SLR camera, 1990s

↑Nikon *D1* digital SLR camera, 1999

↖Nikon *Nuvis S* ultracompact camera, 1996

↑Nikon *Pronea S* autofocus SLR camera, 1996

NINABER, PETERS & KROUWEL (NPK)

FOUNDED 1979
LEIDEN, THE NETHERLANDS

E-Connect 126K3
built-in down-lights
for Wila Leuchten,
1995

Plano hospital bed
and bedside cabinet,
1991

Peter Krouwel (b. 1952) trained as an industrial designer at the Technical University in Delft, graduating in 1978. A year later, together with fellow student Wolfram Peters (b. 1952), he founded the industrial design practice Ingenieorsbureau Peters en Krouwel. In 1985 the office was joined by a third partner, Bruno Ninaber van Eyben, who had previously studied at the Maastricht Art Academy. Renamed Ninaber, Peters & Krouwel Industrial Design, the practice designed a wide range of products for industrial manufacture, including desk accessories for Randstad (1991), the *Yachtboy* world radio receiver for Grundig (1992), light fittings for Wila Leuchten (1995), the *Quin* stacking chair for Kembo (1995) and the *Discovery* child's bicycle seat for Hamax (1995). Ninaber, Peters & Krouwel also produced award-winning "public design" such as road signage for ANWB (the Royal Dutch Touring Club), a bus station in Leiden and a postbox for the Dutch post office (PTT Post). In 1997 the practice was renamed NPK. Today it has some 50 employees and is widely regarded as the leading Dutch industrial design consultancy. The firm's areas of activity include all stages of the design process, from conception through development to pre-production. NPK has received over 100 design awards, including nine Gute Industrieform prizes, and its work is featured in the collections of the Museum of Modern Art, New York and the Stedelijk Musem, Amsterdam.

Sno Family sledge for Hamax AS, 1996

↓Redesign of ANWB (the Royal Dutch Touring Club) signposts, 1997

Nissan assembly
line, 1990s

NISSAN

FOUNDED 1934
YOKOHAMA, JAPAN

After the launch of the original *Dat* car by the Kwaishin-sha Company in 1914, in 1931 the Tobata Casting Co. began producing a new generation of cars which it christened Datsun ("son of Dat"). Two years later, another company, founded by Yoshisuke Aikawa (1880–1967), took over their manufacture. In 1934 this new venture was renamed Nissan Motor Company and its founder, who had a grand scheme of mass-producing 10,000–15,000 cars per annum, began putting his plan into action. The first small-sized Datsun rolled off the assembly line at Nissan's plant in Yokohama just one year later, symbolizing Japan's rapid pre-war industrialization. During the Second World War, production switched to military trucks and engines for aeroplanes and torpedo boats. Nissan resumed production of non-military vehicles in 1945 and two years later recommenced manufacturing Datsun cars. In 1958 Nissan began exporting Datsun cars to the United States and sold its first Datsun compact pick-up truck there a year later. During the 1960s and early 1970s, Nissan continued to make inroads into the American automotive market, with annual sales rising to 255,000 cars in 1971. The oil crisis massively boosted American sales as consumers began opting for smaller, less expensive and more fuel-efficient automobiles such as those exported by Nissan, **Honda** and **Toyota**. By 1973, one million Datsun vehicles had sold in the United States, and two years later the company became the top vehicle exporter to North America. During the 1980s, Nissan established two strategic manufacturing bases in America and Britain

Nissan *Micra*, 1992

and in 1983 began marketing cars under the Nissan name, such as the hugely successful Nissan *Sunny* launched in 1985. By 1987 its cumulative exports had exceeded 20 million units. In the early 1990s, Nissan became one of the first automotive manufacturers to adopt soft design with its introduction of the *Micra* (1992), which was named Car of the Year in Europe in 1993.

MARCELLO NIZZOLI

BORN 1887 BORETTO, ITALY
DIED 1969 CAMOGLI, ITALY

Elettrosumma 14
adding machine
for Olivetti, 1946

Marcello Nizzoli studied painting, architecture and
graphic design at the Scuola di Belle Arti, Parma. In
1914 he exhibited two small paintings and four embroi-
deries with the Futurists in the first "Nuove Tendenze" exhibition in Milan.
After the war, he produced a series of tapestries – an early foray into the applied arts. In 1918 he founded his own design office in Milan and during the
1920s became associated with the Rationalists. His designs for applied art
brought him widespread recognition when they were exhibited at the first
"Esposizione delle Arti Decorative e Industriali Moderne" held in Modena
in 1923. Subsequently, Nizzoli received a diversity of commissions ranging
from the decoration of silk shawls to exhibition architecture. During this
period, Nizzoli also designed advertising posters, most notably for Campari
in 1926. From 1934 to 1936 he worked in partnership with the architect
Edoardo Persico (1900–1936), with whom he designed two stores for **Parker
Pen** in Milan (1934) and several exhibitions, including the 1934 "Aeronauti-

Mirella sewing
machine for Necchi,
1957

Lettera 22 typewriter
for Olivetti, 1950

cal Exhibition" held in Milan. Between 1931 and 1936 Nizzoli also worked
in collaboration with Giuseppe Terragni (1904–1943). From 1931 he worked
freelance for **Olivetti** as a graphic designer, and subsequently joined the
company's advertising office in 1938. While there, Nizzoli became Olivetti's
chief product designer and produced landmark typewriter designs such as
the *Lexikon 80* (1948) and the *Lettera 22* (1950), which won a Compasso d'Oro
award in 1954. He also designed a number of adding machines, including
the *Elettrosumma 14* (1946) and the *Divisumma 14* (1948), as well as workers'
housing for the company. Beyond his work for Olivetti, Nizzoli also produced sculptural designs for Necchi, such as the *BU* sewing machine (c. 1953),
which was awarded a Compasso d'Oro in 1954, and the *Mirella* sewing machine (1957), which won the Premio Compasso d'Oro the same year. In 1959
Nizzoli designed pocket lighters for Ronson and, a year later, petrol pumps
for Agip. As one of the most important post-war Italian designers, Nizzoli's
rational yet sculptural work was immensely influential upon later Italian
product design.

NORTH AMERICAN

FOUNDED 1928
DELAWARE, USA

F-100 Super Sabre,
1953

Having decided in 1934 to concentrate on manufacturing military aircraft, the first North American aeroplane, the *NA-16* trainer, was introduced in 1935. A year later, the company's manufacturing operation was moved to Los Angeles and in 1937 the twin-engine *NA-21 Dragon* bomber was launched. During the Second World War, North American built a staggering 41,000 aircraft and between 1935 and 1967 manufactured more military aeroplanes than any other American company. Among its most famous aircraft were the *P-51 Mustang* fighter (1940), which was described by the US Government as "the most aerodynamically perfect pursuit plane in existence", and the *B-25 Mitchell* medium bomber, which was the first bomber to be deployed in all combat theatres of operation during the Second World War. After the war, North American developed advanced jet fighters such as the *F-86 Sabre* jet (1949), the first swept-wing design in the US fighter inventory, and the *F-100 Super Sabre* (1953), which established the world's first supersonic speed records. North American also produced the legendary *X-15* rocket-propelled research aircraft (1959), which flew beyond the earth's atmosphere and could reach speeds in excess of Mach 6.5.

P-51 Mustang, 1940

ELIOT FETTE NOYES

BORN 1910 BOSTON, MASSACHUSETTS, USA
DIED 1977 NEW CANAAN, CONNECTICUT, USA

Eliot Fette Noyes studied architecture at Harvard University and at the
Harvard Graduate School of Design. In 1939 he joined **Walter Gropius** and
Marcel Breuer's architectural practice. Upon Gropius' recommendation, he
was appointed the first director of industrial design at the Museum of Modern Art, New York – a position he held from 1940 to 1942 and 1945 to 1946.
Between 1946 and 1947 Noyes was design director of **Norman Bel Geddes'**
industrial design practice, which consulted to **IBM**, and in 1947 established
his own office in New Canaan, Connecticut. From 1956 to 1977 he was corporate design director of IBM, during which time he designed several revolutionary products, most notably the *Selectric I* golf-ball typewriter (1961).
Noyes helped establish a strong corporate identity for the company not only
through the integrity of his own product designs, but by commissioning
graphics from Paul Rand (1914–1996) and buildings from Breuer. Noyes
worked as a design consultant to many other companies, including **Westinghouse**, Mobil, **Xerox** and Pan Am. As one of the most influential advocates
of Good Design, Noyes reshaped entire corporations and set new standards
for product design in America.

Executary secretarial
transcribing machine,
Model 212, for IBM,
1961

Selectric 1 golf-ball typewriter for IBM, 1961

Detail showing golf-ball element of the *Selectric 1*, 1961

ANTTI NURMESNIEMI

BORN 1927
HÄMEENLINNA, FINLAND

Antti Nurmesniemi initially worked in a metal workshop and an aircraft factory. Having been inspired by the designs shown at the Museum of Modern Art's 1945 "America Builds" exhibition held in Helsinki, in 1947 he enrolled at Central School of Applied Arts, Helsinki, where the "Nordic Line" approach to design concentrated on social and technical issues. From 1951 to 1956 he worked for the architect Viljo Revell, designing interiors and furniture, including his well-known horseshoe-shaped *Sauna* stool (1951–1952) for the Palace Hotel. Between 1954 and 1955 he worked in Giovanni Romano's design practice in Milan, where he was inspired by functional yet stylish products designed by **Marco Zanuso** and **Roberto Sambonet** among others. In 1956 he established a Helsinki-based design office and a year later designed his well-known enamelled coffeepots for Wärtsilä. He went on to design furniture for Artek, Tecta and Cassina, kitchenware for Arabia and Kymi, trains for the Helsinki metro and the Valmet Oy company. Among his most successful and characteristic designs were the *Antti* and *Yleispuhelin* telephones (both 1984) for Fujitsu, which reflected his belief that, "An object should not be designed simply to fulfil a purpose (not to mention manufacturing), but for the user ... A simply designed object is wrong if its use is complicated ... A designed object does not need to reflect the times. It is much better if it offers a hint of the future."

Yleispuhelin
telephone for
Fujitsu, 1984

OAKLEY

FOUNDED 1975
FOOTHILL RANCH, CALIFORNIA, USA

A Frame snow
goggle, 1999

Founded by Jim Jannard in 1975, Oakley has a worldwide
reputation for the superlative design of its eyewear products. Since its in-
ception, Oakley has defied convention while focusing on new designs, new
materials and new production techniques. In 1984 Oakley came up with the
concept of "sculptural physics" and subsequently introduced designs, such
as the *XYZ Optics*, that exemplified style and performance. Designs such as
these completely redefined the eyewear market, which had been stultified by
a handful of major manufacturers who believed they "knew it all". According
to Oakley, "an idea is born in the depths of our design bunker, given form
with CAD-CAM modelling, brought to life with SLA liquid laser prototyping,
tested with spectrophotometers, environmental simulation chambers and
ANSI Z87.1 impact rigs" before undergoing rigorous field testing by athletes
for whom the designs are originally intended. The form of Oakley's eyewear
is dictated by performance: the *Racing Jacket* (described as "chiselled adren-
aline"), for example, has a vented and hingeless frame that maximizes peripheral vision and provides de-fogging ventilation, while its "nosebomb"
and "earsock" features ensure that the lightweight frame remains securely
in place. Oakley has managed to position itself as a major global brand
and as well as its innovative eyewear – from sunglasses to snow goggles –
it also manufactures sports apparel, footwear and wristwatches.

Racing Jacket
eyewear, 1999

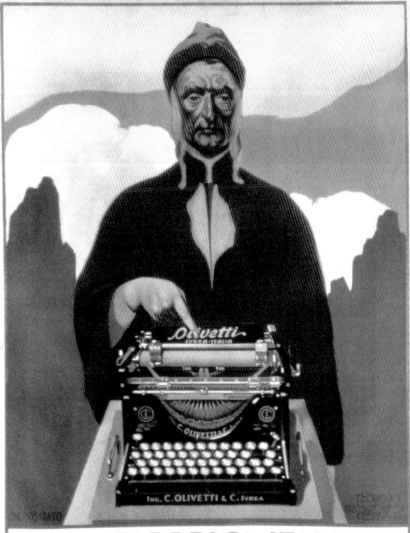

PRIMA FABBRICA ITALIANA
MACCHINE PER SCRIVERE
ING. C.OLIVETTI E Cº IVREA

OLIVETTI

FOUNDED 1908
IVREA, ITALY

Camillo Olivetti

←Poster advertising
Olivetti's *M1*, 1912

Born in Ivrea, Samuel David Camillo Olivetti (1868–1943) studied engineering at the Politecnico di Torino under Galileo Ferraris, who was the discoverer of rotating magnetic fields. After graduating, Olivetti went to London to continue his studies. In 1893 he accompanied Ferraris to the Electricity Congress held in Chicago, and later took a physics course at Stanford University, where he also worked as an assistant electrical engineer. After returning to Italy in 1894, he established a small factory by the name of C. G. S. (Centimetre, Gram, Second) to manufacture electric measuring instruments, which he moved to Milan in 1903. Four years later, he decided to move back to Ivrea and subsequently established Italy's first typewriter factory there in 1908. Aware of American mass production methods, Olivetti began industrially manufacturing typewriters, and in 1909 Ing. C. Olivetti & C. SpA introduced its first typewriter, the *M1*, which was described by a contemporary critic as "robust and elegant" and was praised for its faster carriage speed and smoother key movements. The company grew rapidly, and in the 1920s Camillo's son Adriano Olivetti (1901–1960) was sent to the United States to observe first-hand the latest mass-production techniques, with a view to adopting them back in Ivrea. In 1931 Olivetti launched the *M40* typewriter and the following year unveiled its first portable model, the

Camillo Olivetti, *M1*
typewriter, 1910–1911

MP1. As a moral crusader and social reformer, Camillo Olivetti established a foundation in 1932 that offered his workers social security benefits on a level unprecedented in Italy. Other far-sighted employee incentives were later provided, including seaside summer camps for the workers' children, a cafeteria, a kindergarten, a library and housing. In 1933 Adriano Olivetti was appointed managing director and began diversifying the company's product line, launching its first teleprinter in 1937 and its first adding machine in 1941. He also raised the profile of the company by developing a strong corporate identity through product design, architecture, exhibition design, advertising and graphics. Indeed, during the immediate post-war period, Olivetti was one of only a handful of companies worldwide with a truly Modern image. As a consequence, it was also exceptionally influential upon the design of other firms' corporate identities. As an "intelligence co-ordinator", Adriano Olivetti commissioned leading designers to produce cutting-edge product designs, such as the sculptural *Lettera 22* typewriter (1950) by **Marcello Nizzoli**. He also commissioned leading graphic design-

Marcello Nizzoli,
Lexikon 80 typewriter,
1948

Ettore Sottsass &
Hans von Klier,
Editor 4 typewriter,
1964–1969

Mario Bellini,
Quaderno lap-top
computer, 1989

Michele De Lucchi,
M4–82 Modulo
computer, 1993

ers, such as Giovanni Pintori (b. 1912), to produce eye-catching posters and advertisements. Although Olivetti introduced Italy's first electronic computer in 1959 – the *Elea 9003* designed by **Ettore Sottsass** – the company was obliged for financial reasons to sell its Electronics Division after the death of Adriano Olivetti in 1960. The company continued its research into electronic processing systems, however, and in 1965 it launched the *P101* programmable desktop computer, an innovative forerunner of the personal computer. During the late 1960s and early 1970s the company launched other notable products, including the *Editor 4* typewriter (1964–1969) designed by Ettore Sottsass and Hans von Klier (1934–2000) and the *Divisumma 18* calculator designed by **Mario Bellini** (1973). In 1969 Olivetti also introduced the bright red *Valentine* portable typewriter designed by Ettore Sottsass and Perry A. **King**. A year later, the Olivetti trademark, which is still in use today, was designed by the famous graphic designer Walter Ballmer (b. 1923). While Olivetti experienced financial difficulties in the late 1970s, it nonetheless developed a number of key products, including the company's first electronic typewriter in 1978 and its first personal computer in 1982. During the 1980s Olivetti expanded its operations in the IT field and in the 1990s concentrated on its telecommunications activities. Today, the Olivetti Group is made up of twelve companies operating in either the information technology or telecommunications fields. For Olivetti, "industrial design is a range of activities that not only creates a visual image, but above all contributes to the development of the project as a whole. The designer, in other words, is no longer regarded simply as an expert in aesthetics or style, but as a specialist in the relationship between man and machine."

OLYMPUS

FOUNDED 1919
TOKYO, JAPAN

Olympus μ (mu)
Zoom camera, 1993

Having been founded in 1919, the Takachiho Seisa-kusho company produced the first Japanese microscope in 1920 and a year later adopted the Olympus brand name. In 1930 it launched its first metal-lurgical microscope and five years later opened its Optical Research Centre. It went on to develop the first *Zuiko* photographic lens for its first camera, the bellows-style *Semi-Olympus I* (1936). With increasing demand, the firm introduced a new camera design, the *Olympus 6* (1940). The *Olympus 35* in-troduced in 1948 was the first Japanese 35mm camera with a lens shutter system. In 1949 the company was re-named the Olympus Optical Co. and subsequently unveiled its first medical camera, the *Gastrocamera* (1950). During the late 1950s, the *GTF* gastrocamera with its fibrescope started be-ing used for new medical procedures. Alongside its innovative scientific products, Olympus also released the *Olympus Wide* in the mid-1950s, which inspired a boom for wide-angle lens photography in Japan. In 1969 the com-pany unveiled the world's first microcassette tape recorder, the *Zuiko Pearl-corder*. During the 1960s and early 1970s Olympus pioneered the develop-ment of compact SRL cameras, such as the *Olympus Trip 35* (1968), which was produced for two decades and sold over ten million units, making it one of the most successful 35mm cameras ever. Olympus' compact cameras cleverly incorporated the light-meter cells around the lens, which made their bodies more unified and easier to use. The *Olympus 35-ECR* (c. 1970), for example, had automatic exposure control and focusing that enabled even amateurs to take good quality photographs. As well as continuing to produce state-of-the-art cameras and medical optics, Olympus also manufactures blood-typing systems (80% of all the blood held in North America's blood banks is typed us-ing Olympus machines), along with other diagnostic systems that are used to perform around a quarter of all chemical tests in the US.

Olympus 35-ECR
camera, c. 1970

Opel *Doktorwagen* automobile, 1909

OPEL

FOUNDED 1898
RÜSSELSHEIM, GERMANY

In 1898 the five sons of Adam Opel (1837–1895) converted his sewing-machine and bicycle factory in Rüsselsheim into a car manufacturing plant. Having introduced its first vehicle in 1902 under its own name, Opel AG, the company was beginning to emerge as a major force in the dawning automotive industry when, in 1911, the plant was completely destroyed by fire. Opel took advantage of this calamity by rebuilding and retooling from scratch with the most modern equipment available. In 1913 it introduced the *Rennwagen* racing car and began producing trucks. During this period, Opel also manufactured a distinctive roadster, which was nicknamed the *Laubfrosch* (tree frog) because of its unusual green colour. After the First World War, however, the company's production was restricted by the Allies. Opel nevertheless became one of the first European automotive manufacturers to institute an efficient assembly-line system in 1924, which enabled it to become the largest producer of cars in Europe prior to the Second World War. During the 1920s Opel brought out a number of interesting models, such as the

Opel *Olympia Bi*, 1938

Opel *Corsa*, 1993
(first version
launched in 1982)

small, two-seater *4/16PS* roadster (1928). As a consequence of the hyperinflationary economic conditions of the era, however, the Opel family lost control of the company, and it eventually became a wholly-owned subsidiary of **General Motors**. It later manufactured some notable streamlined cars, such as the *Olympia Bi* (1938) and the *Kapitän* (1953). Although some of Opel's plants were lost to the Soviet Union after the Second World War, the company managed to recover and remains one of the largest European automotive manufacturers, accounting for around 25% of Germany's vehicle production.

PARKER PENS

FOUNDED 1892
JANESVILLE, WISCONSIN, USA

George S. Parker

Advertisement for
the *Jointless Lucky
Curve* fountain pen,
1889

George Parker (1863–1937) was a teacher at the Valentine School of Telegraphy. To supplement his income, he became a part-time representative
of the John Holland Fountain Pen Company and sold pens to his students.
The firm's pens were of poor quality and Parker's customers frequently com-
plained to him about them. He realised that the problems arose from the
design of the feed-shaft, and so with a few simple tools, such as a scroll
saw and a file, he began making shafts for the pens he was selling, which
allowed the air to travel up at a steadier pace and so improve writing performance. Impressed with the re-
sults, Parker decided to manufacture his own pens and with his
meagre savings began purchasing
components from jobbing manufacturers. In 1886 the first Parker
pens were made in the bedroom
of the hotel in which he was living.
He subsequently patented his fountain pen in 1889 and found a firm
to produce them by the gross. In
1892 he established the Parker Pen
Company and designed the first
"jointless" pen, which was patented
in 1899. Advertised as "the greatest
improvement ever made in fountain
pen construction", this design had
"no screw to break, no old-fashioned nozzle, no joints to leak", and
according to the first British advertisement for Parker pens, was retailed by over 6,000 dealers worldwide. 1929 saw the introduction
of the *Duofold*, which could be
converted from a pocket pen to
a desk model with the addition

Kenneth Grange,
Parker 25 pens, 1979

Hollington
Associates, *Frontier*
pens, 1996

of a barrel extension. In 1932 the famous arrow clip was introduced, having
been designed by the New York artist, Joseph Platt. Around 1935, the company launched the *Vacumatic* pen that incorporated an innovative rubber
diaphragm and a mechanism that allowed the ink to be sucked up. The renowned *51* fountain pen, with a sleek profile and Lucite barrel designed to
resist the corrosive effects of the alkali in the ink, was introduced in 1939 –
51 years after the first Parker pen was introduced. Rated as one of the
world's best-designed consumer products, the *51* became one of the company's all-time best sellers and was highly influential upon later pen designs.
Today, Parker retails twelve different product lines, including fountain pens,
rollerballs, ball-point pens and pencils, which range in price from £5.95 to
£5,600.

PENGUIN BOOKS

FOUNDED 1936
LONDON, ENGLAND

Jan Tschichold,
Re-design of the
Penguin logo, 1947

Allen Lane (1902–1970) began his publishing career at Bodley Head, the progressive publishing house owned by his uncle, John Lane. He became a director of the company in 1925 and was appointed its chairman in 1930. In 1935 he started independently publishing paperback reprints and a year later founded Penguin Books – to "put into the hands of people like myself the books they would have read if they had gone to university". Alongside reprints of classics, he also published new titles that were sold at the modest price of 6d (12 US cents) each. Many of these directly promoted the cause of 1930s British Modernism, in particular Anthony Bertram's *Design* (1938) and works by the renowned design historian, Nikolaus Pevsner (1902–1983). Penguin books were extremely well designed and used easy-to-read *Gill Sans* typography. The standardized layouts designed by Jan Tschichold (1902–1974) also gave the imprint a strong visual identity. As the chief pioneer of paperback publishing in Britain, Lane is also credited with having improved printing and binding techniques. Described as having the "noblest list in the history of publishing", Penguin Books attracted a completely new readership who avidly consumed the company's works of quality fiction and non-fiction.

Jan Tschichold,
Penguin book covers,
1946–1949

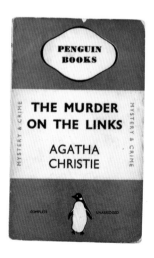

PENTAGRAM

FOUNDED 1972
LONDON, ENGLAND

Kenneth Grange,
Kompakt razor for
Wilkinson Sword,
1998

In 1962 Alan Fletcher (b. 1931), Colin Forbes (b. 1928) and Robert Gill (b. 1931) formed a progressive graphic design partnership based in London. Not wanting to become a fully blown advertising agency, the partners decided to expand into three-dimensional design and in 1965, when Gill left, were joined by the architect Theo Crosby. The office subsequently received two large commissions from Cunard and British Petroleum, the latter of which included the design of petrol stations. For these, they approached the industrial designer **Kenneth Grange** to assist in the design of pump equipment. Eventually, Grange and the graphic designer Mervyn Kurlansky (b. 1936) became officially affiliated with the consultancy and the resulting five-man team was renamed Pentagram in 1972. During the 1970s and early 1980s, John McConnell, David Hillman, Peter Harrison and David Pelham also became partners. Pentagram became noted for its bold and direct corporate identity designs for British Petroleum, Reuters, Faber & Faber, Watneys, the Victoria & Albert Museum, the Tate Gallery, *The Guardian* and Prestel, among others. It also developed typography for many large clients, including **Xerox**, **IBM** and **Nissan**. The Pentagram "house-style" was further perpetuated through its product design, which

Kenneth Grange,
*TT350, TT750 &
TT950* toasters for
Kenwood, 1993

Kenneth Grange,
Power Car Exterior
for the *125* High
Speed Train, British
Railways Board, 1976

was solely undertaken by Kenneth Grange. Under the direction of Crosby, Pentagram has also undertaken exhibition design, including the British Industry Pavilion at Expo '67 in Montreal and the "Environment Game" (1973) and "British Genius" (1977) exhibitions in London. Central to the group's activities is the belief that design should permeate all areas of life, and this idea has been promoted through the publication of several books, including *Pentagram: The Work of Five Designers* (1972), *Living by Design* (1977), *Ideas on Design* (1986) and *The Compendium* (1993). The consultancy has also issued a series of pamphlets, entitled the *Pentagram Papers*, which highlight "curious, entertaining, stimulating, provocative and occasionally controversial points of view". Its consumer product designs are distinguished by a utilitarian directness which is tempered by subtle yet sophisticated styling, and are based on the belief that "product design is not simply a matter of juggling variables, but also a process of discovery where the connections can produce their own surprises." With offices in London, New York and San Francisco, Pentagram has maintained its position as a leading international design consultancy whose partners share "a passion for design, an entrepreneurial spirit, an interest in business, and an interest in the kind of collaborative efforts that we find in a community of designers". Pentagram's longevity and success can be attributed to its collective structure and multi-disciplinary approach to design, which enhances creative synergy and innovative problem solving.

SIGURD PERSSON

BORN 1914 HELSINGBORG, SWEDEN
DIED 2003 STOCKHOLM, SWEDEN

Stainless steel
saucepans for
Kooperativa
Förbundet, 1978

Sigurd Persson trained under his father, Fritiof Persson, as a silversmith in Helsingborg from 1928 to 1937. He gained a journeyman's diploma in 1937 and subsequently studied at the Akademie für Angewandte Kunst, Munich, from 1937 to 1939 and at the Konstfackskolan, Stockholm, from 1939 to 1942. He established his own workshop in Stockholm in 1942 and began designing cutlery and jewellery. In 1949 Persson founded an industrial design practice in Stockholm and subsequently designed sculptural stainless-steel vegetable dishes for Silver & Stål (1953) and the *Jet Line* range of cutlery (1959) for the Kooperativa Förbundet (Co-operative Society of Sweden). In 1959 he won a competition to design in-flight cutlery for the Scandinavian Airlines System (SAS) and later executed matching grey plastic tableware for the airline. Be-

Grey plastic in-flight
service for the airline
SAS, 1959

tween 1968 and 1982 Persson worked as a freelance consultant to the Kosta glassworks, producing brightly coloured, minimally decorated designs. In 1976 he designed uniform badges for the Swedish Army and, with nine other silversmiths, opened the Argentum gallery in Stockholm. In 1978 he designed a range of strong and boldly formed saucepans for the Kooperativa Förbundet which exemplified his approach to the design of practical, everyday objects. Successfully bridging the worlds of handcraftsmanship and industrial design, Persson believed that design had to be truthful in order to be successful. "In every well-designed object", he felt, "it is not just the function that is described in the form. Something else is there, too – we might call it an expression of the times."

PHILCO

FOUNDED 1892
PHILADELPHIA, PENNSYLVANIA, USA

Workers putting the finishing touches to the newly launched *Model 444* broadcast receiver (soon afterwards renamed the *People's Set*) at the Philco radio factory in Perivale, 1936

Philco, *Model 444* broadcast receiver, 1936

Philco was established as the Helios Electric Co. and initially produced batteries. In 1909 it was re-named the Philadelphia Battery Storage Company (Philco) and in 1927 began manufacturing radios. As early as 1929, it was mass-producing radios using assembly-line techniques and rapidly became one of the "big three" radio manufacturers. In 1932 Philco made inroads into the British market by establishing a custom-built manufacturing facility in Perivale. In 1936 Lord Selsden's Ullswater Committee investigated the high price of radios in Britain and challenged the whole industry to produce better and cheaper models. That same year, Philco responded by launching its one-piece Bakelite *Model 444* radio, which sold for just six guineas. Like **Walter Maria Kersting**'s earlier *Volksempfänger*, the radio had a standardized design that was specifically intended for mass production. Initially it did not sell well, but when it was renamed the *People's Set* it became extremely popular and over 500,000 units were sold. During the 1930s, Philco also financed

the researches of the American Philo Taylor Farnsworth (1906–1971), an early pioneer of television. From the late 1940s to the mid-1950s, the company produced a wide range of technically sophisticated televisions housed in historicist cabinets. In the late 1950s, however, Philco began using futuristic styling for its television sets. Known as the Philco *Predictas*, these Space-Age televisions were the most distinctive sets ever designed in America. Despite the use of highly innovative forms and features such as swivelling picture screens, Philco was dogged by the poor picture quality of its televisions and eventually went out of business in 1962.

PHILIPS

FOUNDED 1891
EINDHOVEN, THE NETHERLANDS

Philips *All Transitor*
reel-to-reel tape
recorder, c. 1960

In 1891 Gerard Philips founded a company to "manufacture incandescent lamps and other electric products".
The firm concentrated on the production of carbon-fila-
ment lamps and by 1900 had become one of the largest
manufacturers of its kind in Europe. As a means of stimulating product innovation, in 1914 Philips established a research department that explored
both physical and chemical phenomena. In 1918 the company manufactured
its first X-ray tube and from 1925 began conducting experiments into television. Two years later, Philips launched its first radio and by 1932 it had sold
over one million. The following year, the company produced its 100 millionth radio valve and began manufacturing X-ray equipment in the United
States. Around 1950, having previously pursued an only loose approach to
design, Philips' appointed the architect Rein Veersema to oversee the design
of the company's shavers, radios, televisions and record players. During his
14-year tenure at Philips, Veersema instituted a systematic design strategy
that addressed every aspect of the design process, from ergonomics research to product costings. He also promoted the idea of developing product families so as to project a coherent corporate identity. Many of the designers employed by Philips had worked previously in the **Braun** design department and were thus particularly adept at designing products within the
parameters of a house style. During this period, Philips became a leading

Philips Argenta/TSF
Brevets Français
Philips enamelled
metal sign, c. 1920s

pioneer of sound recording technology and in 1963 introduced its
landmark Compact Audio Cassette,
which was soon after adopted as
an international standard format. In
1972 the company had also developed a laser videodisc, but by the
time this was finally introduced in
1978 the market was already satu-
rated with video tape systems. In
the course of developing the videodisc, however, it was discovered
that it offered significantly better

sound reproduction than conventional vinyl records. This led directly to
the introduction of Philips' revolutionary compact disc (CD) in 1983, which
in a stroke rendered the gramophone record virtually obsolete and seriously
threatened the future existence of audio cassettes. From 1980 Philips' design team was headed by the American industrial designer and architect
Robert Blaich, who espoused the concept of "global design" and championed "product semantics" in an attempt to compete with Japanese companies. Since then, Philips has continued to strengthen the visual identity of
its products with an increasingly "high-tech" aesthetic. Through its commitment to design, Philips has developed an exceptionally strong brand
that is guided by its corporate motto: "Let's Make Things Better".

PIAGGIO

FOUNDED 1884
SESTRI PONENTE, ITALY

Vespa motor scooters launched by Piaggio in 1946

Rinaldo Piaggio (1864–1938) founded the Società Rinaldo Piaggio in 1884. Subsequently renamed Piaggio & C., the company initially specialized in luxury ship fitting but eventually began manufacturing railway carriages, engines, coaches, vans, trams and truck bodies. During the First World War, production included aeroplanes and seaplanes, and the company later manufactured several notable aircraft, including the futuristic-looking *P7* seaplane (1929) and the all-metal *P16* bomber (1935). Piaggio's plants were virtually all destroyed in the Second World War, but the company decided to rebuild and address the pressing need for an inexpensive means of transportation. This led to the development of a scooter, which was designed by the aircraft engineer Corradino D'Ascanio (1891–1981) and whose now famous name was coined by Enrico Piaggio, who when he first saw the new design exclaimed: "It looks just like a wasp" (wasp = *vespa*). After being patented, the *Vespa* went into production and became an immediate success. By 1956, one million *Vespas* had been produced and the scooter became synonymous with the Italian Reconstruction and Latin high-life. Today, the *Vespa* is enjoying an unprecedented revival of popularity as a stylish, practical and affordable form of transport.

Corradino d'Ascanio, Vespa 125 motor scooter, 1951 – original version designed in c. 1945

Lancia *Aprilia Coupé*,
1936–1937

PININFARINA

FOUNDED 1930
TURIN, ITALY

Battista "Pinin" Farina (1893–1966) worked in his brother's bodyshop, the Stabilimenti Farina, in Turin from the age of eleven. In 1920 he visited Detroit in order to meet Henry **Ford** and learn about the new industrial car production techniques which he was pioneering. In 1930 he established his own bodyshop, the Carrozzeria Pinin Farina. His goal was to develop coachbuilding into an independent industry, and with this in mind he instituted an assembly line that was able to produce seven to eight car bodies a day. Throughout the 1930s and 1940s, Farina produced a number of innovative car designs, such as the **Alfa Romeo** *6C* (1931) and the aerodynamic Lancia *Aprilia Coupé* (1936–1937). He also designed the landmark *Cisitalia* (1947), which became the first car to go on permanent display at the Museum of Modern Art in New York as "one of the eight outstanding cars of our time". It was described as the best expression of beauty and simplicity in the automotive field and set new standards for post-war car design. From 1952 Farina began styling exquisite models for **Ferrari**, such as the *250 SWB* (1961) and the *Dino Berlinetta* (1965). In 1958 the Carrozzeria Pinin Farina built a much larger plant to increase production and three years later became known officially as Pininfarina. 1966 saw the opening of the Pininfarina Study and Research Center, which subsequently developed many influential concept and production cars, including the Ferrari *Modulo* (1970) and the **Jaguar** *XJS* (1978). In 1972 Pininfarina constructed the first full-scale

Cisitalia, 1947

Lancia *Florida II*, 1957

Ferrari *250 SWB*, 1961

wind tunnel in Italy for its researches into aerodynamics. During the 1980s, Pininfarina lived up to its reputation as Italy's premier carrozzeria with designs such as the Ferrari *Testarossa* (1984). From 1981 the company was also assembling major components and building complete vehicles – today it produces around 40,000 car bodies per annum. Reviewing Battista Pinin Farina's life and work, the committee of the Gran Premio Nazionale Compasso d'Oro stated in 1957: "After a long period and rigorous elaboration of form, he has created the plasticity of line so influential in industrial production and in the determination of aesthetic standards for the automobile." Today, as throughout its history, Pininfarina's research into new areas of design, aerodynamics, new technologies and new materials, as well as safety and environmental protection, has allowed the company to offer innovative ideas to the automotive industry and to make practical contributions to the evolution of automotive engineering and design.

EZIO PIRALI

BORN 1921 ITALY

Metal table fan for
Zerowatt, 1954

An electromechanical engineer, Ezio Pirali was the
managing director of the Italian electrical appliance
manufacturer, Zerowatt. He designed and developed
numerous products for the company and forged a distinctive house style that relied on sleek, essentialist
forms and precision engineering. In 1954 he received
one of the first Compasso d'Oro awards and won a
silver medal at the Milan Triennale for his *VE505* table
fan (1954). The motor of this unusual fan was cased in highly polished aluminium, and its rubber blades rotated within a chrome-metal wire cage
that also served as a structural support. The *VE505*, which was fabricated
by Elettriche Riunite of Milan for Zerowatt, was characteristic of the smooth
yet rationalist industrial aesthetic of Pirali's work and was produced from
1954 to 1960. Ezio Pirali's product designs were later shown at the 1979
"Design & Design" exhibition in Milan and at the "Design Since 1945" exhibition held in Philadelphia (1983–1984). Pirali's work reflected the sheer in-
ventiveness of post-war Italian design, an attribute born of necessity that
contributed significantly to the country's ensuing economic reconstruction.

Zerowatt *VE505* table
fan for Fabbriche
Elettriche Riunite,
1953

POLAROID

FOUNDED 1937
BOSTON, MASSACHUSETTS, USA

Edwin Land

In 1932, while at Harvard University, Edwin Land (1910–1991) developed the first polarizing filters made from synthetic material. That same year, Land and his physics professor, George Wheelwright, established a company to develop applications for this new polarizing technology, including sunglasses, cameras and car headlamps. The Polaroid trademark was officially registered in 1935 and the Polaroid Corporation founded two years later. The company licensed its polarizing technology to American Optical and Bausch & Lomb, and in 1939 Polaroid sold over one million pairs of its anti-glare sunglasses. Land's revolutionary concept of instant photography was born during a Christmas holiday in Santa Fe in 1943, when his daughter asked: "Daddy, why can't I see the picture now?" Land developed his idea over the next three years and in February 1947 presented the first Instant Photograph to the Optical Society of America. The first Polaroid *Land Camera*, the *Model 95*, was subsequently

Polaroid *Land Camera Model 95*, 1948

launched in 1948, and by 1956 production of this groundbreaking design had reached one million units. With its concertina lens, the camera was nevertheless relatively bulky and could only take black-and-white images. 1957 saw the introduction of the first Polaroid slide-system, which enabled images to be projected within a two-minute waiting period. In 1963 Polaroid introduced the first instant colour film, representing the culmination of 15 years' research and development. By 1964, over five million *Land Cameras* had been manufactured and in 1971 the *MiniPortrait* camera was introduced, designed for taking multiple passport photographs. By incorporating

an automatic ultrasound rangefinder, Land was able to reduce the dimensions of his cameras, which resulted in the compact *SX-70* (1978) designed by **Henry Dreyfuss** Associates. This design served as a blueprint for many subsequent models, such as the *Spectra System* (also known as *Image System*) launched in 1986. This computerized camera used an advanced film that required two different chemicals for its processing. In 1996 Polaroid introduced its first digital camera, the *PDC2000*, and a year later launched the *SpiceCam* instant camera (named after the Spice Girls pop band). These new products have neverthess failed to restore the embattled company's diminishing fortunes, which have suffered considerably from the emergence of digital photographic technology.

GIO PONTI

BORN 1891 MILAN, ITALY
DIED 1979 MILAN, ITALY

→Office chair for
Kardex Italiano, 1938

Coffee machine for
La Pavoni, 1949

Giovanni (Gio) Ponti studied architecture at the Politecnico di Milano and subsequently worked for the architects Emilio Lancia and Mino Fiocchi. From 1923 to 1930 he was art director of the Richard Ginori ceramics factory, where he designed porcelain wares decorated with neo-classical motifs in the Novecento style, some of which were awarded a Grand Prix at the 1925 Paris "Exposition Internationale des Arts Décoratifs et Industriels Modernes". Ponti also designed low-cost furniture for the La Rinascente department store in the mid-1920s, and from 1925 to 1979 was the director of the Monza Biennale exhibitions. In 1925 Ponti designed his first building – his own neo-classical style house on the Via Randaccio in Milan – and in 1926 he established an architectural practice in partnership with Emilio Lancia in

Milan, which continued until 1933. In 1928 Ponti launched *Domus*, the prestigious design journal established to promote the Novecento Movement and counter both "the fake antique" and "the ugly Modern" in architecture and design. Between 1933 and 1945, Ponti worked in partnership with the engineers Antonio Fornaroli and Eugenio Soncini in Milan and during this period undertook many architectural commissions for both private and public buildings. From 1930, he designed lighting and furniture for the Fontana company and in

Series P sanitary fixtures for Ideal Standard, 1953

1933, with Pietro Chiesa (1892–1948), was made artistic co-director of its subsidiary, Fontana Arte. During the 1940s, Ponti contributed extensively to *Stile* magazine, produced sets and costumes for La Scala opera house (1947), created multi-coloured glass bottles, glasses and a chandelier for Venini (1946–1950) and designed the famous La Pavoni coffee machine (1949). During the late 1940s and throughout the 1950s, he collaborated with Piero Fornasetti (1913–1988) on several furniture designs and interiors, including those for the Casino at San Remo (1950). Alongside several prestigious building commissions, such as the second Montecatini Building (1951) and Pirelli Tower (1956), both in Milan, Ponti also designed flatware for Krupp Italiana (1951) and Christofle (1955), sanitary ware for Ideal Standard (1953) and the widely celebrated *Superleggera* chair (1957) for Cassina, which possessed the timeless classicism that was so characteristic of his work. Ponti skilfully balanced the rational with the poetic to create designs that exemplified the Italian Line. Ponti's forward-looking products were extremely influential and reflected his belief that: "Happiness, man's last hope, is a dream. But nothing ever came to pass without being dreamed of first."

PORSCHE

FOUNDED 1931
STUTTGART, GERMANY

Professor Ferdinand Porsche

The design practice that would later evolve into Porsche AG was founded in Stuttgart in 1931 by Ferdinand Porsche (1875–1951), who went on to design the hugely successful **Volkswagen** *Beetle* (1934–1938). During the Second World War, Porsche designed several military vehicles, including the *Tiger* tank. After the war, his son "Ferry" used Volkswagen components and a flat four-cylinder engine to create the *356* roadster (1948), which became the first vehicle to carry the Porsche name. The streamlined body of this sports car was designed by **Erwin Komenda** not only to look beautiful, but to be as functionally efficient as possible. By 1958, over 10,000 *356*s had been manufactured. In 1961 the company began work on a new model with a rear-mounted, air-cooled six-cylinder engine and a body designed by Komeda and Ferry Porsche's son **Ferdinand Alexander Porsche**. This resulted in the immortal *911*, which was first produced in 1964 and went on to become one of the most famous sports cars of all time. Still in production today, numerous evolutions of the *911* were manufactured, including the *911 Carrera RS* (1972), which was the first production car with a front and rear spoiler, and the *911 Turbo* (1974), which was the world's first production car with an exhaust gas turbocharger. Porsche's first front engine transaxle configured car, the *924*, was introduced in 1976 and was followed by the similarly configured *928* in 1977, which was conceived – controversially to purists – as the successor to the *911*. By the mid-1990s, however,

Evolution of the Porsche *911 Carrera*, 1972–1998

Porsche *959*, 1985

Porsche *959*,
1985 – rear view

both these models had been discontinued, despite healthy sales. As well as producing highly acclaimed road cars, such as the limited edition *959* supercar (1985), Porsche has also produced numerous racing cars, including the *917* (1970) and the *956* (1982) – two of the most successful racing cars ever. In 1993 Porsche launched a new concept car at the Detroit Auto Show – the six-cylinder mid-engine *Boxster*. Production of this roadster began in 1996 and it became an immediate success. Through its commitment to design, engineering and technological excellence, Porsche has established one of the strongest brands in Germany and its name remains synonymous worldwide with the sports car. In the company's own words: "For Porsche, the ultimate test is to exhaust all possibilities for excellence even at the cutting edge of technology. We regard every idea as an opportunity. It is this perception that has created our company and what guides us still. It is the very essence of Porsche".

Porsche *Boxster*, 1997–1998

FERDINAND ALEXANDER PORSCHE

BORN 1935
STUTTGART, GERMANY

Kandido Tavolo lamp
for Lucitalia, 1982

Ferdinand Alexander "Butzi" Porsche is the grandson of Professor Ferdinand Porsche (1875–1951) who founded **Porsche** AG. He apprenticed as an engineer at **Bosch** in Stuttgart and from 1957 studied at the Hochschule für Gestaltung in Ulm. The following year, he began working under **Erwin Komenda** in the Porsche design department, which he later headed from 1961 to 1972. In this position, he designed the bodies of several cars, including the famous *911* (1964). In 1972 he established the independent Porsche Design Studio, now based in Zell am See, Austria. The experience he gained in automotive design proved useful in the creation of "lifestyle" products. His sleek and highly engineered products, which balance function, technology and styling, epitomise German product design. The Porsche Design Studio has produced work for a prestigious international clientele, ranging from the *Cobra* motorcycle for Steyr-Puch (1976) to ergonomic high-tech sunglasses for Bausch & Lomb (1997). During the late 1990s, the studio also produced designs for transit systems for the cities of Singapore, Bangkok, and Vienna and developed a proposal for a computer-aided taxi system (CATS). While most of the products designed by the studio are produced and marketed under the brand name of Porsche Design, the creative origin of other products designed for some of the studio's customers is not always made evident.

Singapore *Airport Express* train for Siemens SGP, 1998

Diver's watch for
IWC, 1983

→Motorcycle helmet
for Römer, 1976

Telephone for NEC,
1981

TC 91Lz11d1.5100
coffee machine, *TW
91100* cordless kettle
& *Tt 91100* toaster
for Siemens, 1997

Trend porcelain tea service for Thomas, 1981

QUEENSBERRY-HUNT

FOUNDED 1964
LONDON, ENGLAND

Lord (David) Queensberry (b. 1929) studied ceramics at the Central School of Art & Design, London, and design and technology at North Staffordshire College of Technology. He later studied industrial ceramic design at the Royal College of Art in London before working as a designer in the ceramics industry in Stoke-on-Trent. From 1955, he worked as a freelance for the German company, **Rosenthal**. During the early 1960s, Queensberry designed glassware for Webb Corbett that was distinguished by bold geometric forms and repetitive patterning. In 1960 he was appointed Professor of Ceramics at the Royal College of Art and four years later founded the Queensberry-Hunt Design Group with Martin Hunt (a fellow student at the RCA). Believing in "good design for the mass-market", the practice designed industrially produced ceramics and glassware for Australian Fine China, **Corning** Glass Europe, Curver, Dartington Crystal, Ideal Standard, Le Creuset and Poole Pottery. One of Queensberry-Hunt's most

Kyomi washbasin for Ideal Standard, 1997

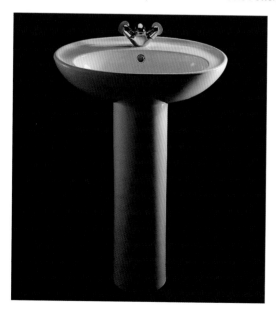

successful designs is the *Trend* dinner service (1981) which is still manufactured by Thomas – a subsidiary of Rosenthal. After many years' experimentation, in 1982 Rosenthal finally perfected the manufacture of "Queensberry Marble", a material that requires the simultaneous processing of two different colours of porcelain or ceramic slip. Queensberry-Hunt's restrained designs, which balance practicality and versatility with attractive yet simple forms, are ideally suited to mass production. The practice was later joined by Robin Levien (b. 1952), who was nominated as a Royal Designer for Industry, and renamed Queensberry Hunt Levien.

DIETER RAMS

BORN 1932
WIESBADEN, GERMANY

Dieter Rams is Germany's most important post-war designer. During his 40-year career at **Braun**, he transformed the way consumer products are designed. Guided by deeply held moral convictions, Rams pioneered an approach to design that was based fundamentally on logic. From an early age, Rams was exposed to construction techniques in his grandfather's cabinet-making workshop in Wiesbaden, which produced well-made undecorated furniture without the use of machinery. He went on to study architecture and interior design at the Werkkunstschule in Wiesbaden from 1947 to 1948. His father, an electrical engineer, insisted that he gain practical experience, however, and so he served a carpentry apprenticeship in Kelkheim from 1948 to 1951. On returning to Wiesbaden, Rams resumed his training under Professor Söder, who taught both architecture and interior design. After graduating in 1953, Rams worked briefly for an architect who designed public sector buildings, which Rams detested for their shoddy quality and lack of honesty. He eventually managed to get a job in the Frankfurt architectural

Phonosuper SK 4 radio-phonograph for Braun, 1956 (co-designed with Hans Gugelot)

practice of leading German Modernist Otto Apel (1906–1966). Apel's office had an affiliation with Skidmore, Owings & Merrill, and Rams became inspired by the "can-do" attitude of the American practice and learnt from it "the simple essence of industrial construction". During this period, Rams applied for a position at **Braun**. He was subsequently summoned to the company for an interview, during which he met Erwin Braun and was shown designs by **Hans Gugelot**, which he found astoundingly Modern. In 1955 Rams began working as an architect and interior designer for Braun and the following year began designing products for the company, most notably the *Phonosuper SK 4* hi-fi system (1956) which he co-designed with Hans Gugelot. Rams also designed other audio equipment for Braun, including the portable *Transistor 1* radio (1956) and the combination phonograph and pocket radio (1959), all of which embodied the practical and ordered approach to oversee the design of that was promoted by the **Bauhaus** and the Hochschule für Gestaltung, Ulm (with which Braun maintained close links). In 1961 Rams was appointed head of the company's design department and went on to oversee the design of the *KM 2* kitchen appliances, the *Sixtant* electric razor, calculators, radios, hifi systems and a cylindrical table lighter. These products were characterized by a pared-down rationalist aesthetic and exemplified the attributes of good design. Also in the 1960s, Rams designed the *606* shelving system and *620* and *601* modular seating ranges

RT 20 radio for
Braun, 1961

for the furniture manufacturer Vitsoe, all of which shared the rational clarity of his product designs for Braun. In 1968 Rams became director of design at Braun and the same year was elected an Honorary Designer for Industry by the Royal Society of Arts, London. By the late 1960s, the idea of "product aesthetics" and in particular the Modern vocabulary of functionalism was coming under increasing attack, as it was seen by many as a promotional sales tool. Rams, however, remained unswayed by such criticism and continued to promote essentialism in design through efficient, well-designed and executed products that incorporated state-of-the-art technology. Rams believes that the central responsibility of designers is to instil order in contemporary life. His approach to design has always been based on his "Ten Commandments on Design" – "Good Design … is innovative, makes a product useful, is aesthetic, helps to understand a product, is unobtrusive, is honest, is durable, is consequential to the last detail, is concerned with environment … Good design is as little design as possible." His rallying cry of "Back to purity, back to simplicity" and his declaration that the "aesthetic requirement of an industrial product is that it should be simple, carefully made, balanced and unobtrusive" have ensured that his work has maintained an unparalleled consistency and quality throughout his career. Rams officially retired from Braun in 1995, but his legacy continues to powerfully influence the ethos of the company.

RCA

(RADIO CORPORATION OF AMERICA)

FOUNDED 1919
NEW YORK, USA

Elridge R. Johnson,
Talking Machine,
1900 – an improved
version of Emil
Berliner's earlier
gramophone

← *Victrola* phono-
graph, 1906

On entering the First World War in 1917, the US government decreed that all American wireless stations were
to be taken over by the US Navy, including **Marconi**'s
operations. After the war, the Assistant Navy Secretary, Franklin D. Roose-
velt, felt that all patents pertaining to broadcasting equipment should remain in America and blocked **General Electric**'s planned sale of equipment
to the British-owned subsidiary of the Marconi Wireless Telegraph Company.
Instead, GE took over Marconi's American assets and established the Radio
Corporation of America (RCA) in 1919. Headed by David Sarnoff (1891–
1971), the company initially marketed radios manufactured by GE and **Westinghouse**. In 1920 RCA obtained the first commercial broadcasting licence
and the following year made the first-ever sports broadcast. The company
subsequently transmitted the first radio photograph across the Atlantic
Ocean in 1924 – an astonishingly early precursor of the facsimile machine.
In 1926 Sarnoff established the National Broadcasting Company (NBC) to
take over responsibility for RCA's radio broadcasting, such as the first coast-
to-coast broadcast of the Rose Bowl game in 1927. Two years later, RCA acquired the Victor Talking Machine Company and began manufacturing its
own radios and phonographs. Victor had been founded by Elridge Johnson
around 1900 and manufactured Emil Berliner's gramophone and the highly
popular *Victrola* phonograph (1906). With the acquisition of Victor, RCA
also obtained its famous "His Master's Voice" (HMV) trademark.

Cathedral-style table-
top radios rolling
down the massive
assembly lines at the
RCA Victor Company's
manufacturing head-
quarters in Camden,
New Jersey, 1934

During the 1930s, RCA became a household name not only through its
ubiquitous cathedral-style table-top radios, but also through its associations

with Hollywood and the construction of Radio City in New York, its
corporate headquarters and the
home of NBC. The company also
developed its first experimental television set, which it unveiled at the
1939 **New York World's Fair**. The development of commercial television
was halted, however, as America
entered the Second World War and

The first factory-fitted automobile radios in America, 1936

RCA Victor experimental television receiver featuring a 21" tri-colour tube, 1951

RCA began manufacturing equipment for military use. After the war, RCA immediately resumed radio production and, having benefited from its wartime research into picture tubes for the US Navy, began manufacturing the black-and-white *Model 630TS* 10" television set in 1946. Although not the first colour television on the market, the company's all-electronic black-and-white-compatible colour television, which was developed in 1953, was recognized as the US Standard. Despite the fact that the 12" set cost a massive $1000 when it was first marketed in 1954, within six years there were 500,000 colour televisions in use across America and 60% of NBC's evening programmes were broadcast in colour. It would be RCA cameras that later transmitted the historic images of **NASA**'s space flights into millions of homes. The company's investment in space technology considerably assisted its development of solid-state colour televisions, which were introduced in 1970. In 1977 the company launched the first-ever four-hour VCR and the following year began marketing the first programmable VCR and colour home recording cameras. In 1979 RCA produced its 100 millionth television picture tube. In 1986 it was acquired by its founder and rival GE, which subsequently sold its consumer electronics interests to Thomson Consumer Electronics the following year. In 1989 RCA manufactured its 50 millionth colour television and in 1994 Thomson entered the digital home entertainment market with the launch of the RCA *Satellite System*, the first high-power direct broadcast satellite system. Featuring DIRECTV programming, the system ushered in the era of HDTV (High-Definition Television) – a technology with vast implications for the future of multimedia. Today, Thomson is the fourth largest producer of consumer electronics in the world and is continuing to expand with the launch of state-of-the-art products marketed under the famous RCA brand.

John Vassos (attributed), portable phonograph *Model M, RCA Victor Special*, c. 1937

REMINGTON

FOUNDED 1816
ILION GORGE, NEW YORK, USA

C. L. Sholes & C.
Glidden, Remington
No.1 typewriter, 1873

In 1816 Eliphalet Remington (1793–1861) fashioned a flintlock rifle that became renowned for its accuracy, and that same year turned his family's forge in Ilion Gorge, New York, into a gun manufacturing business. The venture grew, and in 1828 Remington established a large factory alongside the Erie Canal. Remington and his son Philo subsequently pioneered numerous innovations in arms manufacturing, including a lathe for cutting gun stocks and the reflection method used to straighten the barrels of guns. They also developed the first American practical cast-steel drilled rifle barrel and in 1847 supplied the US Navy with its first breech-loading rifle. As a pioneer of large-scale industrial production, it is not surprising that the company eventually diversified into other areas of manufacture. In 1873 E. Remington & Sons developed the first-ever commercial typewriter, the Remington *No. 1*. This landmark product was designed by Christopher Latham Sholes (1819–1890) and was based on an earlier, cruder model which Scholes had patented in 1868. It was Sholes who first coined the word "typewriter", and his Remington *No. 1* featured the "QWERTY" keyboard layout that is still the standard today. In-

Advertisement for
Remington sewing
machines, c. 1880

deed, many features of this revolutionary machine remained standard for over a century: the cylinder, with its line-spacing and carriage-return mechanism; the escapement mechanism, which moves the carriage along between each letter; the actuation of the typebars by means of key levers and wires; and printing through an inked ribbon. 1878 saw the introduction of the improved Remington *No. 2*, which was the first typewriter to have a shift-key mechanism and was a much better seller than the Remington *No. 1*, which could only print capital letters. Around this time, the company further diversified its product range to include sewing machines. In 1886 E. Remington & Sons sold its typewriter business and became the Remington Arms Company, which supplied a significant number of small arms to the US government during the First and Second World Wars and remains in operation today. Meanwhile, the Remington

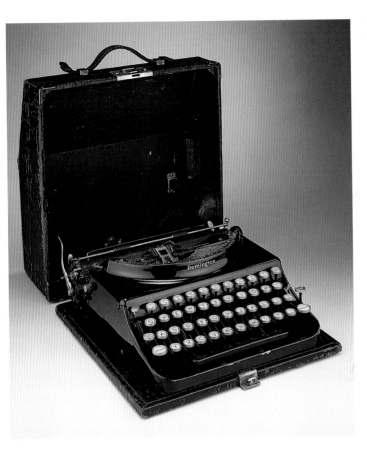

typewriter manufacturing company merged with the Rand Kardex Company
in 1927 and subsequently became a leading supplier of office equipment. In
1936 it established an electric shaver division and the following year launched its innovative *Close Shaver*. At the 1939 **New York World's Fair**, Remington
introduced the idea of dry shaving to literally millions of Americans, and a
year later launched its first dual-headed shaver, the Remington *Dual*. In 1960
Remington brought out the first cordless shaver, the *Lektronic*, employing
rechargeable nickel cadmium energy cells, and in 1975 adopted the *Soft Touch*
flexible foil cutting system. Four years later Victor Kiam formed Remington
Products Inc., stating: "I liked the shaver so much I bought the company".
In 1981 the first Remington retail store was opened and over the succeeding
decades the company has become a leading developer of shaving, grooming, personal care and travelling appliances.

RENAULT

FOUNDED 1898
BOULOGNE-BILLANCOURT, FRANCE

In 1898 Louis Renault (1877–1944) built his first auto-mobile, the *Voiturette*, in a small workshop on his fam-ily's estate in Billancourt. This early design had a direct transmission – an invention that Louis Renault patented. The same year, he founded the So-ciéte Renault Frères with his two brothers, Fernand and Marcel. Together they designed pioneering racing cars which won numerous competitions, but when Marcel was tragically killed during a Paris-Madrid race in 1903, the company decided to concentrate its efforts on production vehicles. During the First World War, Renault manufactured a tank that was used as a troop escort vehicle. After the war, Renault's production capacity was increased and during the 1930s the company manufactured a vast array of vehicles, in-cluding taxis. Although Louis Renault knew how to make successful cars, he was not so politically astute and continued to produce military equipment

during the German occupation of France. Afterwards, he was impris-oned for collaboration and his com-pany was nationalized in 1945. In 1961 the company issued the highly successful Renault *R4* (1961), which in turn became the basis of the landmark Renault 5, introduced in 1972. Designed by Michel Boué (1936–1971), the stylish Renault 5 hatchback was specifically intended for young people and women and created a completely new category of car– the "supermini". By 1975 Renault was producing 1.5 million cars per annum, of which 55 % were exported. Much of this success can be attributed directly to the Renault 5, which became the best-selling French car of all time. A new version of the Renault 5 was launched in

Renault *Espace 2000 TSE*, 1986

Renault *Twingo*, 1992

1982 and in 1986 the company introduced another landmark design, the *Espace*, which similarly created a new market sector – the "people carrier". During the 1990s, the *Twingo* (1992) was one of the first automotive expressions of soft design, while the compact *Scenic* (1996) – voted Car of the Year in 1997 – was the first monospace design to appear in the lower-medium (M1) segment of the market. Today, Renault remains the leading French car manufacturer and continues to be one of the world's most innovative automotive companies.

ROSENTHAL

FOUNDED 1879
SELB, GERMANY

Tapio Wirkkala,
Kurve cutlery, 1963

Raymond Loewy &
Richard Latham,
Form 2000 coffee
service, 1954

During the 1870s, Philipp Rosenthal (1855–1937) worked in a porcelain factory in America. He eventually returned to Bavaria in the late 1870s and in 1879 established Rosenthal Porzellan AG, a workshop in Erkers-
reuth, near Selb, specializing in the decoration of porcelain. Experiencing
difficulties in acquiring blanks, he finally established his own factory in Selb
in 1891. In the early years of the 20th century, the company produced ceramics in a simplified Jugendstil style, which brought it widespread acclaim.
These wares, which included Philipp Rosenthal's *Darmstadt* service (1905)
and *Donatello* service (1907), were influenced by the design reform advocated by the Darmstädter Künstlerkolonie (Darmstadt Artists' Colony) and
the **Deutscher Werkbund**. In his *Isolde* service of 1910, Rosenthal subsequently attempted to break away from conventional forms altogether by inte-
grating the handle into the overall shape. Rosenthal initially manufactured
such ceramics with undecorated finishes in order to emphasize their strong
graphic forms, but they were later made available with simple underglazed
hand-painted decoration, such as small boughs of cherries. The factory also
produced wares based on 18th-century patterns, as well as Art Nouveau-

Jasper Morrison,
Moon porcelain
service, 1997

style figurines, including several of bathing beauties designed by R. Marcuse around 1914. **Wilhelm Wagenfeld** designed a range of Modern ceramics for the company in the 1930s, and in the 1950s **Raymond Loewy** and Richard Latham designed the well-known *Form 2000* service (1954). By commissioning leading designers such as these, the company gained much recognition and as a result became the most prominent German ceramics manufacturer. In 1960 Rosenthal opened its first "Studio-Haus" to retail not only its own products but also those of other progressive factories and studio potteries. All the designs were approved by a panel of advisers that included both artists and design critics, and further Studio Houses were subsequently established in most major European cities. In 1969 Rosenthal introduced **Walter Gropius'** *TAC1* service, which epitomized the company's aesthetic – modern, sculptural and functional. Apart from its porcelain, Rosenthal has also produced glassware as well as silver and stainless steel products, such as Tapio Wirkkala's *Kurve* range of cutlery (1963). Among the many other designers and design practices that have designed products for the company's Studio-Line range are Timo Sarpaneva (b. 1926), Lino Sabbatini (b. 1925), Michael Boehm (b. 1944), **Queensberry-Hunt**, Michael Young (b. 1966) and **Jasper Morrison**. Rosenthal also has its own design studio based in Selb and a subsidiary, Thomas, that produces similar but more utilitarian wares.

RUBBERMAID

FOUNDED 1920
WOOSTER, OHIO, USA

All-plastic dustpan,
1956, and *2001*
spatula, 1957, with
RM.33 dustpan and
brush, 1934

→ *Roughneck* refuse
container, 1990s

Rubbermaid
advertisement,
1950s

The Wooster Rubber Company was founded in 1920 initially to produce Sunshine-brand rubber balloons. Around the same time, James Caldwell (1896–1977), a rubber company executive, noticed that his hanging metal dustpan was chipping the paint off his walls. He solved the problem by inventing the rubber dustpan, and was so impressed by the quality of the product that he set about designing rubber draining boards, soap dishes and sink strainers. Caldwell subsequently christened his new enterprise Rubbermaid. In 1934 the company joined forces with the Wooster Rubber Company to begin the large-scale manufacture of Rubbermaid products, and Caldwell was appointed the firm's president. During the Second World War, rubber was restricted to military use and so the company produced components for self-sealing aircraft tanks as well as life jackets and tourniquets. After the war, Rubbermaid resumed manufacturing housewares, although as colourants were still restricted its products were available initially only in black. In 1956 the company introduced its first all-plastic product, a dishpan (washing-up bowl), and around the same time began running advertisements that appealed to the new generation of homemakers. Over the following decades Rubbermaid expanded rapidly and today operates in numerous sectors, owning major brands such as Curver, Goody, Little Tikes, Rotring and Rolodex. Remarkably, the company manufactures over 5,000 products and introduces on average one new product a day. No other household products company can equal Rubbermaid's 97% brand recognition in the United States nor can claim to have its products in 9 out of 10 American homes. Rubbermaid's success can be directly attributed to its approach to design. The company's high-quality yet affordable utilitarian products, such as its *Roughneck* refuse container, are extremely well conceived in terms of function, durability and suitability to high-volume mass production.

Sixten Sason,
Saab 92, 1947

SAAB

FOUNDED 1937
TROLLHÄTTAN, SWEDEN

The Svenska Aeroplan Aktiebolaget (Swedish Aircraft Company) was founded by the financiers Marcus Wallenberg and Axel Wennergren in response to the Swedish government's re-arming initiative prior to the outbreak of World War Two. Having recruited some 40 engineers from **Boeing**, the company initially manufactured versions of Northrop and **North American** aircraft, before developing its own designs, such as the Saab 17 bomber, the Saab 21, featuring the world's first ejector seat, and the Saab 29, Europe's first swept-wing fighter. In 1945 the Swedish Air Force reduced its purchasing of aircraft, so Saab decided to produce automobiles instead. Its first car, the Saab 92 (1947), was a revolutionary streamlined design styled by **Sixten Sason** and engineered by Gunnar Ljungström. Inspired by aircraft design, it had sweeping aerodynamic lines and was the first small car to be produced in Sweden. The Saab 92 reflected Sason's design philosophy that form should follow function and that products should not be too expensive nor look too utilitarian. The Saab 92 evolved over the following decades into other Saab models, such as the well-known Saab 95 estate car (1959). Sason and Ljungström went on to develop a new car, the Saab 99, which was compact yet spacious inside, aerodynamically efficient and economical to run. Launched in 1969, the Saab 99 became the blueprint for later Saab cars, including the Saab 90 (1985) and the Saab 9–5 SE (1997).

Sixten Sason,
Saab 95, 1959

Saab 9–5 SE, 1997

MARC SADLER

BORN 1945
INNSBRUCK, AUSTRIA

Drop light fixture for
Arteluce, 1993

Marc Sadler studied industrial design at E. N. S. A. D. in Paris and in 1968 became a member of Design Center I, a French organization for designers. The following year he founded Deco Design, an innovative interior design and furnishings company based in the south of France. Sadler subsequently joined the France Design group in 1970 and around this time began studying plastic products and the technologies used to make them. From 1971 he began working with the Italian company, Caber, designing plastic ski boots with patented asymmetrical shells. Spending increasing amounts of time in Italy, Sadler opened a studio equipped with CAD and three-dimensional modelling technology in Asolo, and gathered together a team of model-makers and technicians to assist him in the development of state-of-the-art products for his growing clientele. After opening branches of his studio in New York and Venice, he finally settled in Milan, where he continues to work today. Sadler is most noted for his sports equipment design, such as his motorcyclist's *Bap* back protector (1992) designed for Dainese, which was exhibited in the Museum of Modern Art's 1995 "Mutant Materials" exhibition. This innovative design features an anatomically shaped, expanded polyethylene compo-

Bap back protector
for Dainese, 1992

nent, which conforms to the rider's body and acts as a shock absorber, and an outer shell of flexible polypropylene. His *Drop 1* and *Drop 2* light fixtures (1993) for Arteluce similarly incorporate state-of-the-art materials and production technology, with their soft injection-moulded silicon-elastomer diffusers and rigid injection-moulded Lexan polycarbonate supports. In 1994 Sadler was awarded a Compasso d'Oro for these designs and was named Createur de l'Année for 1997 by the Salon du Meuble in Paris.

Stainless steel
cutlery for
Sambonet, 1959

ROBERTO SAMBONET

BORN 1924 VERCELLI, ITALY
DIED 1995 MILAN, ITALY

Roberto Sambonet studied architecture at the Politecnico di Milano from 1942 to 1945. Between 1948 and 1953 he worked at the Museo de Arte Moderna in São Paolo, collaborating with Pietro Maria Bardi. In 1953 he worked in Alvar Aalto's office, before returning to Italy, where he designed products, graphics and exhibitions for La Rinascente. In 1954 his family firm in Vercelli began manufacturing his stainless steel designs, such as his elegant fish-serving dish *Pesciera* (1956) with a bivalve shell form. With his later flatware (1959) and eight-piece nesting *Center Line* cookware (1964), Sambonet pioneered new forms while expressing an essentialist aesthetic. He won Compasso d'Oro awards in 1956 for a series of oval trays and in 1970 for his series of stainless steel containers (including the fish server and *Center Line* cookware). He was also art director of *Zodiac* magazine from 1957 to 1966 and designed glasses for Baccarat (1971) and porcelain dishes for Richard Ginori (1979). As an advocate of "objective" design, Sambonet distrusted consumer decadence and during the 1980s became interested in the social aspects of town planning.

Center Line cookware
for Sambonet, 1964

SANYO

FOUNDED 1947
OSAKA, JAPAN

14T-183 black-and-white television, 1955

The name Sanyo means "three oceans", which reflects the company's global vision and its management's commitment to "become an indispensable element in the lives of people all over the world". The Sanyo Electric Works was originally founded in Osaka in 1947 by Toshio Iue, the brother-in-law of Konosuke **Matsushita**. Within three years the firm had become Japan's leading manufacturer of generator lamps for bicycles. In 1952 Sanyo began manufacturing and marketing plastic-housed radios and a year later launched its "whirlpool"-action washing machines – an event considered a milestone in Japan's post-war consumer electronics boom. In 1955 the firm began mass-producing televisions and by 1959 it had become Japan's largest exporter of transistor radios. In 1963 the company introduced the Sanyo *Cadnica* battery, which was the first of many batteries developed by Sanyo and which heralded the era of cordless electrical appliances. In response to the oil crisis of the early 1970s, in 1979 Sanyo developed amorphous silicon solar batteries, which it began mass-producing from 1982. In 1990 the company developed a new refrigeration system for air-conditioning units which was based on hydrogen rather than ozone-damaging CFCs. As well as pioneering and commercializing "clean technologies", Sanyo is also one of the world's largest consumer electronics companies. It designs and manufactures a wide array of products – from digital cameras and televisions to videos, audio equipment, mobile phones, refrigerators, vacuum cleaners and microwaves, all of which voice Sanyo's corporate motto: "Precision craftsmanship to be proud of the world over".

MCD X77 portable CD, late 1990s

RICHARD SAPPER

BORN 1932
MUNICH, GERMANY

Richard Sapper studied mechanical engineering and economics from 1952 to 1954 at the University of Munich. From 1956 to 1957 he worked in the car styling department of **Mercedes-Benz** in Stuttgart. He later moved to Italy and worked in the Milan studio of Alberto Rosselli (1921–1976) and **Gio Ponti** from 1957 to 1959. In 1959 he designed his *Static* table clock for Lorenz (which was awarded a Compasso d'Oro in 1960) and from 1959 to 1961 worked for the in-house design department of La Rinascente. Around this time, he also began working in the studio of **Marco Zanuso**. In 1970 Sapper established his own design office in Stuttgart and acted as a design consultant to **Fiat** and Pirelli, among others. He continued collaborating with Zanuso until 1975 and together they produced an impressive series of landmark designs, including the *Lambda* chair for Gavina (1963); the injection-moulded polyethylene *No. 4999/5* stacking child's chair for **Kartell**

Details of *Tizio* lamp
for Artemide, 1972

(1961–1964); the *Doney 14* television (1962); the *TS 502* radio (1964) and the *Algol* and *Black 201* portable televisions for **Brionvega** (1965 and 1969); and the *Grillo* folding telephone for **Siemens** (1966). In 1972 Sapper independently designed the highly successful *Tizio* task lamp for **Artemide**, which with its blatantly High-Tech design rhetoric became a cult object in the 1980s. In 1972, together with Gae Aulenti (b. 1927), he established the Urban Transport Systems Study Group, whose research culminated in an exhibition held at the XVIII Milan Triennale in 1979. Sapper has been a product design consultant to **IBM** since 1980, and since the late 1970s has also created a number of notable designs for **Alessi**, in which his earlier High-Tech style is synthesized with Post-Modernism. Such designs include the *9090* espresso coffee-maker (1979), which was the first Alessi product de-

Soundbook cassette player for Brionvega, 1974

signed for the kitchen since the 1930s; the *9091 Bollitore* whistling kettle (1983), which was Alessi's first "designer" kettle; the *Uri Uri* watch (1988); the elegant *Bandung* automatic teapot (1995); and the *Cobán* range of domestic espresso machines (1997–1998). According to Alberto Alessi, Sapper is a designer who has the "ability to transgress the rules of force in industry", allowing him to open up creative possibilities within the rigid constraints of mass production. His products – which Sapper describes as "transitional objects" – transform the cultural identities of the companies that he works for. He has also designed furniture for Knoll, Unifor, Molteni and Castelli. Sapper's technically sophisticated yet aesthetic design solutions spring from his extraordinary originality of thought. As he himself recognizes: "You cannot learn anywhere how to have an idea."

Richard Sapper &
Marco Zanuso,
Grillo telephone for
Siemens, 1966

Richard Sapper &
Marco Zanuso,
TS 502 radio for
Brionvega, 1964

SIXTEN SASON

BORN 1912 SKOVDE, SWEDEN
DIED 1967 SWEDEN

Design for an
industrial washing
machine for
Electrolux, 1950

→Design for vacuum
cleaner attachments
for Electrolux, 1944

Sixten Sason initially studied with Swedish artists living in Paris and later trained as a pilot. He subsequently worked for Husqvarna making motorcycles and small arms and later worked as a graphic designer and magazine illustrator. During the Second World War, he joined **Saab**, then a fledgling aircraft manufacturing company, and worked as a technical illustrator. Around 1945 the company decided to diversify, and Saab's chief engineer Gunnar Ljungström and his team began developing its first car, the Saab 92 (1947). Inspired by aeronautical design, Sason transformed this front-wheel-drive vehicle into a streamlined vision with integrated side panels, bonnet and boot – an innovation that was later adopted by **Pininfarina** and **Raymond Loewy**. The Saab 92 was Sweden's first small car and became the blueprint for later Saab models, such as the 95, 96 and 99 (all of which were designed by Sason). As one of the first European industrial design consultants, Sason also designed highly influential streamlined products for **Electrolux** and Hasselblad.

Design for *Model
248* vacuum cleaner
for Electrolux, 1943

MARGARETE SCHÜTTE-LIHOTZKY

BORN 1897 VIENNA, AUSTRIA
DIED 2000 VIENNA, AUSTRIA

Margarete Schütte-Lihotzky was the first woman to graduate from the School of Applied Arts in Vienna. Influenced by the work of Heinrich Tessenow (1876–1950), she was dedicated to the improvement of workers' housing and initially collaborated with the Rote Wien group. In 1925 Ernst May (1886–1970), the director of town planning and building in Frankfurt, embarked on a large-scale mass-housing programme which focused on the concept of the "minimum-existence dwelling". For this, Mayer sought highly efficient solutions to housework and asked Schütte-Lihotsky to design the most rational kitchen possible. The resulting standardized *Frankfurt Kitchen* was equipped

The *Frankfurt Kitchen*, 1924

like a scientific laboratory, while its layout and size were designed in accordance with findings gleaned from time and motion studies. Strongly influenced by Christine Frederick's book, *Scientific Management of the Home* (1915 – translated into German in 1922), the kitchen was devised solely for food preparation and its space-saving built-in cupboards, storage units and worktops were similar to those in ship's galleys. The kitchen provided a well-lit and unadorned space and was equipped with a swivelling stool and an adjustable overhead lamp. Schütte-Lihotsky's prefabricated and standardized design was inexpensive to produce and was used for the new housing estates built in Frankfurt during the 1920s. The *Frankfurt Kitchen* played a significant role in the transformation of the art of cooking into the science of home economics.

DOUGLAS SCOTT

BORN 1913 LONDON, ENGLAND
DIED 1990 LONDON, ENGLAND

Routemaster bus for
London Transport,
1954

Pay-on-answer coin
box for the General
Post Office, c. 1960

Douglas Scott trained as a silversmith at the Central School of Art & Design in London and subsequently worked in **Raymond Loewy**'s London office from 1936 to 1939. While there, he worked on designs for vacuum cleaners and refrigerators commissioned by **Electrolux**. After the Second World War, he set up his own design office and was responsible for establishing the industrial design course at the Central School of Arts and Crafts in London. In 1949 he co-founded the Scott-Ashford Associates design office with Fred Ashford. He later designed the classic *Routemaster* bus (1954) for **London Transport**, which is still in use in London today and became a widely recognized and much-loved symbol of the city. The Routemaster (also known as the RM) was the last of a series of buses designed specifically for London and intended to compliment and enhance the look of the capital's streets. Having first entered service in 1956, the *Routemaster* is still considered by many a high-point in public transport design – a triumph of pragmatic engineering that is simple to maintain and operate. With its two-person crew and open back, the *Routemaster* can pick up and set down passengers more quickly than modern driver-only buses. Despite these advantages, the *Routemaster* is unable to comply with recent European health and safety legislation and is slowly being phased out.

Following this successful foray into transport design and engineering, in 1955 Scott established his own independent industrial design consultancy, Douglas Scott Associates. He subsequently designed products for companies such as **Marconi**, Ideal Standard and ITT. Scott believed that there was no place for personal aesthetics in industrial design. His telephone coin box (c. 1960) for the GPO (General Post Office), which combines American **styling** with the traditional restraint of British design, is highly representative of his work.

SIR GILES GILBERT SCOTT

BORN 1880 LONDON, ENGLAND
DIED 1960 LONDON, ENGLAND

Model K6 public telephone boxes, 1936 – also known as Jubilee Kiosks

Sir Giles Gilbert Scott was the grandson of the famous Gothic Revivalist architect, Sir George Gilbert Scott (1811–1878), who designed most notably the Albert Memorial in Kensington Gardens (1863–1872). Following in his grandfather's footsteps, Gilbert Scott trained as an architect and in 1898 articled with Temple Lushington Moore (1856–1920). In 1904, at the age of just 24, he won the competition for Liverpool Anglican Cathedral, defeating such strong contenders as Charles Rennie Mackintosh (1868–1928). His Gothic-style cathedral revealed the bold massing that came to characterize his later buildings, most notably Battersea Power Station (1932–1934) and the colossal Bankside Power Station (1947–1960), which is now the home

of the new Tate Modern art gallery. As well as buildings, Scott also designed two public telephone boxes for the GPO (General Post Office) in 1924 and 1936. The latter was the classic red *Model K6*, which is sometimes referred to as the *Jubilee Kiosk*. This design, which was inspired by the architect Sir John Soane's tomb in London, is as quintessentially British as **Douglas Scott**'s *Routemaster* bus and the famous black London cab. Like Gilbert Scott's buildings, the phone box is a successful synthesis of modernity and tradition. When British Telecom began replacing Gilbert Scott's well-loved phone box with new kiosks in 1988, it prompted a public outcry, with the result that many examples of this classic design can still be seen across Britain today.

Sundberg-Ferar,
automatic washing
machine and drier
manufactured by
Whirlpool for Sears,
Roebuck & Co.,
c. 1950

SEARS, ROEBUCK & CO.

FOUNDED 1893
CHICAGO, ILLINOIS, USA

Beyond design and manufacturing, the success of a product has always hinged on distribution. The concept of distribution by mail order was first pioneered in America by the retailer Montgomery Ward in 1872. Fourteen years later, in Redwood Falls, Minnesota, Richard Warren Sears (1863–1914) established the R. W. Sears Watch Company as a mail-order company selling watches. In 1887 he hired Alvah Roebuck as a watch repairer and moved the business to Chicago. That same year he published his first catalogue, in which he offered watches, diamonds and jewellery with a money-back guarantee. Having become almost immediately successful, the company was sold for $100,000 only two years later. Sears and Roebuck subsequently founded another company, A. C. Roebuck & Co, which became Sears, Roebuck & Company in 1893. The new venture published its first catalogue the same year, running to 196 pages and offering a wide array of products from sewing machines and saddles to bicycles and shoes. By 1894 the catalogue had swollen to 507 pages. Sometimes known as the "Wish Book", the Sears catalogue was often the only book apart from the Bible in many rural North American households. To those living in the American mid-west or on the Canadian prairies, the catalogue offered a powerful link to the cosmopolitan world and inspired many hours' dreaming, which the local general store could not possibly match. Richard Warren Sears, who also wrote the copy for the catalogues, would often only attempt to locate a manufacturer for an item once orders began pouring in – a remarkably early example of "just-in-time" production. By 1914 consumers could mail-order practically anything, from cutlery and furniture suites to complete houses. From 1924 Sears, Roebuck & Company was headed by General Robert Wood, who recognized that the automobile was making retail outlets in urban centres more accessible to consumers in outlying suburbs and rural areas. To exploit this situation and the boom in car ownership, he opened the first Sears retail store in Chicago in 1925. Many other stores

Cover of Sears,
Roebuck & Co.
Consumers Guide
catalogue, *No. 108,*
1899

→Clarence Karstadt,
*Silvertone Model No.
6111* radio, 1938

Raymond Loewy,
Coldspot refrigerator,
1934

were established over the next few years and by 1931, retail store sales had exceeded those of mail order. In 1934 Sears Roebuck launched the *Coldspot* refrigerator designed by **Raymond Loewy**, which was the first domestic appliance to be marketed primarily on its aesthetic appeal. The company also produced and retailed the futuristic white plastic *Silvertone* radio (1938) designed by Clarence Karstadt. During the 1950s, Sears' in-house design division was headed by Carl Bjorncrantz, who designed sleek, streamlined products ranging from household appliances to adding machines. For over 100 years, Sears has been one of the world's largest retailers of general merchandise. Today, its famous catalogues continue to offer a wide range of products, which like the early "Wish Books" provide small and remote communities access to quality goods that are difficult if not virtually impossible to obtain locally.

SEYMOUR POWELL

FOUNDED 1983
LONDON, ENGLAND

Richard Seymour (b. 1953) trained as a graphic designer at the Royal College of Art, London, and subsequently worked in the advertising industry, which gave him a useful insight into the minds of consumers. Dick Powell (b. 1951) also graduated from the Royal College of Art, where he studied industrial design. After working as a freelance designer, Powell teamed up with Seymour in 1983. Three years later, Seymour Powell received widespread recognition for its *Freeline* kettle (1986) for Tefal – the first cordless model on the market. The team has also designed many other consumer products, but it is perhaps best known for its design of the *MZ Skorpion* bike (1993), which was created for the ailing former Eastern Block manufacturer, MuZ. Described by Dick Powell as "like a **BMW** that has had an affair with an Italian **Ducati**", the *MZ Skorpion* was designed to be "sustainable, cheap, simple and easy-to-maintain" and was styled to make an "emotional connection". Seymour Powell's ability to combine the technical with the aes-

Freeline cordless
kettle for Tefal, 1986

thetic is also evident in the digital wristwatches it designed for **Casio**, including the *Tri Chrono* (1992) and *Overland* series (1993) as well as the highly successful *Baby G* series (1995), all of which were less aggressively masculine than Casio's previously launched *G-Shock* wristwatches. The *Vectis GX-4* waterproof camera (1995) for Minolta also reveals a similar "Soft-Tech" approach to design. Seymour Powell has been the subject of two English television series (*Designs On ...* 1998 and *Better by Design* 2000), which explored the duo's approach to redesigning a number of diverse products, from bras to electric cars.

Vectis GX-4
waterproof camera
for Minolta, 1995

MZ Skorpion
motorbike for MuZ,
1993

SHARP

FOUNDED 1912
TOKYO, JAPAN

Sharp QT-12 stereo-radio cassette player, 1973

MD-M11 mini disc player, 1995

Having founded a metalworks in 1912, three years later Tokuji Hayakawa designed the world's first mechanical pencil known as the *Ever-Sharp*, from which the company derives its name. Sharp Corporation began the production of crystal radio sets in 1925 and four years later introduced the *Sharp Dyne*, the first AC vacuum-tube radio set. In 1951 the company developed a prototype television and two years later, to coincide with the first Japanese television broadcasts, began mass-producing black-and-white television sets. These were followed by colour television sets in 1960 and microwave ovens in 1962. In 1964 Sharp produced the world's first all-transistor-diode electronic desktop calculator. During the 1970s, the company continued developing calculators and introduced the revolutionary *EL-805* calculator (1973), which was the first ever calculator to incorporate a liquid crystal display (LCD). Sharp went on to become the world leader in LCD technology. In 1973 it also launched the *QT-12* stereo-radio cassette player, which broke away from the convention of the black box by being offered in either silver or yellow finishes. Sharp introduced numerous consumer electronic prod-

ucts in the 1980s, from the world's first automatic front-loading VCR system (1980) to combination toaster microwave ovens (1986). The company also produced the world's first 4" and 5" colour LCD televisions in 1990 and later formed development partnerships with Intel and **Apple Computer** to produce advanced flash memory products and personal information equipment respectively. The company is dedicated to the promotion of "sharp products" – designs that "help individuals, families and corporate teams connect effortlessly, communicate clearly and unleash creativity."

Sharp *Zaurus PI-7000* (G Mark)
multi-functional fax-internet
personal organiser, 1996

Sharp *MI-10*
multi-functional
fax-internet personal
organiser, 1996

SIEMENS

FOUNDED 1847
BERLIN, GERMANY

T1200 telex,
1983–1986

Siemens is a highly innovative electrical engineering and electronics company that manufactures approximately one million different products and aims to "benefit mankind, help protect the environment and utilize resources responsibly" within its six areas of operation – energy, industry, information and communications, healthcare, transportation and lighting. The company was founded in 1847 by Werner von Siemens (1816–1892) and Johann Georg Halske (1814–1890) as the Siemens & Halske Telegraph Construction Company. The following year, the company began the construction of Europe's first long-distance electrical telegraph line between Frankfurt and Berlin. Between 1853 and 1855 Siemens constructed the Russian state telegraph network. In 1866 Werner Siemens discovered the dynamo-electric principle and invented the self-excited generator, which heralded the era of economical electric power that could be produced in large quantities. Siemens continued laying its telegraph cables, including the Indo-European line from London to Calcutta in 1870 and the first transatlantic line from Ireland to the United States in 1874. The company demonstrated the first electric railway at the 1879 Berlin Trade Fair and two years later built the first electric tramway in Lichterfelde, near Berlin. Like **AEG**, Siemens understood early on the benefits that standardization could offer industrial production, and implemented an extensive program of standardized component interchangeability for its products, including its differential arc lamp produced from 1878 and its wide range of domestic appliances retailed from around the 1890s onwards. 1903 saw the establishment of a sister company, Siemens-Schuckertwerke GmbH, which took over the

Siemens Design,
Dental system
Sirona C1, 1994

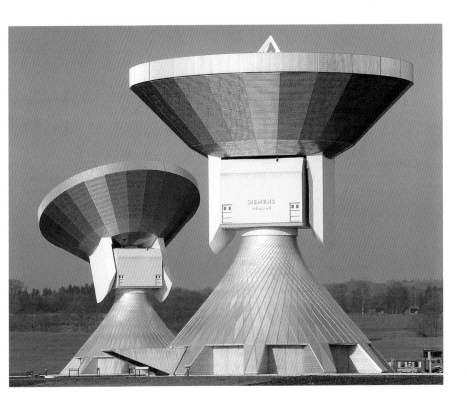

company's engineering activities and in 1905 developed an electrically powered car. That same year, Siemens & Halske developed the tantalum lamp,
which was the first practical metal-filament incandescent light bulb. In 1919
the company, together with two other German incandescent lamp manufacturers, founded Osram GmbH KG for the large-scale mass production of
light bulbs. During the First World War, the Nonnendamm industrial quarter
of Berlin was officially named Siemensstadt (Siemens City), but by the end
of the conflict the company had lost approximately 40% of its plant capacity. In 1932 another subsidiary was founded, Siemens-Reiniger-Werke AG,
which developed and manufactured diagnostic and therapeutic medical
equipment, including microscopes and X-ray equipment. After the Second
World War some 90% of the three companies' manufacturing facilities and
equipment was expropriated, but the firms survived and merged in 1966 to
form Siemens AG with headquarters in Munich. Since then, Siemens has
grown into a global company with over 49,000 employees actively engaged
in product research, design and development.

IGOR SIKORSKY

BORN 1889 KIEV, UKRAINE
DIED 1972 EASTON, CONNECTICUT, USA

Igor Sikorsky studied engineering in Paris and later in Kiev. In 1909 he moved back to Paris to learn as much as he could about this fledgling science, before returning to Russia to develop rotary-wing aircraft. Although his first helicopter design and its successor were complete failures, Sikorsky, undeterred, began designing fixed-wing aircraft, most notably the *S-6-A*. He subsequently worked for the aviation subsidiary of the Russian Baltic Railroad Car Works, where he designed the world's first multi-engined aeroplane, the *S-21* (1913), whose numerous innovations included an enclosed cabin and a lavatory. In 1919 he emigrated to America and four years later founded the Sikorsky Aero Engineering Corp., which built a number of successful flying boats. Turning his attention again to rotary-wing craft, Sikorsky perfected the world's first practical helicopter, the *VS-300* (1939). Following the success of this revolutionary design, United Aircraft (which had previously purchased his company) began full-scale manufacture of the first-ever production helicopter, the *R-4* (1943). Sikorsky subsequently pioneered innumerable "firsts" in the helicopter industry. He was a man guided by a deep religious conviction, and often remarked: "the work of the individual still remains the spark which moves mankind ahead."

VS-300 helicopter, 1939 – the first practical helicopter

CLIVE SINCLAIR

BORN 1940
RICHMOND, SURREY, ENGLAND

Executive (centre) and
Sovereign electronic
calculators, 1972–1975

Clive Sinclair, an early pioneer of consumer electronics, established Sinclair Radionics in 1962 initially selling radios and amplifiers by mail order. This changed in 1972 when Sinclair launched his landmark *Executive* calculator, the world's first pocket calculator and a significant step forward in the miniaturization of electronics. Sinclair produced other pocket calculators during the early 1970s, including the slender *Sovereign*, *Scientific* and *Cambridge* models. In 1976 Sinclair designed a miniature television set, the *Microvision*, which won a **Design Council** Award in 1978. He also pioneered the development of personal computers with his inexpensive and highly successful *ZX 80* (1980) and the later *ZX Spectrum* (1984), which had 48K of RAM. During the mid-1980s these designs sold in their millions, rivalling the sales of both **Apple Computer** and **IBM**. By the end of the decade, however, Sinclair's machines had been surpassed technologically by American computers. In 1985 Sinclair produced the *C5*, an innovative single-person electric vehicle that became a celebrated commercial failure. The following year, Sinclair established a firm devoted solely to research.

ZX Spectrum
microcomputer,
1984

JOSEPH SINEL

BORN 1889 NEW ZEALAND
DIED 1975 USA

Detail of Joseph
Sinel's scales,
c. 1927

Joseph Sinel is reputedly among the first to have used the term "industrial design", which he defined as the art and science involved in the creation of machine-made products. He initially worked as a freelance designer in New Zealand, but moved to Australia in 1913 prior to emigrating to the United States around 1918. In America in 1921, while working for an advertising agency, Sinel began undertaking "product improvement" jobs for his employer's clients. He eventually established his own industrial design consultancy some eight years before designers such as **Raymond Loewy**, **Henry**

Model S scales,
*Height & Weight
Meter* for the
International Ticket
Scale Corporation,
c. 1927

Dreyfuss and **Walter Dorwin Teague** followed suit, and this led him to describe himself in later life as America's first industrial designer. His *Model S* height and weight machine (c. 1927) for the International Ticket Scale Corporation of New York reflected the popularity of the contemporaneous Art Deco style, with its sky-scraper form and machine-polished stainless steel and chromed metal elements. Sinel also worked extensively as a commercial artist, designing trademarks and packaging, and in 1923 wrote *A Book of American Trademarks and Devices*. In 1930 he became a member of the American Union of Decorative Artists and Craftsmen. During the 1970s he was a professor emeritus at the California College of Arts & Crafts, and in 1975, the year of his death, he became a founding member of the American Society of Industrial Design (ASID) – a field of design in which he was an important pioneer.

SINGER

FOUNDED 1851
NEW YORK, USA

Isaac Singer's first
sewing machine, 1851

Isaac Merit Singer (1811–1875) served an apprenticeship as a machinist and in 1839 patented a machine for drilling rock. While working in a Boston machine shop in 1851, he was asked to repair a Lerow & Blodgett sewing machine. Seeing a way in which the design could be improved, he set about developing his own model. Within eleven days he had built his first sewing machine. Unlike **Elias Howe**'s earlier design, Singer's machine could be operated by a treadle, allowing both hands to be kept free, and had a presser foot which kept the material in place. Both these features had been used previously by Barthélemy Thimonnier in France, but it was Singer who was the first to incorporate them successfully into a practical sewing machine. In 1851 Singer patented his design and established I. M. Singer & Company to manufacture the machine, which from 1855 was sold around the world. By 1860 Singer had become the world's largest producer of sewing machines, and three years later the firm was incorporated as the Singer Manufacturing Company. In 1885 the company unveiled its first elec-

Singer Manufacturing
Company's factory at
Kilbowie, Scotland,
1867

Singer sewing machine, c. 1900

XL-1000 top-of-the-range sewing machine, late 1990s

trically powered sewing machine, although it would be another fifteen years before the first mass-produced electric sewing machines went on sale. Isaac Singer also pioneered the use of instalment credit plans, which had an enormous impact on consumer sales in modern society. In later years, the Singer Manufacturing Company diversified its product range to include furniture, power tools and floor-care products.

HANS ERICH SLANY

BORN 1926
ESSLINGEN, GERMANY

BR Trike for Kärcher, 1990s

Hans Erich Slany studied engineering in Eger (Hungary) and in Esslingen. Between 1948 and 1955 he developed products for Ritter Aluminium, before establishing his own design studio, Slany Design, in 1956. Around this time he collaborated with Heinrich Löffelhardt, co-designing the *Ikonette* compact camera for Zeiss-Ikon (1956). In 1959 Slany was one of the founders of the Verband Deutscher Industriedesigner (German Association of Industrial Designers), and that same year designed his first electric drill for **Bosch**. Utilizing ergonomics data and shockproof and heat-resistant thermoplastics, Slany went on to design numerous other power tools for Bosch, including an electric screwdriver (c. 1960), the *Combi-E* range of DIY tools (1966) and the first electro-pneumatic rotary hammer weighing less than 2.5 kg (1981). Since 1962 Slany has designed over 200 high-pressure cleaning products for Alfred Kärcher GmbH and has also designed a wide variety of other products, ranging from medical equipment to electronic rocketry components. He was made an honorary professor at the Berlin Hochschule der Künste in 1985. Over its 40-year history, Slany's office – which in 1997 renamed itself Teams Design – has won more than 900 national and international design awards, more than any other design consultancy in the world. Slany believes that "individual product personality" is essential for success in today's increasingly homogenous marketplace.

GBH 24 VRE drill for Bosch, c. 1992

SNAP-ON

FOUNDED 1920
MILWAUKEE, MINNESOTA, USA

Earliest socket
wrench set, 1920

Joseph Johnson worked for a company that manufactured traditional socket wrenches – tools that were constructed as one-piece units with the socket permanently attached to the
handle. While there, Johnson questioned the logic of having to buy a handle
in order to have the use of a single socket and wondered if it would be possible to have, for example, five handles and ten sockets that could "snap-on"
to provide fifty different combinations. A colleague, William Seidemann, saw
the potential of "interchangeable sockets" and together they developed a
sample socket wrench set that was marketed with the slogan "Five Do the
Work of Fifty". Johnson and Seideman established the Snap-on Wrench
Company in 1920 and enlisted representatives to demonstrate the benefits
of the Snap-on system directly to potential customers. In 1923 the first company catalogue featured over 50 designs, including open-ended wrenches,
and by 1935 sales had exceeded $1 million. The Great Depression proved a
difficult trading period, however, and so Snap-on introduced T. P. (time pay-
ment) selling, which allowed customers to purchase their "dream orders"
and pay over time. This was an astute sales strategy that also bred much
goodwill and customer loyalty. From 1941 Snap-on tools were retailed over-
the-counter and the company continued developing new products, such as

Joseph Johnson &
William Seidemann
with a display of the
"original tool set"
on Snap-on's 25th
anniversay, 1945

the patented *Flank Drive* wrenching
system (1945) – a revolutionary design developed for the US Navy,
which prevented the corners being
sheared off aircraft fasteners under
high torque conditions. The system
was later incorporated into most
Snap-on wrenches. In 1990 another
innovative feature was pioneered,
the *Flank Drive Plus* wrench, which
has serrated jaws to increase the
grip of open-ended wrenches. By
1993 Snap-on's sales had surpassed
$1 billion and today the company
remains a leader in its field.

Flank Drive
wrenching system,
1945

Flank Drive Plus
wrench, 1990

SONY

FOUNDED 1946
TOKYO, JAPAN

TR-55 transistor
radio, 1955 – Japan's
first transistor radio

→ *TC50* compact
cassette recorder,
c. 1969

TV8–301 television,
launched in 1959 –
the world first's
portable television

In the period immediately following the Second World War, Japan saw a huge surge in demand for radios, fuelled by a populace desperate for news from around the world. In September 1945 the engineer Masaru Ibuka responded to this opportunity by opening a small electrical repair shop in Tokyo. Called the Tokyo Tsushin Kenkyujo (Tokyo Telecommunications Research Institute), the business repaired war-damaged radios and shortwave units and made its own short-wave adapters and converters that could turn short-wave radios into all-wave receivers. Ibuka was joined by his friend, the physicist Akio Morita (1921–1999), and together they founded the Tokyo Tsushin Kogyo Kabushiki Kaisha (Tokyo Telecommunications Engineering Corporation) in May 1946. Although the company's best-selling product was initially an electrically heated cushion, in 1950 – by now in larger premises – it introduced the first Japanese magnetic tape, the *Soni-Tape*, and also began marketing Japan's first reel-to-reel tape recorder, the *G-Type*. Later, the company developed the *H-type*, a less bulky and easier-to-use tape recorder in a case, which was more suitable for home use and was especially popular for educational purposes. In 1952 it devel-

oped a stereophonic audio system for the first Japanese stereo broadcast by NHK. That same year, Ibuka visited the United States and while there discovered that Western Electric was planning to release rights to its transistor patent to companies prepared to pay royalties – the transistor having been previously developed in 1948 by scientists at **Bell** Laboratories. Ibuka's company managed to obtain a licence to manufacture the transistors in 1954 and the following year introduced the first Japanese transistor radio, the *TR-55*. This was followed by the

Voyager Watchman, 1982

← Walkman, first launched in 1979

AIBO robotic dog, 1999

world's first pocket-sized transistor radio, the *TR-63* (1957). In 1958, aware of the need to appeal not just to Japanese consumers but to a global audience, the company changed its name to Sony Corporation, which sounded more Western. Around this time Ibuka noted: "The days of radio are over. The future lies in television", and in 1959 the company launched the first-ever transistor television, the *TV8-301*. Two years later, Sony entered into a contract with Paramount Pictures that involved the studio providing "technical assistance in the production of a chromatron tube and colour television utilizing it". This agreement led to the development of the small but revolutionary *Trinitron* colour televisions, which were first introduced in 1968. They included the *TV5-303*, the smallest and lightest micro-television in the world, which created a boom in America for micro-TVs. Sony's pioneering miniaturization of electronic technology continued with the introduction of the first *Walkman* personal stereo in 1979. With excellent and unwavering sound reproduction, the light and highly portable *Walkman* was an instant success and initially Sony could not keep up with consumer demand. Unlike conventional stereos with large speakers, the *Walkman* needed only a small amount of battery power because the sound was directed straight into the listener's ears. Sony has subsequently produced many versions of this landmark design, including models suitable for sporting activities. It went on to launch a flat-screened portable television, the *Watchman*, in 1982, and a portable compact disc player, the *Discman*, in 1990. More recently, Sony has been developing various digital technology products, including its miniature *Memory Stick*, which provides compact portable storage for digital data, and the sound-responsive *AIBO* robotic dog, which has infra-red eyes and can be trained to perform tricks. Sony predicts: "The Eighties was the age of the PC and the Nineties was the age of the Internet, the 2000s will be the age of the robot."

ETTORE SOTTSASS

BORN 1917
INNSBRUCK, AUSTRIA

Valentine typewriter
for Olivetti, 1969
(co-designed with
Perry King)

Ettore Sottsass studied architecture at the Politecnico di Torino and in 1958 began working as a design consultant to **Olivetti**. He designed some of the company's best-known products, including the *Logos 27* calculator (1963), the *Praxis 48* typewriter (1964), the *Valentine* typewriter (with Perry **King**, 1969) and the *Synthesis* office system (1973). His most remarkable project for Olivetti, however, was the design of the *Elea 9003* main-frame computer (1958), for which he was awarded a Compasso d'Oro in 1959. During the 1960s and 1970s, Sottsass became a prominent member of the Radical Design movement. In 1979 he collaborated with Studio Alchimia and in 1981 established the Memphis design group. His bold and colourful Memphis designs heralded a new direction in the decorative arts – Post-Modernism. As the most influential propagandist of Post-Modernism in the 1980s, Sottsass showed that it was possible and in many circumstances desirable to break away from the rational constraints traditionally imposed on the design of industrially manufactured products, in favour of a new language of design concerned with the articulation of creative expression and in particular, emotion.

Elea 9003 main-
frame computer for
Olivetti, 1959

SPRINGFIELD ARMORY

FOUNDED 1794
SPRINGFIELD, MASSACHUSETTS, USA

Detail of *Model 1865* rifle

In 1777, during the American Revolution, an arsenal was built in Springfield to produce cartridges and gun carriages. Later, the new Federal Government decided to begin manufacturing its own muskets so that it would no longer have to rely on imported arms. To this end, it established a new Armory in Springfield in 1794 and a year later the production of muskets began, with 40 workers producing 245 flintlock muskets per month. In 1819 Thomas Blanchard invented a special lathe which enabled rifle stocks to be mass-produced to a standardized design. In the period running up to the outbreak of the American Civil War, the Armory became a beacon of the Industrial Revolution. During the 1840s, the flintlock was replaced by a percussion cap and hammer, an innovation which led to better weather performance and faster firing. The development of rifling – cut spiral grooves in a barrel – to spin projectiles also ensured greater range and accuracy. The Union Army's subsequent victory at Gettysburg in 1863, and indeed in the Civil War as a whole, can to some extent be attributed to the *Springfield* rifle (1861) and the unprecedented scale of its mass production: between 1863 and 1864, the Springfield Armory boosted its productivity to a phenomenal 25,000 weapons per month using rationalized manufacturing techniques such as the division of labour, mechanization where possible and standardization. In testimony to the influence of the Springfield Armory and other pioneering small arms manufacturers such as **Eli Whitney**, mass production in America was initially referred to as "armory practice" when it was first employed by automobile manufacturers in Detroit. The Springfield Armory also produced the well-known *M1* rifle (1926), also called the *Garand* Rifle after its inventor John Garand, of which 4.5 million were made. Today, in recognition of its contribution to both American history and the evolution of industrial manufacture, the Springfield Armory has been designated a National Historic Site.

Springfield rifle with telescopic sight, 1903

STANLEY

FOUNDED 1843
NEW BRITAIN, CONNECTICUT, USA

Frederick T. Stanley

In 1843 Frederick Stanley established The Stanley Works, a small company manufacturing door hardware, including bolts and hinges. His goal was to create a hardware company that would be unequalled in its product innovation anywhere in the world. With the early success of this mission and an increasing demand for his high-quality goods, Stanley began exporting his products in the 1870s. In the early years of the 20th century, the company opened the first of many plants outside the United States, thereby assisting in the realization of Frederick Stanley's global vision. Shortly afterwards, The Stanley Works acquired America's largest manufacturer of hand tools, Stanley Rule & Level, which had been established by a relative in 1857. By the 1920s, Stanley hardware and tools, including its famous high-quality planes, were available on every continent. In accordance with its founder's belief in expansion and diversification, Stanley today manufactures some 50,000 different products that are sold throughout the world. In recognition of its design innovation, the Industrial Designers Society of America (IDSA) awarded Stanley two silver Industrial Design Excellence Awards (IDEA) in 1999. That same year saw the launch of the *MaxGrip* line of hand tools, including ergonomically designed self-adjusting pliers and a locking wrench described by Stanley's product manager as "another step forward in the evolution of the wrench", as it eliminates the tendency of the wrench to slip from the fastener. Today, Stanley works closely with the Ergonomic Technology Center at the University of Connecticut to ensure that all its products are designed to have an "ergonomic advantage."

Stanley plane, early
20th century

MaxGrip line of hand
tools, 1999

PHILIPPE STARCK

BORN 1949
PARIS, FRANCE

Philippe Starck has the highest profile of any designer working today, not only because of his multitude of highly successful products but also because of his compelling showmanship, which has helped to promote the cause of design to new audiences. Starck studied at the École Nissim de Camondo and later became art director of the Pierre Cardin studio, where he produced 65 furniture designs. During the 1970s he worked as an independent designer, most notably designing nightclub interiors. In 1980 he established his own manufacturing and distribution company, Starck Products, to commercialize his earlier designs such as the *Easy Light* (1977) and *Von Vogelsang* sofa (1978) and the *Francesca Spanish* chair (1979–1980). During the 1980s, Starck became the leading "superstar of design" and worked prolifically on numerous projects. He designed elegant and sumptuous interiors, most notably for the Royalton Hotel, New York (1988) and Paramount Hotel,

Moto 6,5 motorbike
for Aprilia, 1995

Axor twin-handled basin mixer for Hansgrohe, 1998

↖ *Axor Starck* single-lever basin mixer for Hansgrohe, 1994

New York (1990), which incorporated his own designs for furniture, lighting, door handles, vases and other objects. He also became celebrated for his numerous furniture designs, from the three-legged *Costes* chair (1982) and the plastic injection-moulded *Dr. Glob* chair (1990) to the elegant *Lord Yo* tub chair (1994) and collapsible *Miss Trip* chair (1996), created for manufacturers such as Vitra, Disform, Driade, Baleri, XO and Idée. Like his furniture, Starck's lighting and product designs were also personalized with characterful names and sensual, appealing forms. Amongst the most commercially successful of these were the *Arà* table lamp for **Flos** (1988) and the *Juicy Salif* lemon squeezer (1990–1991), *Max le Chinois* colander (1990–1991) and *Hot Bertaa* kettle (1990–1991) for **Alessi**. During the 1990s, Starck designed consumer electronic products for Thomson, Saba and Telefunken that attempted to humanize technology. His *Jim Nature* television (1994) for Saba is particularly innovative with its high-density chipboard casing. Starck also designed the *Motó 6,5* motorcycle (1995) as well as the *Lama* scooter (prototype 1994) for **Aprilia** with seductive rounded forms. Today, Starck acknowledges that much of the design produced during the 1980s and early 1990s, including some of his own, was narcissistic "over-design" driven by novelty and fashion. He now promotes product durability (longevity) and believes that morality, honesty and objectivity must be an integral part of the design process. He has also stated that "the designer can and should participate in the search for meaning, in the construction of a civilized world" and predicts that "the 21st century will be immaterial and human."

← *Lama* scooter (prototype) for Aprilia, 1992

STEELCASE

FOUNDED 1912
GRAND RAPIDS, MICHIGAN, USA

David Hunting (far
right) and colleagues
demonstrating the
strength of the Metal
Office Furniture
Company's filing
cabinet, c. 1919

→Steelcase adver-
tisement, 1935

Peter M. Wege and eleven other stockholders founded the Metal Office
Furniture Company in 1912. The company initially manufactured only "steel
cases" (ie. metal office safes) and it was not until 1914 that it began diversifying with the introduction of the *Victor* fireproof steel wastebasket. David
Hunting was recruited as sales and marketing vice-president that same year,
and he subsequently steered the firm into desk manufacturing. Having
learned that the Boston Customs House required non-flammable furniture,
it was obvious to him that steel was the ideal material for the job, so he bid
for the contract even though the company was not yet making desks. The
company managed to produce a prototype desk within 90 days and eventually won the contract to supply 200 fireproof metal desks. In 1919 the company introduced its first steel filing cabinets, which were advertised on the
merits of their superior strength. Around 1936, the company manufactured
one of the first Modern office systems, the revolutionary steel workstation
and office chair designed by Frank Lloyd Wright (1867–1959) for the S. C.
Johnson & Son Administration Building. In the late 1940s the company introduced the industry's first standard desk sizes, based on multiples of 15
inches, and in 1953 launched the first office furniture available in different

Steelcase *Co-
ordinated* files,
c. 1962

Frank Lloyd Wright, workstation and office chair, 1936

Archival photograph showing Frank Lloyd Wright's office system in use

colours (previously grey had been the only option). In 1954 the company was renamed Steelcase Inc. and over the succeeding decades continued developing advanced office systems. In 1974 Steelcase became the largest furniture manufacturing company in the world – a title it retains to this day thanks to its innovative product designs, such as the *Pathways* system, which demonstrates the company's pursuit of product integration so as to provide the most responsive office environments possible. With a strong global culture, Steelcase is highly committed to continuous environmental improvement – from the recycling of its workstations and the revamping of certain manufacturing processes in order to reduce pollution, to the minimization of waste wherever possible.

↗Office system, 1947

→ *Pathways* office system, 1998

STELTON

FOUNDED 1960
HELLERUP, DENMARK

Erik Magnussen,
No. 140 water kettle,
1988

→Erik Magnussen,
No. 460 cake knife,
1980, *No. 462* pizza
server, 1990 and
No. 463 fish server,
1990

Arne Jacobsen,
Cylinda-Line, 1967

A/S Stelton was founded in 1960 to sell the stainless steel hollowware – sauceboats, platters, salad bowls etc. – manufactured by the Dansk Rustfrit company based in Fåre vejle, Stationby, a little town in Northwest Zealand. The products, which were of mediocre design, were intended by Stelton to form the basis of a range of contemporary housewares aimed primarily at the Danish market. Salesman Peter Holmblad was hired to market the line and in 1964, with a view to raising the company's profile, he convinced the well-known architect and designer **Arne Jacobsen** to design a range of hollowware for the firm. Jacobsen subsequently produced some sketches of cylindrical forms that were "terse, logical and functional", but the technology needed to translate his ideas into three-dimensional products did not yet exist – stainless steel being an extremely demanding material to work, as its surface can become easily distorted if it is not handled in exactly the right way. Stelton first had to develop new machines and welding techniques, therefore, before it could produce

Jacobsen's *Cylinda-Line*, which was eventually introduced in 1967. In recognition of the superior quality of its design, the *Cylinda-Line* was awarded an ID-prize by the Danish Society of Industrial Design in 1967 and the International Design Award by the American Institute of Interior Designers in 1968. The *Cylinda-Line* initially comprised around 18 pieces, but Jacobsen continued to add new designs up to 1971, while others were introduced posthumously from drawings. Today, the range comprises 34 pieces in total. Described as "products which live up to the demand for a beautiful, functional, and practical hollowware range in contemporary design", the success of the *Cylinda-Line* can be largely attributed to Jacobsen's ability to express the inherent qualities of stainless steel through highly rational yet aesthetically pleasing, timeless forms. From 1976, **Erik Magnussen** also worked for Stelton and designed a number of well-known table pieces in both stainless steel and plastic, including his classic three-piece vacuum jug set of 1977. The company's range today also includes designs by Peter Holmblad, such as his *No. 100–15* houseplant watering can, which has a similar rational aesthetic to the *Cylinda-Line*. Stelton's high-quality products exemplify Danish design and are represented in the permanent collections of museums around the world, including the Museum of Modern Art, New York, and the Victoria & Albert Museum, London.

GEORGE & ROBERT STEPHENSON

BORN 1781 WYLAM, NORTHUMBER-
LAND, ENGLAND
DIED 1848 CHESTERFIELD, DERBYSHIRE,
ENGLAND

BORN 1803 WILLINGTON QUAY,
NORTHUMBERLAND, ENGLAND
DIED 1859 LONDON, ENGLAND

George Stephenson's father was a mechanic who operated an atmospheric-
steam engine, designed by Thomas Newcomen (1663–1729) and used to
pump water out of coal mines. By the age of 19, the younger Stephenson
was operating one of these revolutionary engines himself and the skill he
developed led, in 1812, to his appointment as enginewright (chief mechanic)
at the Killingworth High Pit colliery. A year later, he visited the neighbouring
Middleton Colliery to appraise a "steam boiler on wheels" conceived for
hauling coal trucks out of the mines. This double-cylinder locomotive had
been devised by John Blenkinsop (1783–1831) and employed a patented system of toothed wheels that fitted into a racked rail. After seeing Blenkinsop's
design, Stephenson believed he could produce an improved model and in
1814 built his first locomotive, the *Blucher*. This design did not meet with
much success, however, until he added a funnel for the waste steam (as Richard Trevithick had done on his earlier engines), which effectively doubled
the engine's power. The following year, Stephenson designed a miner's
safety lamp (contemporaneously with **Humphrey Davy**'s better-known lamp)
and built the *Killingworth* engine, which incorporated an innovative sprocket
and chain drive. In 1817 the Duke of Portland commissioned Stephenson
to construct one of his engines for use on the Kilmarnock & Troon line. In
1821 he was commissioned to build a proper steam locomotive for the new

The *Rocket*
locomotive, 1829

Stockton to Darlington line, a project which culminated on 27 September 1825 with Stephenson's
Active engine pulling the world's
first public passenger train – an
achievement that heralded the age
of steam railways. The *Active* engine
later became known as the *Locomo-
tion* and could carry 450 passengers
at a speed of 15 mph. Stephenson
was subsequently asked to survey
and construct a new line running
from Liverpool to Manchester. In
1829, shortly before construction

LOCOMOTIVE ENGINE.

CONSTRUCTED BY MESS™ R. STEPHENSON & C™.

NEWCASTLE UPON TYNE.

CYLINDER 8 DIAMETER. STROKE 17.

1828.

of the line was completed, the famous Rainhill Trials were staged in Merseyside, with a prize of £500 being offered for the best locomotive. For the competition, Stephenson built a new engine – his now legendary *Rocket* – featuring single drive wheels on both sides which were coupled to a steam cylinder on either side of the firebox. It won with a top speed of 36 mph, and when the line was eventually opened in September 1830, eight modified locomotives, which had been built at Stephenson's works in Newcastle, were put into operation. Almost immediately, the *Rocket* came to symbolize the Industrial Revolution. Known as "the father of the railways", Stephenson greatly influenced the rapid transition from horse power to steam power. He went on to build other railroads in the Midlands, as well as acting as a consultant engineer on numerous railway projects both at home and abroad. His son, Robert Stephenson, was also a celebrated engineer and assisted his father during the 1830s with improved designs for locomotives. He too constructed railways and was acclaimed for his design of long-span rail bridges, such as the structure linking Anglesey to mainland Wales. The success of this unique tubular design led to his building of other innovative tubular bridges both in England and overseas.

Coloured engraving dating from 1836 showing the *Rocket* as rebuilt in 1831

Access, 1997/1998

SWATCH

FOUNDED 1983
BIEL-BIENNE, SWITZERLAND

Unable to compete with the plethora of cheap electronic watches being mass-produced in the Far East, and in particular Japan, by the mid-1970s the Swiss watch industry was deep in crisis. Against this background, Nicolas Hayek snr. (b. 1928) masterminded a four-year reorganization of the ASUAG and SSIH watch companies (founded in 1931 and 1930 respectively), which eventually merged in 1983 into the SMH (Swiss Corporation for Microelectronics and Watchmaking Industries). Hayek also made the strategic decision to develop a new watch brand that would rival the sales of Japanese models and rescue the embattled Swiss watch industry. The resulting slim plastic watch was based on a prototypical design by Ernst Thonke, Jacques Müller and Elmar Mock of Hayek Engineering AG. Comprising only 51 components (compared to the 90 or more components normally required in a conventional watch), it was very well suited to large-scale mass production. By combining high technology with affordability and artistic, emotional styling, the *Swatch* became a must-have fashion accessory following its

Range of early
Swatch watches,
1983

Nam June Paik,
*Zapping, Swatch Art
Special*, 1996

launch in 1983. Its success was boosted by the fact
that the watch was being marketed along the same
lines as *haute couture*, with new "collections" launched
every season. By producing limited editions for collectors as well as a range of "classic" watches in every possible style and colour, Swatch ensured that it had a design to suit almost every taste. Thanks to the remarkable success of the *Swatch*, the Swiss watch industry
regained its leading position in the sector in 1984. The
following year, Hayek and a number of Swiss investors
gained control of the Swatch Group, which included
other brands such as Blancpain, Omega, Longines,
Rado, Tissot, Certina, Mido, Hamilton, Balmain, Calvin
Klein, Lanco and the children's range Flik Flak. Today,
the company is the world's largest manufacturer of finished watches. Some 200 million *Swatch* watches have
been sold to date, ranging from the standard plastic
model to the metal cased *Irony*. The Swatch Group was
also highly instrumental in the development of the
diminutive *Smart* car in the late 1990s.

ROGER TALLON

BORN 1929
PARIS, FRANCE

Roger Tallon is one of France's foremost industrial designers. He studied electrical engineering in Paris from 1947 to 1950, before joining the design consultancy Technès in 1953. While there, he designed numerous innovative products, including cameras for SEM (1957 and 1961), a typewriter for Japy (1960), a portable television for Téléavia (1963), the *Module 400* furniture range (1964), drinking glasses for Daum (1970) and the *Chronograph X* watch for LIP (1973). He also worked as a consultant, designing Frigidaire refrigerators for **General Motors** from 1957 to 1964. In 1973 he established the multi-disciplinary design consultancy, Design Programmes SA, and went on to gain an international reputation for his transportation design, which included the *Corail* locomotive (1977) for SNCF. In 1983 Tallon founded the design partnership ASDA + Partners with Pierre Paulin (b. 1927) and Michel Schreiber (b. 1950). He has continued designing trains, including the high-speed *TGV-Atlantique* (1988) for SNCF and the *Trans-Euro-Star* shuttle for Euro-Tunnel (1987). In 1973 he was elected an Honorary Royal Designer for Industry in London and in 1985 was awarded the National Grand Prix for industrial design.

*Chronograph X
2230.194* wristwatch
for Lip-France, 1975

WALTER DORWIN TEAGUE

BORN 1883 DECATUR, INDIANA, USA
DIED 1960 FLEMINGTON, NEW JERSEY, USA

Cash register for
the National Cash
Register Company,
c. 1937

Walter Dorwin Teague was one of the great pioneers of industrial design consulting in America. Having moved to New York in 1903, he subsequently attended evening classes at the renowned Art Students League. He later worked as an illustrator for a mail-order catalogue and also for the Hampton Advertising Agency. In 1912 he established his own studio in New York, working as a typographer and graphic designer, and in the mid-1920s also began designing packaging. In 1926 he founded Walter Dorwin Teague Associates, which was one of the earliest industrial design consultancies. Teague's first client was Eastman **Kodak**, for which he undertook a comprehensive design programme that included design research and product development. His first camera, the *Vanity Kodak* (1928), was designed specifically for the female market and was produced in various colours with matching silk-lined cases. Teague later received widespread acclaim for his *Baby Brownie* camera (1933), which was one of the first consumer prod-

Bantam Special
camera for Kodak
Company, 1933–1936

←Gas refrigerator for Servel Inc., c. 1950

→Automatic accounting machine for the National Cash Register Company, c. 1950

↘*Edgemount Combination Barometer* for Taylor Instrument Companies, c. 1950

Label design for soft-drink bottles for Nehi Corporation, c. 1950

ucts made of plastic. His best-known camera, however, was the distinctively styled *Bantam Special* (1936), which was more compact and user-friendly than earlier Kodak models. In 1930 Teague designed the streamlined body of the Marmon *Model 16* car, which was among the most aerodynamically efficient automobiles of its time. He also designed other streamlined products, including glassware for **Corning**, kitchenware for Pyrex, pens and lighters for Scripto, lamps for **Polaroid**, mimeographs for A. B. Dick, radios for Sparton and the *Centennial* piano for Steinway. As well as consumer prod-

WALTER DORWIN TEAGUE . 479

Mock-up of an interior for the Boeing *707*, 1956

ucts, Teague also designed a plastic truck body for UPS, supermarkets for Colonial Stores, interiors for the **Boeing** *707* airliner, United States pavilions at various international trade fairs, exhibition interiors for **Ford**, service stations for Texaco and various exhibits at the 1939 **New York World's Fair**, including a gigantic cash register, which recorded visitor numbers and was based on his earlier design for the National Cash Register Company. He also wrote *Design This Day – The Technique of Order in The Machine Age* (1940), which celebrated the potential of machines and the "new and thrilling style" of the Modern era. Now headquartered in Redmond, Washington, with a regional office in California, Walter Dorwin Teague Associates Inc. employs a staff of 200 professionals and specializes in transportation design, systems engineering, human factors and corporate identity.

→ *Bluebird* radio for the Sparton Corporation, 1934–1936

TETRA PAK

FOUNDED 1951
LUND, SWEDEN

Ruben Rausing

Detail of *Tetra Brik*
Aseptic ReCap
packaging, 1994

Dr. Ruben Rausing (1895–1983), the visionary founder of Tetra Pak, studied at the Stockholm School of Economics and later at Columbia University, New York. While in America, he was exposed for the first time to "self-service" stores and was convinced that the concept would eventually be adopted in European countries, which would in turn increase the demand for pre-packaged foodstuffs. On his return to Sweden, he founded, with Erik Wallenberg, the first Scandinavian specialist packaging company, based on Rausing's belief that "A package should save more than it costs". In 1943 the company began researching and developing a form of milk packaging which would require the minimum amount of material while providing the maximum hygiene. These early studies ascertained that the optimum shape for a milk carton was a tetrahedron. The following year saw the invention of new techniques for coating paper with plastics and for sealing packages. In 1951 Rausing and Wallenberg established AB Tetra Pak as a subsidiary of Åkerlund & Rausing, and the same year launched their new packaging system based on the tetrahedron-shaped carton. A year later, the Lundaortens Mejerifsrening dairy in Lund became the first to use a *Tetra Pak* machine to package cream into 100ml cartons. Several other Swedish dairies introduced the revolutionary packaging system in 1953, and that same year the company began using polyethylene as a coating for its paperboard. In 1954 the first *Tetra Pak* machine was exported to Germany and the half-litre pack-

aging machine was introduced. The commercial packaging of milk in 300ml cartons subsequently began the following year. The company eventually opened a much larger manufacturing plant in Lund and commenced the development of an aseptic packaging system (launched in 1961). By 1959 the factory was producing one billion cartons per annum and began working on a new brick-shaped design, the *Tetra Brik* (introduced in 1963).

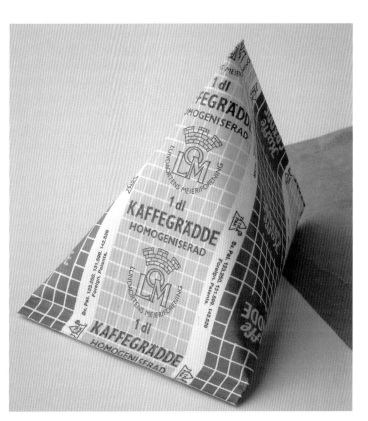

Over the next few years, the business expanded its production operations
into Mexico, the United States, Lebanon and Italy. In 1965 Åkerlund &
Rausing was sold, but Ruben Rausing retained its subsidiary, AB Tetra Pak,
and with the funds from the sale was able to invest more resources into
the research and development of even better packaging designs, such as
the ubiquitous gable-top package known as *Tetra Rex* (1965). During the
1970s, the company exported its packaging machinery all over the world,
including to the USSR, Japan, China, Australia and Iran. By the beginning
of the 1980s, over 30 billion packages were being sold per annum for the
storage of milk, cream, fruit juice and wine. In 1991 AB Tetra Pak merged
with Alfa-Laval and subsequently diversified its product line to include PET
bottling equipment. Today, Tetra Pak remains the undisputed world leader
in the field of packaging, selling over 200 million products a day in some
150 countries. Its simple yet ingenious designs have revolutionized food
retailing and hygiene.

TEXAS INSTRUMENTS

FOUNDED 1951
DALLAS, TEXAS, USA

Jack St. Clair Kilby holding up an early integrated circuit, c. 1959

Texas Instruments originated from a company, Geophysical Service, established in 1930 by Clarence "Doc" Karcher and Eugene McDermott. Geophysical Service specialized in the reflection seismograph method of exploration and, later, submarine-detection equipment and airborne radar systems. In 1951 it became a wholly-owned subsidiary of the newly-founded Texas Instruments. In 1952 Texas Instruments acquired a licence from the Western Electric Company to manufacture transistors, and in so doing entered the semiconductor industry. In 1958 Texas Instruments' in-house inventor, Jack St. Clair Kilby, demonstrated the first-ever integrated circuit (IC) and three years later the company delivered the first integrated-circuit computer to the US Air Force. 1964 saw the introduction of the first consumer product to incorporate an integrated circuit – a hearing aid – and the same year the company began mass-producing the first plastic-packaged ICs. Over the next decade, Texas Instruments was at the forefront of electronic technology, developing the first electronic hand-held calculator (1967) and the first laser-guided missile system (1969). In 1971 the firm pioneered the single-chip

The first electronic hand-held calculator, developed by Jack St. Clair Kilby, Jerry D. Merryman and James Van Tassel, 1967

microprocessor and the single-chip microcomputer, and a year later entered the consumer market with its *Datamath* hand-held calculator. In 1974
it launched the *TMS1000* one-chip microcomputer and the following year
introduced three-dimensional data-processing technology. The first single-
chip speech synthesiser was unveiled in 1978, as was the first consumer
product to incorporate low-cost speech synthesis technology, the *Speak &
Spell* learning aid. In 1987 the company pioneered the first single-chip 32-bit
artificial-intelligence microprocessor, and four years later became the first
American manufacturer of semiconductors to establish a research and development facility in Japan. Over the years, Texas Instruments has won numerous awards for manufacturing excellence and its many inventions, which
have driven completely new industrial design typologies.

Advertisement for
Thermos flask, 1909

→Thermos flask
Model 65, 1928

Advertisement for
Thermos flask, 1909

THERMOS

FOUNDED 1904
BERLIN, GERMANY

The philosopher and scientist, Sir James Dewar (1842–1923), was a pioneer of vacuum technology and in 1892 devised an innovative flask for the storage of liquefied gases that worked on the principle of temperature retention. This ingenious design comprised two glass bottles, sealed together one inside the other, with the air in the cavity between them pumped out so as to produce a vacuum. Later, Dewar silvered the glass with mercury to increase its insulating performance. Dewar managed to create a stronger, more resilient version of his flask with the assistance of a German glass blower, Reinhold Burger. While developing these flasks for scientific experiments, Dewar and Burger came up with the idea of making a flask with a protective metal casing for domestic use. The resulting "Thermos" was patented in Germany in 1903 and a year later the Thermos GmbH manufacturing company was subsequently established in Berlin. The flasks were retailed under the Thermos tradename from 1904 and by 1905 they were being marketed in the United States. Thermos Ltd. was founded in Britain two years later and a factory was opened in London shortly afterwards. The early Thermos flasks were time-consumingly handblown and cost an expensive one guinea. By 1911, however, production had become mechanized and the price of the vacuum bottles plummeted to 2s 6d. The Thermos flask became very popular with both workers and picnickers in the 1920s and 1930s, and during the Second World War Thermoses were used extensively by troops. With its double-walled design that isolates heat, the Thermos flask works equally well at keeping liquids hot or cold for many hours. Although other companies have manufactured insulating vacuum flasks, the name Thermos has become a generic term and the company continues to export its products throughout the world.

Michael Thonet,
Boppard Chair I,
1836–1840

MICHAEL THONET

BORN 1796 BOPPARD, GERMANY
DIED 1871 VIENNA, AUSTRIA

Michael Thonet established a furniture workshop in
Boppard am Rhein in 1819. From around 1830, he began
experimenting with laminated wood and produced a
number of innovative chairs in the Biedermeier style.
These designs prompted the Austrian chancellor, Prince
Metternich, to invite Thonet to Vienna, where he was
granted a patent for his new process for bending wood
laminates in 1842. Having secured the necessary financial backing, Thonet
moved to Vienna and in 1849 established a furniture workshop in Gumpen-
dorf, a suburb of Vienna, with his sons, Franz, Michael, August and Joseph.
For the next two years, the Thonet family concentrated on developing techniques for mass-producing furniture, including the steam bending of solid
wood. In 1851 Thonet exhibited its new furniture designs at the **Great Exhibition** in London and won a gold medal. By 1853 Gebrüder Thonet, as the
company was now known, had moved into larger premises and was fully

Page from Thonet
catalogue, 1904

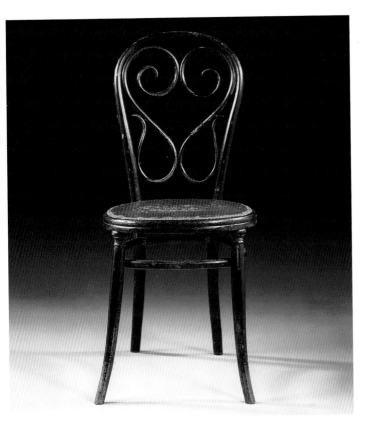

mass-producing chairs distinguished by a reduction of elements and the elimination of extraneous ornament. In 1856 the company opened its first factory in Koritschan, Moravia. Gebrüder Thonet's remarkable success was due to its adherence to mechanized methods of production, which allowed it to sell its products at very competitive prices. In 1860, for example, the firm's best-known model, the *No. 14* chair, cost less than a bottle of wine, and by 1891 a staggering 7,300,000 of these ubiquitous café chairs had been sold. In the early 1900s several leading Viennese architects, including Josef Hoffmann (1870–1956), began designing furniture for Gebrüder Thonet in the Secessionist style. In 1929 a French subsidiary was established, Thonet Frères, which produced progressive tubular metal furniture designed by **Marcel Breuer** and others. The production of this furniture was later moved to Frankenberg, Germany. Thonet continues to operate and manufactures re-issues of its earlier seating as well as contemporary designs.

TOSHIBA

FOUNDED 1939
TOKYO, JAPAN

Osamu Kondo,
Libretto lap-top
computer, 1995

In 1875 Hisashige Tanaka (1799–1881) founded Japan's first telegraphic equipment manufacturing company, the Tanaka Seizo-sho (Tanaka Engineering Works). Later renamed the Shibaura Seisaku-sho (Shibaura Engineering Works), his company became one of the largest Japanese manufacturers of heavy electrical apparatus. Meanwhile, another company – Hakunetsusha & Co. – was established in 1890 to manufacture the first Japanese electric incandescent lamps. Renamed Tokyo Denki (Tokyo Electric Company) in 1909, this firm became a notable manufacturer of consumer electronic products. Between them, the two companies were responsible for producing the first Japanese induction motors (1895), the first Japanese X-ray tubes (1915) and the first Japanese washing machines and refrigerators (1930). In 1939 the two companies merged to form the Tokyo Shibaura Electric Co. (officially renamed Toshiba in 1978). A year later, this new venture produced the first Japanese fluorescent lamps and in 1942 constructed Japan's first radar equipment. In 1954 the company introduced the first Japanese digital computer, *TAC*, which was designed specifically for the University of Tokyo, launched the first practical electric rice cooker in 1955 and developed the first Japanese transistor televisions and microwave ovens in 1959. Since then, Toshiba has been a leading pioneer of electrical and electronic products, from computers and home appliances to medical diagnostic equipment, elevators and escalators. Toshiba's commitment to "universal design" is powerfully expressed in the company's user-friendly products and its

↘Toshiyuki
Yamanouchi,
PDR-M4 digital
still camera, 1998

Toshiyuki
Yamanouchi,
TLP651 LCD data
projector, 1999

Yoshiharu Iwata,
ER-4 electric rice
cooker, 1958

pioneering design for disability, such as Braille operating panels on washing machines for the visually impaired. The company's design centre runs a special programme in which its engineers and designers are made to wear adapted goggles that lets them experience what it is like for a person living with impaired sight. This type of "human-centric" design also involves a systemized approach to ergonomics. Toshiba views "universal design" as a means of "harmonizing social responsibility and business requirements."

TOYOTA

FOUNDED 1937
KARIYA, HONSHU, JAPAN

Kiichiro Toyoda

Japan's most famous inventor, Sakichi Toyoda (1867–1930), designed the first Japanese power loom in 1896 and in 1924 devised the first-ever automatic loom capable of replenishing the shuttle during continuous operation. In 1926 he went on to found the Toyoda Automatic Loom Works, which was one of the world's leading manufacturers of weaving machinery. Three years later Toyoda signed a patent royalty agreement with a British firm, Platt Brother & Co., for the manufacture of one of his weaving machines. This provided sufficient funds for Toyoda's eldest son, Kiichiro, to begin the research and development of an automobile. A steel manufacturing division was also established to develop a "speciality steel" that could be used for car engines and bodies, and in 1933 an automobile department was formed. This produced its first *Type A* engine in 1934 and its first *Model A1* car prototype in 1935. Serial production of the *Model AA* car commenced in 1936 and the Toyota Motor Co. was officially founded the following year. In 1947 Toyota launched its first small car, the *Model SA*, and from 1959 began building a vast network of plants outside Japan in the belief that "localizing operations provides customers with the products they need where they need them". Today Toyota is the largest car manufacturer in Japan and the third largest worldwide. In 1996 the company's Tokyo Design Research & Laboratory division was centralized so that the entire design process, from

Toyota *AA*
automobile, 1936

Toyota *2000 GT*, 1966

Toyota *Yaris*, 1998

concept sketches to prototyping, could be undertaken at a single site. Since then, Toyota has produced award-winning cars such as the fuel-efficient *Yaris* (European Car of the Year 2000) and the world's first production hybrid car, the *Prius* (Japan's Car of the Year 1997–1998), which runs on both gasoline and electric power. With its innovative *RAV4* electric car, Toyota became the first automobile manufacturer to comply with the State of California's zero-emissions legislation.

TUPPERWARE

FOUNDED 1938
LEOMINSTER, MASSACHUSETTS, USA

Drawing showing
the Tupper Seal,
patented in 1947

Single-impression
mould used in the
injection-moulding
of the *Mix-n-store*
container, 1965

Earl Tupper (1907–1983) gained his love of invention from his father, who used to construct labour-saving devices on the family farm. In 1936 Tupper met Bernard Doyle, who had developed a plastic known as Viscoloid, and he subsequently took a job in the plastics manufacturing division of DuPont. Although he only stayed there a year, Tupper later recalled that it was where "my education really began". In 1938 Tupper established his own plastics manufacturing company in Leominster, Massachusetts. The company initially undertook subcontracted work for DuPont, but subsequently won its own contracts for military products such as gas masks and signalling lamps. During the immediate postwar years, Tupper turned his attention to the design of plastic consumer products, such as sandwich picks, bathroom tumblers and cigarette cases, which firms could give away to their customers as promotional sales tools. Plastics in the 1940s still had numerous unpleasant properties, ranging from brittleness and odour to greasiness. To overcome these problems,

You can Christmas shop for nearly everybody at a Tupperware party.

Even dad and the children.

Oyster, multi-purpose storage containers, 1990s

Tupper developed a technique for purifying black polyethylene slag (a by-product of the oil industry) into a strong, flexible, non-porous material that was translucent and non-greasy. Around the same time, he also patented his famous airtight and watertight Tupper seal, the design of which was based on the lids used for paint containers. Tupper combined these two innovations to create a range of airtight plastic containers, but found that the product line "fell flat on its face" in retail stores. He realised that his sealing system had to be demonstrated before consumers would buy into it. In the late 1940s **Stanley** was successfully selling its household products through local distributors, who arranged demonstration home parties, and in 1948 Tupper met several of these distributors. Amongst them was Brownie Wise, who expanded this novel retailing concept for Tupper so that his food storage containers were only available through such demonstrations. From 1951 to 1958 Wise was vice-president of the company, which was then called Tupperware Home Parties. As a skilful saleswoman and motivator of people, Wise understood the importance of her party hostesses, declaring: "If we build the people, they'll build the business." With her large sales force, she transformed the company's fortunes and Tupperware parties became as famous as Tupperware products. Today Tupperware produces a vast range of high-quality plastic housewares, from kitchen equipment and multi-purpose domestic storage containers to organizing products. A Tupperware demonstration party is held on average every two seconds somewhere in the world.

←Advertisement showing a Tupperware party scene, 1960s

Cifra 3 clock for
Solari & C., 1966

GINO VALLE

BORN 1923 UDINE, ITALY
DIED 2003 UDINE, ITALY

Gino Valle trained at the Università di Architettura in
Venice until 1948 and then worked in the architecture
practice founded by his father, Provino Valle. He completed his studies at
the Harvard Graduate School of Design in Cambridge, Massachusetts, graduating in 1951. Valle and his brother, Nani, subsequently took over their father's practice in Udine and designed numerous commercial and industrial
buildings in northern Italy. In 1954 Lino **Zanussi** commissioned Valle to design a range of household appliances. Valle's resulting product system was
based on the concept that appliances should be viewed as integral components of the kitchen. His rational style was also translated into his designs
for automated schedule boards for train stations and airports, a large public-
area clock (1966), and his well-known domestic *Cifra* range of clocks for
Solari & C. (1956–1966), which incorporated a similar modular flap-down
numeral system. He was awarded a Compasso d'Oro for his *Cifra 5* clock
(co-designed with Nani Valle) in 1956 and another in 1962 for his indicator
board, which is still widely used in railway stations and airports throughout
the world.

Large metal public-
area clock for Solari
& C., 1963

solari&c. udine, italy

HAROLD VAN DOREN

BORN 1895 CHICAGO, ILLINOIS, USA
DIED 1957 PHILADELPHIA, PENNSYLVANIA, USA

Harold van Doren was a pioneer of industrial design consulting and streamlining who designed products with a clean, contemporary aesthetic. He first
studied languages before moving to Paris, where he worked at the Louvre.
After returning to the United States, he worked as an assistant to the director of the Minneapolis Institute of Arts, but resigned from this position
when he was given the chance to work in the fledgling field of industrial design. One of his first commissions came from the Toledo Scale Company,
who asked him to design some commercial weighing scales. Van Doren's
lightweight and innovative solution was one of the first products to incorporate a large-scale plastics moulding. Having based his practice in Philadelphia, he went on to design many streamlined products for Maytag,
Goodyear, Ergy, Swartzbaugh and DeVilbiss. Together with John Gordon
Rideout, he designed the widely acclaimed green skyscraper-shaped plastic

First range
manufactured
by Philco, 1950

radio for Air-King (1930–1931) and a child's scooter (1936). In 1940 van Doren wrote *Industrial Design: A Practical Guide to Product Design and Development*, and nine years later published an article in *Design* magazine entitled "Streamlining Fad or Function?". Here he argued that the curved lines common to refrigerators at the time were the result of a manufacturing process involving a metal press known as a "bulldozer", and that the formal language of the product was thus "imposed on the designer by the necessity of obtaining low cost through high-speed production". Van Doren was particularly adept at re-styling products without altering their existing layout, as illustrated by his cooking range (1950) for the **Philco** Corporation. This streamlined design was a re-modelled version of an earlier range produced by Electromaster Inc. that "even experienced men in the industry believed to be entirely new". For Philco, van Doren also designed a streamlined re-frigerator (c. 1950) whose door was emblazoned with the words "Advanced Design" – an early example of the use of design as a marketing tool.

Magnalite tea kettle for Wagner Ware, 1940 (co-designed with J. G. Rideout)

←Photograph show-ing Harold van Doren (left) and Hubert Bennett (right) with van Doren's redesigned scales for the Toledo Scale Company, 1930s

VICTORINOX

FOUNDED 1884
IBACH, SWITZERLAND

Karl Elsener

Karl Elsener (1860–1918) spent a year as a journeyman cutler before setting up a small business close to Schwyz in 1884. In order to boost the local industry, in 1891 he founded the Swiss Cutlery Guild (later re-named Victorinox) to produce soldiers' knives for the Swiss Army, which had hitherto sourced its knives in Germany. Elsener delivered his first batch of knives to the Swiss Army that same year, and soon after began manufacturing penknives such as the *Student Knife*, the *Cadet Knife* and the *Farmer's Knife*. While the *Soldier's Knife* was extremely robust, its corkscrew, punch, can opener, screwdriver and second blade also made it quite heavy, prompting Elsener to design the lighter and more elegant *Officer's Knife*. This hugely successful model, which incorporated just two springs for its six multi-purpose tools, was registered in 1897. Over the succeeding years, numerous other tools were added to the knife, and today this classic design is sold all over the world. Victorinox currently produces 77,000 Swiss Army knives a day, amongst them its top-of-the-range *Swiss-Champ*, which although weighing only 185 grams has 31 different features.

Swiss Army knife,
original model
designed c. 1891

MASSIMO VIGNELLI

BORN 1931
MILAN, ITALY

Massimo Vignelli studied architecture at the Politecnico di Milano from
1950 to 1953 and later trained at the Università di Architettura, Venice. From
1956 he designed glassware for Venini and from 1958 to 1960 taught at the
Institute of Design, Chicago, while his wife, Lella Vignelli (b. 1936), worked
for the architects Skidmore, Owings and Merrill. In 1960 the couple returned to Italy and opened a studio in Milan. Four years later Massimo Vignelli
began working for the Container Corporation of America, Chicago, and designed its logo. In 1965, together with Bob Noorda (b. 1927), Jay Doblin
(b. 1920) and **Reinhold Weiss**, Vignelli founded the design consultancy Unimark International in Milan. The Vignellis moved permanently to America
that same year and in 1966 Unimark opened a New York office, specializing
in corporate identity work. In 1971 the couple established Vignelli Associates
and subsequently designed: corporate identity programmes for Knoll, American Airlines, Bloomingdale's, **Xerox**, Cinzano, **Ford** and Lancia; signage
for the subway systems in Washington DC and New York City; furniture for
Poltronova, Sunar, **Rosenthal** and Morphos; glassware for Venini, Steuben
and Sasaki; melamine tableware for Heller; and showrooms for **Artemide**
and Hauserman.

ⵜStacking cups
for Heller, 1970

Stacking dinnerware
for Heller, 1967

VOLKSWAGEN

FOUNDED 1938
WOLFSBURG, GERMANY

Ferdinand Porsche, original Volkswagen *Beetle*, 1938 – first designed in 1934

In January 1934 the automotive designer Professor Ferdinand **Porsche** submitted a design proposal to the new German Reich government for a revolutionary "car for the masses", called the *Volkswagen* (People's Car). Soon afterwards he signed a contract with the Reichsverband der Automobilindustrie (Automobile Industry Association), which provided him with a development budget on the understanding that a prototype of the car would be completed within ten months. In his pursuit of technological excellence, Porsche introduced innovations into the chassis, engine and transmission, regardless of cost; indeed, despite the availability of cheaper alternatives, he selected the more expensive air-cooled, horizontally-opposed engine for the car because of its full-throttle endurance. The first *Volkswagen* prototype was unveiled in October 1935, and by the following spring successive prototypes were undergoing extensive testing. The earliest cars, with their modern all-steel bodywork, were later improved with the addition of bumpers and running boards. **Erwin Komenda**, the acclaimed aerodynamicist, designed a distinctive streamlined body for the car, which, although compact, could easily accommodate five passengers. In 1938, in advance of the full-scale mass-production of the *Volkswagen*, the Gesellschaft zur

Volkswagen-Werk GmbH brochure, c. 1937, designed by Thomas Abeking

Volkswagen *Beetle*,
model of the late
1970s
(first re-designed
in 1945)

Volkswagen
New Beetle,
1998

Vorbereitung des Volkswagens (Company for the Preparation of the People's
Car) was established, later renamed Volkswagen-Werk GmbH. When a prototype of the *People's Car* was shown to the press later the same year, it was
immediately dubbed the "Beetle" by the *New York Times*. The German government, on the other hand, preferred the propagandist "Kraft durch Freude"
(Strength through Joy) name for the new car, which was initially intended to
be sold through a savings stamp scheme. Work began on the construction
of the world's largest car plant, where the cars were to be manufactured,
and a town for workers was also built (the Town of Strength through Joy,

Volkswagen *Transporter* from the 1970s (first designed in 1950)

Giorgetto Giugiaro, Volkswagen *Golf Mk 1*, 1974

today Wolfsburg). In 1940 the still incomplete factory started producing armaments, and *Volkswagen* production took a back seat while Porsche converted the car into two military jeep-style vehicles called the *Kübelwagen* and the *Schwimmwagen*. A year later, serial production of the *Volkswagen* was initiated with the manufacture of 41 cars, which were mainly used by the Nazi Party for propaganda purposes. After the war, *Volkswagen* production accelerated considerably, with the British Military Authorities placing an order for 20,000 vehicles in September 1946. Some 100,000 *Volkswagens* had been manufactured by 1950, the year in which another "classic" design was launched – the VW *Transporter*. This famous camper van was developed from a rough sketch by a Dutch VW importer who was inspired by the "flat vehicles" used in factories. Functioning as either a mini-bus or a commercial

Volkswagen *Passat V6 TDI*, 2000

vehicle, the *Transporter* was hugely successful, and by 1951 Volkswagen was producing some 12,000 units per annum. The *Karmann Ghia Coupé* was launched in 1955 and two years later a sophisticated cabriolet version was unveiled at the Frankfurt International Motor Show. By 1962 Volkswagen was producing 1,000,000 cars per annum. In 1964 the company built a huge research and development complex in Wolfsburg, which included a state-of-the-art wind tunnel opened in 1965. With this facility, VW was able to increasingly use technological advances as the basis for its future designs. In 1972 the *Beetle* became the world's most produced car. Volkswagen launched the *Passat* in 1973, followed in 1974 by the *Golf* designed by **Giorgetto Giugiaro**. Both cars served as blueprints for subsequent generations of Volkswagen models. The *Golf* – the intended successor of the *Beetle* – proved one of Volkswagen's greatest successes: one million of these "classless quality cars" were manufactured within the first 31 months and in 1977 the *Golf GLS* received a "Gute Form" award sponsored by the Federal Ministry of Economic Affairs. By 1988 an impressive ten million *Golfs* had been produced. Volkswagen's highly successful *New Beetle* was launched in 1998 and immediately received numerous awards and accolades, including being named "The Best Design of 1998" by *Time* Magazine. Based on the new *Golf* chassis, this superb design – like all other Volkswagens – "expresses the German engineering passion for designing and building cars". The Volkswagen Group now owns **Audi**, **Bugatti**, **Lamborghini**, Seat and Skoda and is one of the largest car producers in the world.

VOLVO

FOUNDED 1927
GOTENBURG, SWEDEN

Volvo was the first car manufacturer in the world to offer safety belts as standard in 1959

Sweden in the 1920s was heavily reliant on imported cars, and so Gustaf Larson and Assar Gabrielson decided to produce their own automobiles. In 1926 they came up with the idea of designing high quality car components that would be manufactured to their exacting specifications by selected suppliers, and which they would then assemble with the assistance of experienced car builders. Their first series-manufactured car, the *OV4*, left their factory in Gotenburg in April 1927 – marking a turning point in Swedish industrial design history. In 1929 Volvo introduced its first six-cylinder model, the *PV651*, and in 1935 the company launched the streamlined *PV36* designed by Ivan Örnberg, who had spent the whole of his career working in the US car industry. With its integrated headlamps, this sleek design was undoubtedly inspired by **Carl Breer**'s *Airflow* of 1934. During the Second World War, Volvo produced several military vehicles, including a jeep, and also developed a number of civilian prototypes. 1944 saw the launch of the *PV444*, which combined American styling with an unconventional unitized all-steel body. Developed by Helmer Petterson, the *PV444* was the first Volvo that "could be afforded by the average-man-in-the-street". In 1956 the *120* (known as the *Amazon* in Scandinavia) was introduced, featuring numerous innovative safety features including the world's first three point safety belt and a padded dashboard. With its emphasis on safety, in 1959 Volvo became the first manufacturer in the world to equip its cars with

Volvo *Amazon 120*, 1956

safety belts as standard. During the 1960s and 1970s, models such as the *144* sedan (1966), *145* estate (1967) and *240/260 series* (1974) became best-sellers while pioneering other innovative safety features such as disc brakes all round, split steering columns and outer rear-seat three-point safety belts. In 1976 Volvo presented the world's first catalytic exhaust emission control and by 1988 70% of its new cars were fitted with catalytic converters. The highly successful *700 Series* was launched in 1982, with the subsequent *740/760* estate models (1985) totally dominating their market segment. During the 1990s the company attempted to shake off its conservative image with the introduction of a number of new cars, such as the *S80* (1998), which featured smoother styling yet retained a strong Volvo identity. As a pioneer of revolutionary safety features, such as the safety belt tensioner, the integrated child booster seat, the self-adjusting belt reel and the integrated side impact protection system including side airbags, Volvo has had a significant impact on car design around the world.

WILHELM WAGENFELD

BORN 1900 BREMEN, GERMANY
DIED 1990 STUTTGART, GERMANY

←Table lamp for the
Dessau Bauhaus,
1924

Heat-resistant glass
tea set for Schott &
Gen. Jenaer Glas-
werke, 1930–1934

Wilhelm Wagenfeld trained with the silverware manufacturer Koch & Berg-feld and attended classes at the Kunstgewerbeschule in Bremen from 1916 to 1918. From 1919 to 1922 he trained at the Zeichenakademie (Drawing Academy) in Hanau, where he executed expressionist woodcuts. In 1923 he enrolled on the preliminary course at the Weimar **Bauhaus** and subsequently apprenticed under László Moholy-Nagy (1895–1946) in the school's metal workshop. While there, Wagenfeld co-designed with Karl J. Jucker (1902–1997) the famous "Bauhaus" table lamp (1923–1924), which was put into serial production by the workshop. After completing his journeyman's exam, he remained in Weimar and became an assistant to Richard Winkelmayer in the metal workshop of the Staatliche Bauhochschule. In 1927 he began independently producing designs for industrial production, including metal household items for Walter & Wagner, Schliez and door handles for S. A. Loewy, Berlin. A year later he took over the directorship of Bauhochschule's metal workshop, where he continued designing functional objects such as his *M15* tea caddy, which incorporated geometric forms that were less severe than those of his earlier Bauhaus designs. Although strongly connected with the Bauhaus, Wagenfeld rejected the institution's design doctrine as being too theoretical and self-centred. He shared its belief, however, that function was a prerequisite of good design and that industrial design practice necessitated close cooperation between designer and manufacturer. From 1929 Wagenfeld also worked as a freelance designer for the Schott & Gen. glassworks in Jena and in 1930 established his own design studio. He redesigned Schott & Gen.'s domestic glassware range, which included his famous glass tea set (c. 1930). Unlike so many of his former Bauhaus colleagues, Wagenfeld did not emigrate from Germany to escape Nazi persecution – his unquestionable skill in designing for industrial production being put to use during this period. Having become a professor at the Staatliche Kunsthochschule Berlin in 1931, he gave up teaching in 1935 when he was appointed artistic director of the Vereinigte Lausitzer Glaswerke, for whom he designed glass products that were highly suited to industrial mass-production, including utilitarian pressed-glass ranges of restaurant wine and beer glasses, as well as commercial-use bottles and jars. His best-known design from this period was his famous Kubus range (1938) – a modular system of stacking storage containers for kitchen use. During the 1930s Wagenfeld produced

designs for ceramics, most notably a porcelain service (1934) for Fürsten-
berg and the *Daphne* service (1938) for **Rosenthal**. Wagenfeld's refusal to
join the Nazi party eventually led to him being sent to the Russian Front,
where he was captured and made a prisoner-of-war. After the war and his
release, he outlined his functionalist approach to design in articles published in several journals, including *Die Form*, between 1947 and 1949.
During this period, he was also professor of industrial design at the Hochschule für Bildende Künste in Berlin and continued working as an independent designer, most notably designing cutlery for the Württemburgische
Metallwarenfabrik and electric light fittings, including graduated-sized plastic globe lamps, for Lindner of Bamberg. In 1954 he established the Wagenfeld Workshop in Stuttgart to develop products for industrial manufacture,
including metalware for WMF and a melamine in-flight meal tray for Lufthansa (1955). Wagenfeld's teachings stressed the moral, social and political
obligations of designers and were enormously influential upon the younger
generation of design practitioners in Germany. Today, as a celebrated Bauhaus figure, Wagenfeld's functional and aesthetically restrained industrial
designs are seen to epitomize the aspirations of the institution.

Waterman's **Ideal** Fountain Pen

WATERMAN

FOUNDED 1887
NEW YORK, USA

Lewis Edson Waterman (1837–1901) worked as an insurance broker in New York. According to legend, he was on the verge of selling a valuable policy when he loaned the client a newly-purchased fountain pen to sign the document. The pen failed to work and, to make matters worse, it made a large blot on the paper. While Waterman hurried back to his office to get another copy of the policy, a rival broker signed up the client. Furious at this turn of events, Waterman decided to dedicate himself to the development of an improved fountain pen. Whether this story is apocryphal or not, Waterman did invent the multi-channel feed in the 1870s. His first pen design to incorporate this feature was the *Regular*, patented in 1884. This forerunner of the modern fountain pen used a capillary system of small grooves under the nib to ensure a smooth flow of ink, which was contained in a reservoir filled using an eyedropper. Waterman subsequently established a small workshop and began manufacturing his innovative *Ideal* fountain pen, which was cased in hard rubber. Guaranteed for five years, the *Ideal* was the first truly reliable and leak-proof fountain pen and as such was the first pen to be a large-scale commercial success – Waterman produced around 500 of these pens during his first year, and in 1900 opened a factory in Montreal. After Waterman's death in 1901, his nephew headed the company and in 1905 introduced a revolutionary feature – the pen clip. Around 1914, Jules Fagard obtained the French manufacturing licence and later established Jif-Waterman, which produced exquisitely decorated pens. In 1954 Waterman's American operation ceased trading, after which Jif-Waterman became the largest Waterman concern. Now a subsidiary of the **Gillette** Group, Waterman is the largest manufacturer of fountain pens in Europe and the second largest worldwide.

Early advertisement for Waterman's *Ideal* fountain pen

Patent drawing for Lewis Waterman's *Regular* fountain pen, c. 1884

JAMES WATT

BORN 1736 GREENOCK, RENFREWSHIRE, SCOTLAND
DIED 1819 HEATHFIELD HALL, NEAR BIRMINGHAM, ENGLAND

C. F. von Breda,
James Watt, 1792

At the age of 17, James Watt trained as a mathematical instrument maker and studied under a master in London for one year before ill-health forced him to return to Scotland. In 1757 he subsequently established a shop within the precincts of Glasgow University (then one of the epicentres of the scientific community in Britain) and produced various mathematical instruments, such as quadrants, compasses and scales. While there, Watt met many leading scientists, including the professor of chemistry, Joseph Black (1728–1799), whose investigations into latent heat formed the basis of modern thermal science and inspired Watt's later improvements to the steam engine. In 1764 Watt was asked to repair a model of the atmospheric steam engines designed by Thomas Newcomen (1663–1729). In the course of his work on the engine, Watt realised that it wasted steam, and therefore power.

Patent drawing for
James Watt's steam
engine, 1781

He also identified the biggest design flaw in Newcomen's engine, namely its loss of latent heat (the heat used to alter the state of a substance), a problem which Watt calculated could be solved by adding a separate condenser connected to the cylinder by a pipe. Soon after this discovery, Watt entered into partnership with Dr. John Roebuck (1718–1794), who owned a coal mine in Linlithgowshire that could not be drained with one of Newcomen's engine-driven pumps. With Roebuck's financial assistance, Watt constructed an (unsuccessful) experimental engine in 1768 and the following year patented his famous "New Invented Method of Lessening the Consumption of Steam and Fuel in Fire Engines". Unable to find financing to develop his steam engine further, Watt worked as a surveyor for several years before entering into another partnership in 1775 with Matthew Boulton (1728–1809), who owned the Soho Engineering Works in Birmingham. With the necessary backing secured, Watt was able to improve his engine and an example was built to power a section of Boulton's works. By 1776 two Boulton & Watt engines were in commercial operation – one pumping water at a Staffordshire colliery, the other blowing air into the furnaces at John Wilkinson's ironworks. Between 1776 and 1781 Watt installed a number of his pumping engines in Cornish tin and copper mines. Boulton recognized the enormous potential of steam power for milling and persuaded Watt to design a rotative engine to replace his earlier model, which used a reciprocating action. The

Drawing of single-
action Watt engine
showing condenser,
1788

resulting design of c. 1780 incorporated a crank driven by two pistons, but this element had already been patented by James Pickard. Watt got around this problem in 1781 by devising the novel "Sun-and-Planet" wheel, and a year later harnessed the power derived from the expansiveness of steam to create a double-stroke engine. As well as continuing to improve upon his steam engines for powering machinery, Watt also independently discovered the composition of water (H_2O). As an important pioneer of the practical steam engine, Watt made an immense contribution to the Industrial Revolution.

JOSIAH WEDGWOOD

BAPTIZED 1730 BURSLEM, STAFFORDSHIRE, ENGLAND
DIED 1795 ETRURIA, STAFFORDSHIRE, ENGLAND

Sir Joshua Reynolds,
Josiah Wedgwood,
1783

The youngest of twelve children, Josiah Wedgwood came from a family whose members had been potters since the 17th century. After the death of his father in 1739, he worked for the family pottery, which had been inherited by his eldest brother, Thomas. After a five-year apprenticeship there, in 1754 Wedgwood entered into partnership with an established potter, Thomas Whieldon. Around this time, he began carefully recording his experiments, which included his highly successful formula for green glaze. In 1759 he established his own business at Ivy House Works, where he produced "a species of earthenware for the table, quite new in appearance, covered with a rich and brilliant glaze". A tea and coffee service of this warm cream-coloured earthenware was ordered by Queen Charlotte and in 1765 Wedgwood was given permission to re-christen it *Queen's ware* and to call himself "Potter to Her Majesty". Well finished and clean in appearance, *Queen's ware* was extremely versatile, in that it could be left plain, incised with decoration, hand-painted or transfer printed. Its durability and its serviceable

Wedgwood's Etruria
factory opened on
13 June 1769

Page from Josiah Wedgwood's first pattern book – border designs from c. 1775–1780

Large *Queen's ware (Royal Shape)* dish with *Pink Antique* border, c. 1775 forms also made it highly practical, and it became the standard domestic pottery selling successfully around the world. The success of *Queen's ware* was not only the result of the increasing demand for pottery fuelled by the growing popularity of tea-drinking, but also due to its simple forms, which were utterly in tune with the emerging Neo-Classical style. To meet the demand for his high quality wares, in 1764 Wedgwood moved his business to the larger Brick House Works in Burslem. Here he pioneered the logical division of labour, employed rationalized production methods and used innovative marketing techniques, while the forms of his cream-ware became less and less ornate so as to be better suited to mass-production. Wedgwood also discovered that cream-coloured earthenware could be successfully coloured with oxides to imitate stones, most notably agate, porphyry, granite and Blue John. In 1768 he entered into partnership with Thomas Bentley, a merchant from Liverpool, to manufacture ornamental wares using this type of coloured earthenware in the Neo-Classical style. A year later, they established the famous canal-side Etruria factory to produce these

Pair of white *Jasper ware* vases with green and lilac decoration, 1862 (large) and 1871 (small)

Keith Murray, coffee service, c. 1934

new moulded wares. The pieces were so popular that the company's London agent was, according to Wedgwood, "mad as a March hare for Etruscan vases". After literally thousands of experiments, in 1774 Josiah Wedgwood finally perfected his cameo-like *Jasper ware*. Many leading artists, including George Stubbs (1724–1806) and John Flaxman (1755–1826), designed relief decorations for *Jasper ware*, which was unashamedly Neo-Classical in style. Wedgwood explored every kind of shape and function possible in ceramics. He also investigated new industrial methods of production and in 1782 his Etruria factory became the first to install a steam-powered engine for the mass production of his moulded wares. During the 19th century, the Wedgwood factory continued manufacturing both domestic and ornamental wares. In the 1930s and 1940s, the factory began producing Modern wares designed by Keith Murray (1892–1981) and Eric Ravilious (1903–1942), among others. Alongside its more traditional wares, Wedgwood today continues to produce designs that combine simple forms with modern decoration.

REINHOLD WEISS

BORN 1934
STUTTGART, GERMANY

Reinhold Weiss studied at the Hochschule für Gestaltung in Ulm, where he trained under the industrial designer **Hans Gugelot**. In 1959 Weiss became an in-house designer for **Braun** and three years later was appointed deputy director of its design department. While at Braun, Weiss was also executive designer

HE 1 electric kettle
for Braun, 1962

for the appliance division and designed numerous household products, including the well-known *HL 1* desk fan (1961), which was awarded a "Gute Form" Bundespreis in 1970. This relatively austere design exemplified the highly rational and systematic approach to design that was advocated at Ulm and promoted by Braun. Weiss's *HT 1* toaster (1961), *HE 1* electric kettle (1962) and *KMS 1/11* coffee grinder (1967) similarly articulated this design vocabulary. In 1967 Weiss emigrated to the United States and together with **Massimo Vignelli** opened the New York office of Unimark, a design consultancy specializing in corporate identity. Three years later Weiss established his own design office in Chicago, which concentrated on graphics and product design. In 1972 Weiss received an award from the American Institute of Graphic Arts for a packaging design and in 1982 was awarded

KMS 1/11 coffee
grinder, 1967

a gold medal at the Industry Fair in Brno. Reinhold Weiss currently lives in Evanston, Illinois, and continues to run a design office.

Revere Ware sauce-pans for the Revere Copper & Brass Company, 1938

W. ARCHIBALD WELDEN

AMERICAN, ACTIVE 1920S–1950S

The cooking pot is one of humankind's oldest designed artefacts, with an ancestry that can be traced back to prehistoric times. One of the most innovative designs along its evolutionary path was W. Archibald Welden's copper-bottomed stainless steel *Revere Ware* cooking pans (1938) produced by the Revere Copper & Brass Company. Founded in 1892, the Revere Copper & Brass Company had over many decades attempted to develop "a pot that would combine the thermal efficiency of copper with easier cleaning and maintenance". While stainless steel appeared to offer the solution, it had poor thermal conductivity. It was not until 1936 that a manager of the company, J. M. Kennedy, came up with the idea of depositing a copper coating on the outer base of a stainless steel pan so as to unite the heat conductivity of copper with the durability and imperviousness of stainless steel. After literally hundreds of tests and experiments, the company eventually arrived at a revolutionary low-cost process for the electrodeposition of a heavy, hard and adherent copper coating on stainless steel. In the meantime Welden, who was working as a design con-

Revere Ware sauce-pan for the Revere Copper & Brass Company, 1938 sultant to the company, determined that the fundamental criteria for a cooking pot were heating efficiency, ease of carrying and handling, especially with liquid contents, and ease of cleaning. His resulting designs were entirely logic-driven – the pots were carefully dimensioned in accordance with the design and size of gas, coal, oil and electric range burners, a non-geometric curve was developed for the inside base of the pans that conformed to the shape of the fingers so as to facilitate easier cleaning, while the form of the moulded plastic handles was guided by numerous ergonomic considerations. Revere eventually began retailing Welden's *Revere Ware* in 1939 and in testimony to the timeless functional simplicity of the design, the range remains in production today with only a few minor modifications. For Revere, Welden also designed the first successful stainless-steel pressure cooker and later an institutional range of pans (1954).

WESTINGHOUSE

FOUNDED 1886
PITTSBURGH, PENNSYLVANIA, USA

In 1865 George Westinghouse (1846–1914) filed his first patent for a rotative steam engine. Although this invention proved impractical, he later applied the engineering principles on which it was based to the successful design of a water meter. Also in 1865, he designed a device for placing derailed freight carriages back onto their tracks. His first major invention, however, was the air brake, which he patented in 1869. Westinghouse became an early advocate of standardization when he realised that there were significant advantages to standardizing the design of his air brakes, so that they would be compatible between different rail cars on different lines. Apart from his important contribution to rail transport safety, Westinghouse was also responsible for the widespread acceptance of electrical alternating current in America, which during the 1880s was still employing the direct current system.

Westinghouse advertisement for washing machine, 1950s

... of course, it's electric!

Just Plug It In!

**SAVE INSTALLATION COST
WITH NEW WESTINGHOUSE**

EASY LOADING ... NO STOOPING
No lifting or carrying of heavy wet
clothes. No lines to string or tighten.

Europe, on the other hand, was developing a number of alternating current systems utilizing the transformers designed by Lucien Gaulard and John Gibbs. In 1885 Westinghouse imported a Gaulard-Gibbs transformer and began adapting its design. With this device, Westinghouse was able to "transform" the voltage of alternating current so that electricity could be carried over considerable distances at high voltage, before it was stepped down to the proper voltage for its intended use. In March 1886 the redesigned transformer was demonstrated in Great Barrington, Massachusetts, where people marvelled at the electrically powered lights "that produce no odour, heat or danger of fire". This first Westinghouse transformer heralded the golden age of electricity and its basic features have since remained virtually unchanged. Having founded the Westinghouse Electric Company the same year, by 1890 the company had constructed over 300 power stations and achieved an annual sales figure of $4 million. As well as being a major power provider, the company also became a leading manufacturer of electrical appliances, introducing the first fully automatic electric cooking range in 1917 and the first practical automatic electric iron in 1924. During the Second World War, the company designed the first long-range ground warning radar system (1939) and built the first American-designed and American-built jet engine (1941). In the 1950s Westinghouse also became a pioneer of nuclear power technology. Although its large range of domestic appliances made it a trusted household name, it was never able to rival the **General Electric** Company and in 1975 ceased production of all household appliances, apart from light bulbs. Today, Westinghouse remains a leading developer and manufacturer of defence electronics and a major player in the commercial nuclear power industry.

ELI WHITNEY

BORN 1765 WESTBORO, MASSACHUSETTS, USA
DIED 1825 NEW HAVEN, CONNECTICUT, USA

One of the great pioneers of the Industrial Revolution in America, Eli Whitney studied science and technology at Yale College (now Yale University). After graduating in 1792, he accepted a teaching post in South Carolina, but was shipwrecked on his way there and ended up in New York, where he met Phineas Miller and Catherine Greene. Travelling south with them, he stayed at Greene's plantation, Mulberry Grove, where he learned of the need for a new machine that would help remove seeds from the short-fibred cotton that grew in the South, and which unlike long-fibred cotton could not be easily cleaned. Within six months, Whitney had constructed a working model of his revolutionary cotton gin. "This machine may be turned by water or with a horse, with the greatest of ease", he declared, "and one man and a horse will do more than fifty men with the old machines". The device pulled the cotton through hundreds of closely set teeth mounted on a revolving cylinder that combed the seeds out. The fibre was then passed through slots in an iron breastwork that were narrow enough not to let seeds through. Whitney's cotton gin had an enormous impact on the South's economy, which became almost completely cotton-based. This simple yet ingenious design was patented in 1794. Whitney subsequently

Engraving of Eli
Whitney's cotton
gin, designed in 1793

went into business with Miller to begin mass-producing the cotton gin
using purpose-built machinery. The patent rights to the device, however,
were widely infringed by planters who made their own copies, which resulted in Miller & Whitney going out of business in 1797. Having learned from
this experience, Whitney decided to turn his attention to the manufacture of
small arms, as the US Government, fearing the outbreak of war with France,
was attempting to solicit 40,000 muskets from private contractors. Muskets
at this stage were still made almost entirely by hand and individually, so that
each one differed slightly from the next. Whitney resolved to produce small
arms using specially-designed machinery allowing for the mass-manufacture
of interchangeable parts. This novel and far-reaching concept of standardization was completely revolutionary and had enormous implications for
later industrially produced designs. Standardized interchangeable parts enabled faster assembly and much easier replacement of broken elements.
Funded by the US government, Whitney established an armoury in New
Haven to produce his famous muskets. There he almost single-handedly
pioneered what became known as the "American System" of mass-production, which by the late 1900s was being used for the large-scale manufacture
of numerous products, from sewing machines and clocks to automobiles.

WILBUR & ORVILLE WRIGHT

BORN 1867 NEAR MILLVILLE,INDIANA, USA
DIED 1912 DAYTON, OHIO, USA

BORN 1871 DAYTON, OHIO, USA
DIED 1948 DAYTON, OHIO, USA

Wilbur (top) & Orville Wright

Having grown up in a home where "there was always much encouragement to children to pursue intellectual interests; to investigate whatever aroused curiosity", the Wright brothers initially established a printing shop where they designed and built their own presses. In 1892 they opened a shop for the sale and repair of bicycles and four years later began constructing bicycles to their own designs. They used the profits from these two ventures to fund aeronautical experiments, whereby the experience gained from the design and construction of the presses and bicycles, and their subsequent understanding of the inherent characteristics of materials such as tubular metal, wood and wire, was of enormous benefit to their research. After corresponding with the French civil engineer and leading authority on aviation, Octave Chanute, in 1900, the Wright brothers realized that a successful flying machine would need wings to generate lift, a propulsion system and a control system. Of these, control posed the greatest problem. Having determined that the best means of control would be through the precise manipulation of the centre of pressure on the wings, they devised a method of mechanically inducing a helical twist across the wings in either direction. This provided lift on one side and decreased lift on the other so that the pilot could raise and lower either wing tip at will. This concept of "wing-warping" was used to control the Wright's first glider of 1900, which was mainly flown as a kite without a pilot on board. Having constructed and flown another glider in 1901, they subsequently built a small wind tunnel in which they tested the performance of over 100 wing designs. Their researches led to the very successful 1902 glider with an improved control system comprising a forward monoplane elevator and a moveable rudder linked to the wing-warping system. First tested in late 1902 at Kill Devil Hills, near the village of Kitty Hawk, North Carolina, the 1902 Wright glider demonstrated that the brothers had overcome the key problems of heavier-than-air flight. So important was this aircraft that the Wright's later basic patent featured a glider

Drawing of Wright brothers' basic patent, 1904

design with the control system of the 1902 model, rather than a powered
aircraft. Having made considerable technical progress with their gliders,
the brothers moved on to powered machines and built their 1903 *Flyer*.
This aircraft was powered by a 12.5 horsepower four-cylinder engine of the
Wright's own design, which was linked through a chain-drive transmission
to twin counter-rotating pusher propellers. On 17 December 1903 at Kill
Devil Hills, with Orville as pilot, this aircraft achieved the first ever controlled, powered flight, flying 120 feet in 12 seconds. Over the next couple
of years the Wrights refined their designs so that by 1905 their *Flyer* could
remain in the air for 39 minutes. Three years later, the Wrights were contracted by the US government to supply a military aeroplane capable of
flying for more than one hour at an average speed of 40 mph. In 1909 the
US Army conducted trials of the resulting *Model A*, which was the world's
first military aeroplane. That same year, the Wright Company was incorporated, an aeroplane factory was established in Dayton and a flying school
was founded at Huffman Prairie. In tribute to the Wright brothers' historic
achievements, the label beside their famous 1903 *Flyer* on display in the
Smithsonian Institute reads: "By original scientific research, the Wright
brothers discovered the principles of human flight. As inventors, builders
and flyers, they further developed the aeroplane, taught man to fly, and
opened the era of aviation."

XEROX

FOUNDED 1906
ROCHESTER, NEW YORK, USA

British patent
drawing, 1952 –
the American
patent was
originally filed
in 1940

Chester Carlson (1906–1968) briefly worked for the **Bell Telephone Company**, prior to joining the New York-based electronics firm P. R. Mallory Company. While working there as a patent attorney, Carlson became increasingly frustrated by the difficulties he experienced in obtaining copies of patent drawings and specifications. From 1934, he began exploring techniques for copying both text and drawings using electrostatic methods rather than photographic or chemical methods, which were already being researched by a number of large corporations. Having set up a makeshift laboratory in Queens, New York, he developed a prototype photocopier that produced the first xerographic image in 1939. Carlson then spent several years trying to sell his invention to various large corporations, including **General Electric** and **IBM**, but was met with what he described as "an enthusiastic lack of interest". The reasons for this were manifold – Carlson's prototype copier was both bulky and messy to use, for one, and many business people felt that carbon copy paper remained perfectly adequate for their needs. Eventually the Battelle Memorial Institute in Colum-

Chester F. Carlson
with early photo-
copier

Xerox copier,
mid-1970s

*Model 1075 photo-
copiers, late 1980s/
early 1990s.*

bus, Ohio, contracted Carlson to refine his "electrophotography" process in 1944. The Haloid Corporation, a photographic paper manufacturer and retailer founded in 1906, subsequently purchased the rights to manufacture and market a copying machine based on Carlson's invention. Later, the company obtained all the rights pertaining to Carlson's pioneering technology and, with him, agreed that a new and shorter title for the process was needed. The name "Xerography" was eventually coined and the word "Xerox" was trademarked in 1948. The following year saw the introduction of the first xerographic copier, the *Model A*. In 1958 the Haloid Company changed its name to Haloid Xerox Inc. (becoming the Xerox Corporation in 1961). The first automatic plain-paper office copier, the *Xerox 914*, was launched in 1959 and remained in production until 1976. This design was so successful and influential that the company had to battle to prevent the name "Xerox" from becoming a generic term. Since then, Xerox has been a world leader in the photocopying field and has diversified into other product areas, including word-processors, facsimile machines and the highly successful office communications network *Ethernet* (1979). Since the early 1990s, Xerox has been committed to the design of "waste-free" and "sustainable" products that "optimize resource utilization and minimize environmental impact".

YAMAHA

FOUNDED 1897
HAMAMATSU CITY, JAPAN

GK Design Group,
A-vot audio device,
1987

Torakusa Yamaha produced his first organ in 1887 and ten years later founded the Nippon Gakki Company (renamed the Yamaha Corporation in 1987 to mark the company's centenary). In 1900 the company began manufacturing pianos and concentrated on the production of keyboard instruments until 1954, when it began diversifying and developed its first hi-fi player and first motorcycle, the 125cc *YA-1*. The subsidiary Yamaha Motor Company was founded in 1955 and in 1960 began production of fibreglass powerboats and outboard boat motors. Since the 1960s, Yamaha has developed a wide range of mainly sports vehicles that utilize the company's expertise in engine and fibreglass technology. These include the *SL350* snowmobile (1968), Yamaha's first golf cart (1975), and the *MT500T WaveRunner* personal watercraft (1986). Yamaha's motorcycle business proved particularly successful – in 1977 motorcycle production reached the ten million mark. Many Yamaha Motor Company products have been designed by the **GK Design Group**, including the company's highly successful all-terrain vehicles (ATVs). Yamaha continues to design and manufacture high-quality audio equipment and keyboard instruments, and like many other Japanese companies is also actively involved in the design of robots for automated production lines.

GK Design Group,
Morph II motorcycle,
1991

MARCO ZANUSO

BORN 1916 MILAN, ITALY
DIED 2001 MILAN, ITALY

One of Italy's leading postwar industrial designers, Marco Zanuso is celebrated for his innovative applications of state-of-the-art materials and technology in the design of forward-looking office equipment, consumer electronics and furniture. Having studied architecture at the Politecnico di Milano,
graduating in 1939, in 1945 he established his own Milan-based practice
and executed product and furniture designs as well as commissions in the
fields of architecture and town planning. From 1946 to 1947 Zanuso edited
the journal *Domus* jointly with Ernesto Rogers (1909–1969), and also edited
Casabella magazine from 1947 to 1949. In 1948 he was commissioned by
Pirelli to explore the potential of latex foam as an upholstering material. His
subsequent *Antropus* chair (1949) was the first chair to exploit this material
and was produced by Arflex, a manufacturing company set up by Pirelli. It
was followed by several other latex foam upholstered seating designs, including the *Lady* chair (1951) and the *Triennale* sofa (1951). These two pieces

Duna cutlery for
Alessi, 1995

were first exhibited at the IX Milan
Triennale of 1951, where Zanuso
was awarded a Grand Prix and two
gold medals. In 1956 he was awarded a Compasso d'Oro for his *Model
1100/2* sewing machine for Borletti
– a design that epitomised his work
through its rational yet sculptural
form. Zanuso also undertook several architectural commissions,
including the **Olivetti** manufacturing plants in San Paolo (1955) and
Buenos Aires (1955–1957) and the
Necchi factory in Pavia (1961–1962),
which revealed his interest in product architecture and prefabricated
structures. From 1958 to 1977 Zanuso collaborated with **Richard Sapper**
on numerous landmark product
and furniture designs, including

the enamelled light-weight stamped steel *Lambda* chair for Gavina (1959–1964); the *No. 4999/5* stacking child's chair for **Kartell** (1961–1964), which was the first seating design to be produced in injection-moulded polyethylene; the *TS 502* radio (1964) and *Doney 14* (1964) and *Black 201* (1969) televisions for **Brionvega**, which were remarkable for their miniaturization of technology and synthetic Pop aesthetic; the *Grillo* telephone for Italtel/**Siemens** (1966) with its innovative folding form; and the *2000* kitchen scales for Terraillon (1970). In 1956 Zanuso became a member of CIAM (Congrès Internationaux d'Architecture Moderne) and the INU (Instituto Nazionale Urbanista). During the same year, he also co-founded the Associazione di Disegno Industriale (ADI). He was a city councillor in Milan from 1956 to 1960 and in 1961 became a member of the city's planning commission. Having experimented with new materials, manufacturing methods and types of construction throughout his career, Zanuso often produced product designs that redefined the formal potential of existing types. From 1966 to 1968 he was a professor of materials applications at the Politecnico di Milano, becoming Professor of Industrial Design there in 1976.

ZANUSSI

FOUNDED 1916
PORDENONE, ITALY

Gastone Zanello &
Zanussi Design
Center, *Series 700*
mini cooker, 1961

In 1916 Antonio Zanussi opened a small workshop to make stoves and wood-burning ovens. Four years later, the Officina Fumistera Antonio Zanussi launched its first export design – a wood-burning stove with a cast-iron hotplate. In 1946 Zanussi's son Lino took over the company and began a programme of modernization by adopting efficient industrial production processes. He was also responsible for establishing the Zanussi Design Center in 1954 and for commissioning **Gino Valle** to design household appliances for the firm. With their crisp modern lines, Valle's product designs helped Zanussi gain a reputation for innovation. From 1958 to 1981 the company's design department was headed by Gastone Zanello, who introduced the concepts of modularity and dimensional co-ordination, thus heralding the advent of built-in models. Between 1976 and 1989, the freelance designer Andries Van Onck also produced designs for the company that were based on the concept of "product families", which served to aesthetically differentiate the various brands within the group. In 1982 Roberto Pezzetta

A. Van Onck &
Zanussi Design
Center, *"A"* washing
machine, 1980

succeeded Zanello as director of the Zanussi Design Center and two years later the company became part of the **Electrolux** group. This provided it with the necessary funding to re-establish its reputation for ground-breaking designs, such as the *Jetsystem* washing machine (1985), which automatically adjusted its energy consumption to the size of the load. As part of the Electrolux Design Families Strategy, in 1987 Pezzetta designed the post-modern *Wizard's Collection*, which was considered "a cornerstone in household appliances design history". During the 1990s, Pezzetta also designed several experimental prototypes for "a new generation" of "bio-design" appliances that combine "emotion, style, ergonomy", including the award-winning *Oz* refrigerator and the *Zoe* washing machine. Today, around five million Zanussi appliances are sold every year.

↘Roberto Pezzetta, *Colours* oven, 1998

Roberto Pezzetta & Zanussi Design Center, *Colours* refrigerator, 1999

LZ 126 airship being repaired, c. 1924

FERDINAND GRAF VON ZEPPELIN

BORN 1838 KONSTANZ, BADEN, GERMANY
DIED 1917 CHARLOTTENBURG, BERLIN, GERMANY

In 1863 Ferdinand Graf von Zeppelin made his first balloon ascent in St. Paul, Minnesota, while acting as a volunteer military observer for the Union Army during the American Civil War. He became fascinated by balloon technology and after retiring from the military in 1890 dedicated his energies to the design of rigid dirigible airships. After ten years of research, Zeppelin's *LZ-1* airship made its first test flight. 128 metres long, it had a lightweight aluminium frame and could achieve speeds of up to 32 kph. The *LZ-1* was the first of many lighter-than-air airships which Zeppelin developed, and which performed better than Germany's fledgling aeroplanes. After a Zeppelin airship made a 24-hour flight in 1906, the German government placed an order for an entire fleet. An airship passenger service was launched four years later, by the Deutsche-Luftschiffahrts AG (German Airship Travel Company). Over one hundred Zeppelin airships were used for military purposes during the First World War, after which commercial production resumed under the supervision of Hugo Eckner (1868–1954), who piloted the legendary *Graf*

The *Hindenburg* disaster – the airship crashed and exploded while landing at Lakehurst, New Jersey, on 6 May 1937

Zeppelin on its 21-day around-the-world flight in 1929. When the Nazi Party came to power in 1933, the company was compelled to design an even bigger and better airship that was intended to reflect the superiority of the Third Reich. The result was the *LZ-129*, which made its first flight in 1936. Better known as the *Hindenburg*, this airship famously crashed and exploded while landing at Lakehurst, New Jersey, in 1937. Following this tragedy, airships were deemed too dangerous and all but disappeared from the skies.

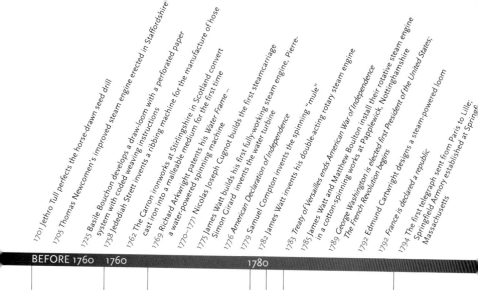

1701 Jethro Tull perfects the horse-drawn seed drill

1705 Thomas Newcomen's improved steam engine erected in Staffordshire

1725 Basile Bouchon develops a draw-loom with a perforated paper system with coded weaving instructions

1758 Jedediah Strett invents a ribbing machine for the manufacture of hose

1762 The Carron ironworks at Stirlingshire in Scotland convert cast iron into a malleable medium for the first time

1769 Richard Arkwright patents his Water Frame – a water-powered spinning machine

1770–1771 Nicolas Joseph Cugnot builds the first steam-carriage

1775 James Watt builds his first fully-working steam engine, Pierre. Simon Girard invents the water turbine

1776 American Declaration of Independence

1779 Samuel Compton invents the spinning "mule"

1782 James Watt invents his double-acting rotary steam engine

1783 Treaty of Versailles ends American War of Independence

1785 James Watt and Matthew Boulton install their rotative steam engine in a cotton-spinning works at Papplewick, Nottinghamshire

1789 George Washington is elected first President of the United States; The French Revolution begins

1792 Edmund Cartwright designs a steam-powered loom

1792 France is declared a republic

1794 The first telegraph sent from Paris to Lille; Springfield Armory established at Springf...

BEFORE 1760   1760   1780

1698
Thomas Savery patents the first steam engine

1759
Josiah Wedgwood establishs his own pottery

1779
Abraham Darby III constructs the first-ever cast-iron bridge in Coalbrookdale, England

1793
Eli Whitney designs the cotton gin

1764
James Hargreaves develops the *Spinning Jenny*

1781
James Watt designs "Sun-and-Plant" wheel for his steamengine

1783
Joseph-Michel and Jacques-Étienne Montgolfier demonstrate their hot-air balloon

1800 Richard Trevithick constructs the first light pressure steam-engine; Eli Whitney develops a musket with interchangeable components

1801 Robert Fulton develops the first submarine, "Nautilus"

1803 Henry Shrapnel invents the first munitions shell in England

1810 Krupp factory established at Essen, Germany for the manufacture of arms

1811 Blenkinsop's steam locomotive with tooth-rack rail propulsion

1812 Philippe Girard develops a machine for spinning flax

1814 George Stephenson builds the first practical steam locomotive at Killingworth Colliery near Newcastle.

1815 Humphrey Davy designs the first safety lamp for miners

1822 First iron railway bridge constructed by George Stephenson for the Stockton-to-Darlington line

1829 William Burt granted the first US patent for a typewriter

1830 Edwin Budding designs first lawn-mower

1832 Charles Babbage builds his Difference Engine (an early precusor of the computer); Isambard Kingdom Brunel appointed chief engineer of the Great Western Railway

1834 Cyrus Hall McCormick patents his revolutionary mechanical reaper

1837 Launching of Isambard Kingdom Brunel's ship, the *Great Western*; John Deere designs the first commercially successful self-scouring steel plough

1839 Charles Goodyear develops a rubber "vulcanization" process

1800   **1820**   **1840**

1801
Richard Trevithick
constructs a steam-
powered carriage

1819
Michael Thonet
establishes a furni-
ture workshop in
Boppard am Rhein,
Germany

1821
Michael Faraday
builds his first dyna-
mo

1832
Samuel Morse
starts to develop
an electric
telegraph

1844
Morse's telegraph is
transmitted for the first
time between Baltimore
and Washington

1805
Joseph-Marie
Jacquard invents
a mechanised
loom

1829
George Stephenson's landmark
engine, *The Rocket* is awarded a
prize of £500 in the Rainhill
Trials

1835
Samuel Colt designs a revolver with interchangable com-
ponents

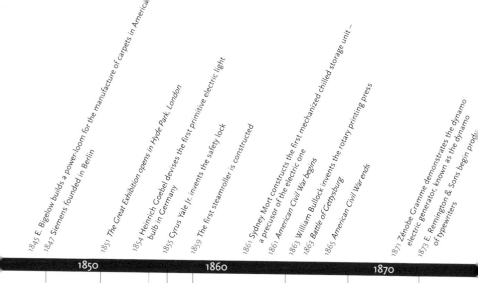

1845 E. Bigelow builds a power-loom for the manufacture of carpets in America

1847 Siemens founded in Berlin

1851 The Great Exhibition opens in Hyde Park, London

1854 Heinrich Goebel devises the first primitive electric light bulb in Germany

1855 Cyrus Yale Jr. invents the safety lock

1859 The first steamroller is constructed

1861 Sydney Mort constructs the first mechanized chilled storage unit – a precusor of the electric one

1861 American Civil War begins

1863 William Bullock invents the rotary printing press

1863 Battle of Gettysburg

1865 American Civil War ends

1871 Zénobe Gramme demonstrates the dynamo electric generator, known as the dynamo

1873 E. Remington & Sons begin produ...

1850   1860   1870

1851
Isaac Singer patents his improved sewing machine

1859
Michael Thonet designs the *Model No. 14* chair

1862
Richard Gatling patented multi-barrel machine gun

1873
C.L. Shoales and C.Glidden develop their improved typewriter

1858
Stapellauf der *Great Eastern* von Isambard Kingdom Brunel

1846
Elias Howe patents the first practical sewing machine

1855
General locomotive built by Rogers for Western & Atlantic Railroad

1867
Nikolaus Otto and Eugen Langen design their first internal combustion engine

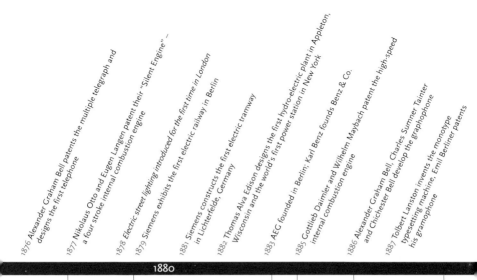

1880

1876 Alexander Graham Bell patents the multiple telegraph and designs the first telephone

1877 Nikolaus Otto and Eugen Langen patent their "Silent Engine" – a four stroke internal combustion engine

1878 *Electric street lighting introduced for the first time in London*

1879 Siemens exhibits the first electric railway in Berlin

1881 Siemens constructs the first electric tramway in Lichterfelde, Germany

1882 Thomas Alva Edison designs the first hydro-electric plant in Appleton, Wisconsin and the world's first power station in New York

1883 AEG founded in Berlin; Karl Benz founds Benz & Co.

1885 Gottlieb Daimler and Wilhelm Maybach patent the high-speed internal combustion engine

1886 Alexander Graham Bell, Charles Sumner Tainter and Chichester Bell develop the graphophone

1887 Tolbert Lanston invents the monotype typesetting machine; Emil Berliner patents his gramophone

1879
Thomas Alva Edison designs the incandescent electric light

1884
Lewis Edison Waterman patents his improved fountain pen

1886
Karl Benz launches the three-wheeled *Motorwagen* – the world's first practical car powered by an internal combustion engine

1888
J. K. Starley baut sein Rover »Sicher-heitsfahrrad«

1877
Thomas Alva Edison designs his phonograph

1885
George Eastman starts manufacturing coated photographic paper

1888
John Boyd Dunlop develops the first pneumatic tyre

1889 Gustave Eiffel constructs the Eiffel Tower

1891 Thomas Alva Edison patents the first motion picture camera; Philips founded in Eindhoven, Netherlands

1892 General Electric founded in Albany, New York

1893 Rudolf Diesel designs a prototype diesel engine; Benjamin Holt patents first "Caterpillar" track-type tractor

1894 Karl Benz launched his first series production car – the Benz Velo

1895 Konstantin Tsiolkovsky develops a system of refrigeration through compression; Auguste and Louis Lumière patent their cinématographe, a motion picture camera and projector combined

1896 Guglielmo Marconi patents radio telegraphy

1897 Karl Ferdinand Braun invents the cathode ray tube

1900 "Exposition Universelle et Internationale" staged in Paris

1901 Guglielmo Marconi transmits the first transatlantic telegraphic radio messages from Cornwall to Newfoundland; the first Mercedes car is manufactured

1904 Sir John Ambrose Fleming invents the diode

1906 Victor Talking Machine Company launches the Victrola ph

1907 Peter Behrens becomes artistic consultant of

1890   **1900**

1889
Linotype machine introduced for composing type

1892
Lingner Werke designs the *Odol* bottle

1892
Sir James Dewar invents the first vacuum flask

1898
Count Ferdinand von Zeppelin constructs his first airship

1893
Sears, Roebuck & Co. is founded in Chicago and publishes its first mail-order catalogue

1900
George Eastman introduces the Kodak *Brownie* camera

1903
The Wright Brothers succeed in flying the first powered aircraft, the *Flyer No. 3*

1907
Deutscher Werkbund founded in Munich

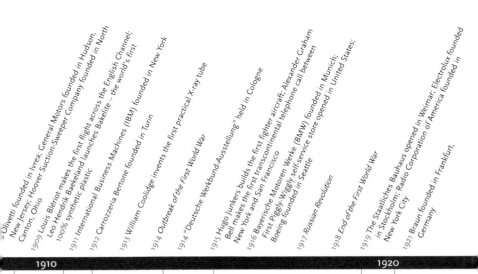

Olivetti founded in Ivrea; General Motors founded in Hudson, New Jersey; Hoover Suction-Sweeper Company founded in North Canton, Ohio

1909 Louis Blériot makes the first flight across the English Channel; Leo Hendrik Baekeland launches Bakelite – the world's first 100% synthetic plastic

1911 International Business Machines (IBM) founded in New York

1912 Carrozzeria Bertone founded in Turin

1913 William Coolidge invents the first practical X-ray tube

1914 Outbreak of the First World War

1914 "Deutsche Werkbund-Ausstellung" held in Cologne

1915 Hugo Junkers builds the first fighter aircraft; Alexander Graham Bell makes the first transcontinental telephone call between New York and San Francisco

1916 Bayerische Motoren Werke (BMW) founded in Munich; First Piggly-Wiggly self-service store opened in United States; Boeing founded in Seattle

1917 Russian Revolution

1918 End of the First World War

1919 The Staatliches Bauhaus opened in Weimar; Electrolux founded in Stockholm; Radio Corporation of America founded in New York City

1921 Braun founded in Frankfurt, Germany

1910

1920

1912
Pyrex developed by
Corning Glass Works

1915
Major Wilson and Sir William
Tritton design the *Big Willie*
tank

1920
Snap-on Wrench Company
founded in Milwaukee

1911
International Business
Machine (IBM) founded
in New York

1914
Bell Telephone Company
introduces the candle-
stick telephone

1919
Citroën launches its
Type A automobile

1908
The Ford Motor Company
produces the first *Model T*
car

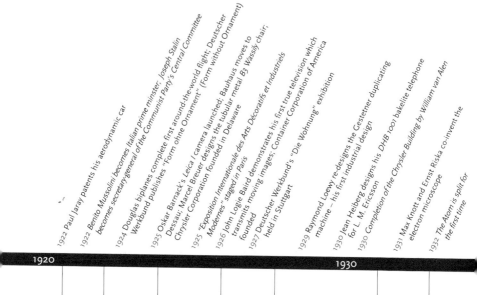

1922 Paul Jaray patents his aerodynamic car

1922 Benito Mussolini becomes Italian prime minister; Joseph Stalin becomes secretary-general of the Communist Party's Central Committee

1924 Douglas biplanes complete first around-the-world flight; Deutscher Werkbund publishes "Form ohne Ornament" (Form without Ornament)

1925 Oskar Barnack's *Leica I* camera launched; Bauhaus moves to Dessau; Marcel Breuer designs the tubular metal *B3 Wassily* chair; Chrysler Corporation founded in Delaware

1925 "Exposition Internationale des Arts Décoratifs et Industriels Modernes" staged in Paris

1926 John Logie Baird demonstrates his first true television which transmits moving images; Container Corporation of America founded

1927 Deutscher Werkbund's "Die Wohnung" exhibition held in Stuttgart

1929 Raymond Loewy re-designs the Gestetner duplicating machine – his first industrial design

1930 Jean Heiberg designs his *DHB 1001* bakelite telephone for L. M. Ericsson

1930 Completion of the Chrysler Building by William van Alen

1931 Max Knott and Ernst Riska co-invent the electron microscope

1932 *The Atom* is split for the first time

1920

1930

1922
Gustav Dalén
designs the *Aga*
cooker

1922
Austin *Seven* launched

1925
Bang & Olufsen
founded in Quistrup,
Denmark

1926
Mart Stam constructs
the first cantilevered
metal chair

1928
Walter Maria Kersting
develops the first version
of his *Volksempfänger*
radio

1931
Hermann Gretsch
designs his *1382*
service for Arzberg

1932
Aino Marsio-Aalto
designs pitcher and
glasses for Karhula

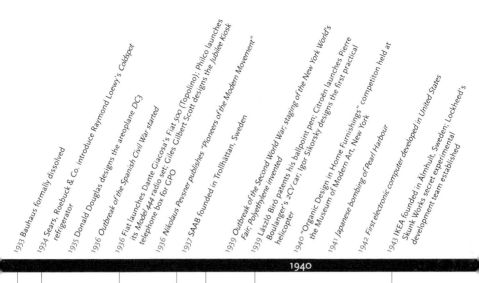

1933 Bauhaus formally dissolved

1934 Sears, Roebuck & Co. introduce Raymond Loewy's *Coldspot* refrigerator

1935 Donald Douglas designs the aeroplane DC3

1936 Outbreak of the Spanish Civil War started

1936 Fiat launches Dante Giacosa's Fiat 500 (Topolino); Philco launches its Model 444 radio set; Giles Gilbert Scott designs the *Jubilee Kiosk* telephone box for GPO

1936 Nikolaus Pevsner publishes "Pioneers of the Modern Movement"

1937 SAAB founded in Trollhättan, Sweden

1939 Outbreak of the Second World War; staging of the New York World's Fair; Polyethylene invented

1939 László Biró patents his ballpoint pen; Citroën launches Pierre Boulanger's 2CV car; Igor Sikorsky designs the first practical helicopter

1940 "Organic Design in Home Furnishings" competition held at the Museum of Modern Art, New York

1941 Japanese bombing of Pearl Harbour

1942 First electronic computer developed in United States

1943 IKEA founded in Älmhult, Sweden; Lockheed's Skunk Works secret experimental development team established

1940

1934
Carl Breer's *Airflow* car
for Chrysler is launched

1939
Chester Carlson produces
first xerographic image
with his prototype
photocopier

1936
Reginald Mitchell
designs a prototype of
the *Spitfire* monoplane

1943
Sixten Sason designs his
bullet-shaped *Model 248*
vacuum cleaner for
Electrolux

1938
Volkswagen *Beetle*
launched

1933
Wilhelm Kåge designs
his *Praktika* tableware
for Gustavsberg

1937
Raymond Loewy designed
his *S-1* locomotive for
Pennsylvania Railroad

1945 First atomic bombs dropped on Hiroshima and Nagasaki; Germany unconditionally surrenders

1946 Tokyo Tsushin Kenkyujo (later renamed Sony) founded in Tokyo; Microwave oven invented by Percy Spencer

1947 John Bardeen, Walter Brattain and William Shockley invent the transistor at the Bell Telephone Laboratories

1948 Alec Issigonis' Morris Minor launched; Polaroid launches its Land Camera; Porsche launches the 356 roadster; Earl Tupper establishes his plastics manufacturing company in Leominster, Massachusett

1949 Richard Buckminster Fuller invents the Geodesic Dome; Xerox launches the first xerographic copier

1949 Chinese People's Republic proclaimed by Mao Tse-Tung

1950 Marcello Nizzoli designs the Lettera 22 typewriter for Olivetti

1950 Start of Korean War

1951 Festival of Britain

1953 Hochschule für Gestaltung opened in Ulm, Germany

1955 Arne Jacobsen designs his Series 7 chairs for Fritz Hansen; Lockheed launched the U-2 Spy Plane; Lufthansa introduces the first transatlantic scheduled flight service using Lockheed Super Constellations; Isetta micro-car launched by BMW

1956 Dieter Rams and Hans Gugelot design the Phonosuper SK 4 for Braun

1957 Launch of Sputnik satellite

1957 Sony launches the Type 63 transitor radio the first "shirt pocket" radio

1950

1949
Henry Dreyfuss designs the *Model 500* telephone for Bell Telephone Company

1948
Porsche launches its first car, the *356* roadster

1946
Piaggio launches the *Vespa* scooter

1950
Charles & Ray Eames' fibreglass shell chair put into production

1951
Tetra Pak founded in Lund, Sweden and first Tetra Pak packaging launched

1952
Boeing begins production of the *B52 Stratofortress*

1954
Douglas Scott designs his *Routemaster* bus for London Transport

1957
Max Bill designs a wall clock for Junghans

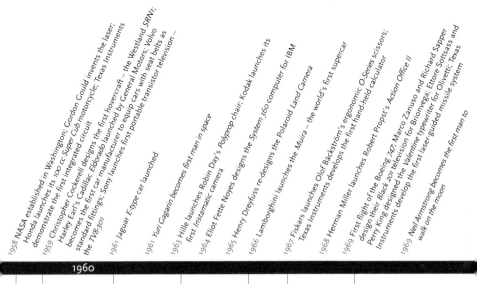

1958 NASA established in Washington; Gordon Gould invents the laser; Honda launches its 100 cc Super Cub motorcycle; Texas Instruments demonstrate the first integrated circuit

1959 Christopher Cockerell designs the first hovercraft – the Westland SRN1; Harley Earl's Cadillac Eldorado launched by General Motors; Volvo becomes the first car manufacturer to equip cars with seat belts as standard fittings; Sony launches first portable transistor television – the TV8-301

1961 Jaguar E-type car launched

1961 Yuri Gagarin becomes first man in space

1963 Hille launches Robin Day's Polyprop chair; Kodak launches its first Instamatic camera

1964 Eliot Fette Noyes designs the System 360 computer for IBM

1965 Henry Dreyfuss re-designs the Polaroid Land Camera

1966 Lamborghini launches the Muira – the world's first supercar

1967 Fiskars launches Olof Bäckström's ergonomic O-Series scissors; Texas Instruments develops the first hand-held calculator

1968 Herman Miller launches Robert Propst's Action Office II

1969 First flight of the Boeing 747; Marco Zanuso and Richard Sapper design their Black 201 television for Brionvega; Ettore Sottsass and Perry King designed the Valentine typewriter for Olivetti; Texas Instruments develop the first laser-guided missile system

1969 Neil Armstrong becomes the first man to walk on the moon

1960

1969
First flight of Concorde –
the world's first supersonic
commercial airliner

1961
Eliot Fette Noyes designs
Selectric "golfball"
typewriter for IBM

1963
Porsche launches the 911 car

1959
Alec Issigonis' Morris
Mini launched

1967
Stelton introduces Arne Jacobsen's
Cylinda-Line stainless-steel hollow-ware
range

1958
Ettore Sottsass designs
the Elea 9003 main frame
computer for Olivetti

1966
Marco Zanuso and Richard
Sapper design their Grillo
telephone for Siemens

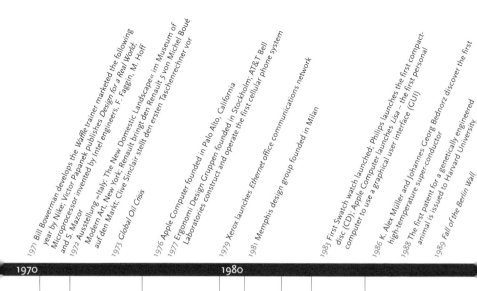

1971 Bill Bowerman develops the *Waffle* trainer marketed the following year by Nike; Victor Papanek publishes *Design for a Real World*; Microprocessor invented by Intel engineers; F. Faggin, M. Hoff and S. Mazor

1972 Ausstellung »Italy: The New Domestic Landscape« im Museum of Modern Art, New York; Renault bringt den Renault 5 von Michel Boué auf den Markt; Clive Sinclair stellt den ersten Taschenrechner vor

1973 Global Oil Crisis

1976 Apple Computer founded in Palo Alto, California

1977 Ergonomi Design Gruppen founded in Stockholm; AT&T Bell Laboratories construct and operate the first cellular phone system

1979 Xerox launches Ethernet office communications network

1981 Memphis design group founded in Milan

1983 First Swatch watch launched; Philips launches the first compact-disc (CD); Apple Computer launches *Lisa* – the first personal computer to use a graphical user interface (GUI)

1986 K. Alex Müller and Johannes Georg Bednorz discover the first high-temperature super-conductor

1988 The first patent for a genetically engineered animal is issued to Harvard University

1989 Fall of the Berlin Wall

1970 **1980**

1974
Volkswagen launches Giorgetto Giugiaro's *Golf Mk I*

1982
Sony launches the *Watchman* micro-television

1972
Jakob Jensen designs the *Beogram 4000* for Bang & Olufsen

1979
Sony launches the first *Walkman* personal stereo

1981
NASA launches the first Space Shuttle

1984
Apple Computer launches the *Macintosh I* personal computer

1971
Lamborghini *Countach* launched

1980
Nikon launches Giorgetto Giugiaro's *F3* camera

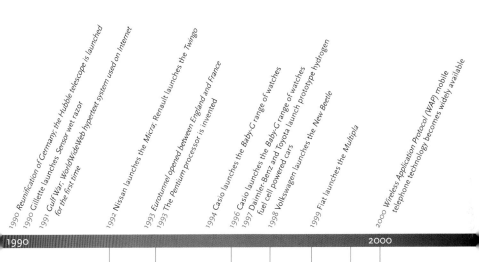

1990 Reunification of Germany; the Hubble telescope is launched

1990 Gillette launches Sensor wet razor

1991 Gulf War; WorldWideWeb hypertext system used on Internet for the first time

1992 Nissan launches the Micra; Renault launches the Twingo

1993 Eurotunnel opened between England and France

1993 The Pentium processor is invented

1994 Casio launches the Baby-G range of watches

1996 Casio launches the Baby-G range of watches

1997 Daimler-Benz and Toyota launch prototype hydrogen fuel cell powered cars

1998 Volkswagen launches the New Beetle

1999 Fiat launches the Multipla

2000 Wireless Application Protocol (WAP) mobile telephone technology becomes widely available

1990   **2000**

1999
Sony launches its
AIBO Robotic dog

2000
Hanover tram TW2000 for
Expo' 2000 launched

1993
James Dyson's
DC01 vacuum
cleaner launched

1999
Audi TT Coupé Quattro
launched

1998
Apple Computer
launches the iMac
computer

1992
Bill Stumpf and Don
Chadwick design their
Aeron office chair
for Herman Miller

1996
Trevor's Baylis' Freeplay
clockwork radio wins the
Product Category of BBC
Design Awards

Acknowledgements

We would like to take this opportunity to thank all at Taschen – especially our editors Susanne Husemann, Uta Hoffmann and Karen Williams – for the successful realization of yet another *Klotz* project. We also wish to thank Nick Bell, Sacha Davison and Christopher Brawn of UNA for their wonderful graphic design work and for their steady nerve while under fire. Lastly – and as ever – we thank our daughters Emelia and Clementine for their good humour and tolerance of publishing deadlines.

We are also enormously grateful to the many individuals, manufacturers, distributors, design offices, auction houses, public institutions and picture libraries who have lent their assistance and provided images. Thanks must also go to Paul Chave for the new photography generated specially for this project.

Special thanks to:

A&E Design, Stockholm
Advanced Vehicle Design, Altrincham
AEG Aktiengesellschaft, Frankfurt
Airstream, Jackson Center
Alessi, Crusinallo
Alex Moulton Ltd., Bradford-on-Avon
Alfa Romeo Museo, Arese
Emilio Ambasz, New York
Animal, Wareham
Apple Computer, Cupertino
Aprilia SpA, Novale
Archivio Storico Olivetti, Ivrea
Artemide, Milan
Arzberg (Winterling AG), Kirchenlamitz
Atomic, Altenmarkt
Authentics, Holzgerlingen
Automobili Lamborghini, S. Agata Bolognese
Aviation Picture Library, London
Barry Friedman Ltd., New York
Bang & Olufsen, Struer
Bauhaus-Archiv, Berlin
Stephen Bayley, London
Caroline & Denis Bellessort, London
Bertone, Turin
Biro Bic Ltd., London
Black & Decker, Slough
B-Line Srl, Grisignano di Zocco
BMW Museum, Munich
Boeing Company, Seattle
Bombardier Inc., Montreal
Braun GmbH, Kronberg im Taunus
BT Archives, London
Burton Snowboards, Burlington
Canon Europa NV, Amstelveen
Callaway Golf Europe, Chessington
Case Corp., Racine
Caterpillar Inc., Peoria
Christies Images, London
Citroën AG, Cologne
Cooper-Hewitt Museum, New York
Corning Inc., Corning
Daimler-Chrysler, Detroit
Danese, Grumello del Monte
JC Decaux, Neuilly
Deere & Co., Moline
Design Council, London
Design Council Archives, Brighton
Design Research Unit, London
Di Palma, Milan
Ducati, Bologna
Dyson Appliances Ltd., Malmesbury
Ecco Design, New York
Electrolux, Stockholm
Erco, London
Ergonomi Design Gruppen, Bromma
L. M. Ericsson, Stockholm
Euclid-Hitachi Inc., Cleveland, Ohio
Fiat UK, Slough
Fiskars, Billnäs
Flos UK Ltd., London

Ford Motor Company, Brentwood
Franz Schneider Brakel GmbH, Brakel
Frazer Designers, London
Freeplay Energy, Cirencester
Frogdesign, Altensteig
General Electric Co., Fairfield
General Motors Corp., Detroit
Ghia, Turin
Gillette Company, Boston
Giugiaro Design, Moncalieri
GK Design Group Inc., Toyko
Gustavsberg Fabrike, Gustavsberg
Hansgrohe, Schiltach
Harley-Davidson Motor Company, Milwaukee
Haslam & Whiteway, London
Keith Helfet, Coventry
Henry Dreyfuss Associates, Woodridge
Herman Miller Inc., Zeeland
Hollington Associates, London
Knud Holscher, Copenhagen
Honda UK Ltd., Reading
Hoover Company, North Canton
IBM Corp., Armonk
IDEO, Palo Alto
IDEO, Tokyo
Iittala Glass Museum, Iittala
Ikea, Älmhult
Imperial War Museum, London
InterDesign, Stockholm
Jaguar Ltd., Coventry
JCB Inc., Rochester
Junghans Uhren GmbH, Schramberg
Kartell, Milan
Kawasaki Motors UK, Bourne End
Kompan A/S, Ringe
Leica Camera AG, Solms
Lewis Moberley, London
London Transport Museum, London
Louis Poulsen, Copenhagen
Ross & Miska Lovegrove, London
MAG Instruments Inc., Ontario Cal.
Master Cutlery Corp., Chiba
Alberto Meda, Milan
Minale Tattersfield, London
Jasper Morrison, London
Motivation, Bristol
National Motor Museum, Beaulieu
Nautilus, Independence
Necchi, Pavia
Die Neue Sammlung, Munich
Nike UK Ltd., London
Nikon UK Ltd., Kingston-upon-Thames
Nissan Motor Co., Tokyo
NPK Industrial Design, Leiden
Oakley Inc., Foothill Ranch
Opel AG, Rüsselsheim
Parker Pen Company, Newhaven
Pentagram Design, London
Pethick & Money Ltd., London
Philips NV, Eindhoven

Pi Design, Lucerne
Pininfarina SpA, Turin
Polaroid Corporation, Cambridge, Mass.
Porsche Cars (GB) Ltd., Reading
Porsche Design, Zell am See
Queensberry Hunt Levien, London
Radio Corporation of America, New York
Raymond Loewy International, London
Remarkable Pencils Ltd., London
Renault Presse, Billancourt
Robert Bosch GmbH, Stuttgart
Robert Krups GmbH, Solingen
Rosenthal AG, Selb
Rubbermaid Inc., Wooster
Saab UK Ltd., Marlow
Sadler Design, Milan
Sanyo UK Ltd., Watford
Science & Society Picture Library, London
Seymour Powell, London
Shimano Europe, Nunspeet
Siemens AG, Munich
Silhouette International, Linz
Smart Cars UK, London
Snap-on Inc., Kenosha
Solari, Udine
Sony Corporation, Tokyo
Springfield Armory Museum, Springfield
Stanley Tools, New Britain
Steelcase Inc., Grand Rapids
Stelton, Hellerup
Sunseeker International Ltd., Poole
Swatch International, Zurich
TAG Heuer, Marin
Tangerine, London
Team Design, Esslingen
Tetra Pak Ltd., Uxbridge
Texas Instruments, Dallas
Torsten Bröhan GmbH, Düsseldorf
Toshiba Corp., Tokyo
Toyota Motor Corp., Tokyo
Tupperware Europe, Aalst
Vent Design, Campbell
Victorinox, Ibach
Vignelli Associates, London
Vignelli Associates, New York
Vin-Mag, London
Vitrac (Pool) Design, Paris
Volkswagen AG/Audi, Wolfsburg
Volvo Corp., Gotenburg
The Wedgwood Museum, Barlaston
Robert Welch, Chipping Campden
Werkbund-Archiv, Berlin
Weston Medical Ltd., Eye
The Wolfsonian (FIU), Miami
Xerox Corp., Stamford
Zanussi Design Center, Pordenone

We are immensely grateful to those individuls and institutions that have allowed us to reproduce images. We regret in some cases it has not been possible to trace the original copyright holders of photographs from earlier publications. We would also like to thank the numerous designers, manufacturers and institutions that have kindly supplied portrait images. The publisher has endeavoured to respect the rights of third parties and if any rights have been overlooked in individual cases, the mistake will be correspondingly amended where possible. The majority of historic images were sourced from the design archives of Fiell International Ltd., London and TASCHEN GmbH, Cologne.

A&E Design 14 (top), 14 (bottom), 15 Advanced Vehicle Design 118 (top), 118 (bottom), 119 AEG Aktiengesellschaft 10 (top), 10 (bottom), 11, 12 (top left), 12 (top right), 13, 58 (top), 58 (bottom), 59 (top), 59 (bottom), 60, 63 Aga UK Ltd 158 Airstream 16 (top), 16 (bottom), 17 Alessi 18 (top), 18 (bottom), 19, 20, 21, 49 (bottom), 340 (bottom), 534 (bottom) Alex Moulton Ltd. 356 (bottom) Alfa Romeo Museo 22 (top), 22 (bottom left), 22 (bottom right), 23 (top), 23 (bottom), 24 (top), 24 (bottom), 25, 71 (top) Emilio Ambasz 26 (top), 26 (bottom), 27 Apple Computer 28 (top), 28 (bottom), 29 (top), 29 (bottom), 30, 31, 285 (top), 285 (bottom), 286 Aprilia SpA 32 (top), 32 (bottom), 33, 463 (bottom), 464 Archivio Storico Olivetti 68 (all images), 69 (all images), 373 (top), 380, 381 (top), 381 (bottom), 382 (all images), 383, 460 (top), 461 (bottom) Artemide 37 (top), 37 (bottom), 163 (bottom), 431 (bottom left), 431 (bottom right) Arzberg (Winterling AG) 54 (top), 54 (bottom), 55 Ateliè Sundahl AB 39 (bottom) Audi 39 (top), 39 (bottom), 40, 41 (top), 41 (bottom), 42, 292 (top) Authentics 43 (top), 43 (bottom) Automobili Lamborghini 310 (top), 310 (bottom) Aviation Picture Library (aviationpictures.com) 88 (bottom), 354 (top), 354 (bottom) Bang & Olufsen 46 (top), 46 (bottom), 47, 77 (bottom), 296 (bottom) Barry Friedman Ltd. 51, 281 (top), 336 (bottom), 508 Bauhaus-Archiv 50, 53, 96 (all photos: Gunter Lepkowski) Bell Helicopter Textron 66 (top), 66 (bottom) Bernice Jackson Fine Art 115 (top) Bertone 71 (bottom), 72 Bildarchiv Photo Marburg 436 (bottom) Bildarchiv Österreichische Nationalbibliothek 38 (top) Biro Bic Ltd. 75 (top), 75 (bottom), 76 Black & Decker 79 (top), 79 (bottom) B-Line Srl 151 BMW Museum 80 (top), 80 (bottom), 81 (top), 81 (bottom), 82 Boeing Company 85 (top), 85 (bottom), 86 (top), 86 (bottom), 87, 88 (top), 343 (top), 344 (bottom), 345 (all images), 375 (top), 375 (bottom) Bombardier Inc. 89 (top), 89 (bottom), 90, 162 (top), 162 (bottom), 312 (top), 312 (bottom) Braun GmbH 97 (top), 97 (bottom), 98, 99, 101, 102, 332 (top), 333 (bottom), 357 (top), 357 (bottom), 358, 411 (top), 412, 413, 520 (bottom) Brionvega 108 (top) BT Archives 204, 437 (bottom) Buckminster Fuller Archives 226 (top), 226 (bottom), 227 Burton Snowboards 120 (top), 120 (bottom), 121 Canon Europa NV 123 (top), 123 (bottom), 124, 125 (top), 125 (bottom) Case Corp. 128 (top), 128 (bottom), 129 (top), 129 (bottom) Casio UK Ltd. 130 (top), 131 (all images) Caterpillar Inc. 136 (top), 136 (bottom), 137 (top), 137 (bottom) Christies Images 48, 106, 108 (bottom), 109, 127, 132 (bottom), 133, 144, 228 (top), 228 (bottom), 254 (bottom), 301 (bottom), 351 (bottom), 398 (top), 398 (bottom), 432, 489, 496 (bottom) Citroën AG 142 (top), 142 (bottom), 143 (top) Colombo Design 347 (bottom) Cooper-Hewitt Museum 155 (bottom), 179 (bottom), 433 (top) Corning Inc 154 (top), 154 (bottom), 155 (top), 192 (top) Culver Pictures Inc. 365 (top), 365 (bottom) Daimler-Chrysler 104, 103 (top), 138 (top), 138 (bottom), 139, 140 (top), 140 (bottom), 141, 348 (top), 348 (bottom), 349 (all images), 350 Danese 159 (top), 159 (bottom) JC Decaux 164 (top), 164 (bottom), 165 Deere & Co. 166 (top), 166 (bottom), 167 (all images) Design Council (Millennium Products) 220 Design Council Archives (University of Brighton) 169 (bottom) Design Research Unit 78 (top), 78 (bottom), 170 (top), 170 (bottom), 171 Ducati 182 (top), 182 (bottom left), 182 (bottom right), 183 (top), 183 (bottom) Dyson Appliances Ltd. 122 (top) Duracell 122 (bottom) Ecco Design 190 (top), 190 (bottom), 191 Electrolux 200 (top), 200 (bottom), 201, 319, 320, 321, 434 (photo: Hans Thorwid – Nationalmuseum) (top), 434 (photo: Hans Thorwid – Nationalmuseum) (bottom), 435 (photo: Hans Thorwid – Nationalmuseum) Eli Whitney Museum 525 (top), 525 (bottom), 526 Erco 202 (top), 202 (bottom) Ergonomi Design Gruppen 203 (top), 203 (bottom) L. M. Ericsson 204 (top), 205, 259 (top), 259 (bottom) Euclid-Hitachi Heavy Equipment Inc. 208 (top), 208 (bottom) Ferrari 209 (top), 209 (bottom), 210 (top) Fiat UK 211 (top), 211 (bottom), 212 (top), 212 (bottom), 213, 242 (top), 242 (bottom) Fiell International Limited (all photos: Paul Chave) 100, 130 (bottom), 161 (bottom), 186 (bottom), 288, 313 (top), 313 (bottom), 333 (top), 374, 388 (bottom), 438 (bottom) Fiskars 45 (top), 38 (bottom), 214 (top), 214 (bottom) Flos UK Ltd. 135 (all images), 215 (top), 215 (bottom) Fondation pour l'Architecture 70 (bottom) Ford Motor Company 216 (top), 216 (bottom), 217 (all images), 218 (all images), 219 Frazer Designers 221 (top), 221 (bottom) Freeplay Energy 56 (top), 56 (bottom), 57 Frog 222 (top), 222 (bottom) Franz Schneider Brakel GmbH 223 (top), 223 (bottom) General Electric Co. 230 (top), 230 (bottom), 231, 232 (top), 232 (bottom), 233 General Motors Corp. 188 (top), 188 (bottom), 189, 234 (top), 234 (bottom), 235 (top), 235 (bottom), 236 (top), 236 (bottom), 237–238, 239 (top), 239 (bottom) Ghia 240 (top), 240 (bottom), 241 (top), 241 (bottom) Gillette Company 243 (top), 243 (bottom left), 235 (bottom right), 236 (all images) Giugiaro Design 108 (bottom), 237 (top), 237 (bottom), 238 GK Design Group Inc. 190 (top), 198 (bottom), 199, 247 (top), 247 (bottom), 248 (top), 248 (bottom), 533 (bottom) Gustavsberg Fabrike 297 (top), 297 (bottom) Hansgrohe 465 (left), 465 (right) Harley-Davidson Motor Company 258 (top), 258 (bottom), 259 (top), 259 (bottom) Haslam & Whiteway 176 (bottom), 177, 353 (bottom), 488 (top) Henrik Kam Photography 206 (bottom), 207 (top) Henry Dreyfuss Associates 67 (top), 175 (bottom), 178 (top), 179 (top), 180 (top), 180 (bottom), 181, 401 (bottom) Herman Miller Inc. 261 (top), 261 (bottom), 262 Hollington Associates 263 (top), 263 (bottom), 387 (bottom) Knud Holscher 264 (top), 264 (bottom) Honda UK) Ltd. 265 (all images), 266 Hoover Company 267 (all images) IBM Corp. 270 (top), 270 (bottom), 271 (top), 271 (bottom), 372, 376 (bottom), 377 Ideal Standard 404 (top), 404 (bottom) IDEO 224 (top), 224 (bottom), 225, 273 (top), 273 (bottom), 274, littala Glass Museum 276 (top), 276 (bottom) Ikea of Sweden 277 (top), 277 (bottom), 278 Imperial War Museum 351 (top) InterDesign 280 (top), 280 (bottom) Jaguar Ltd. 289 (top), 291 (top), 291 (bottom) JCB Limited 293 (top), 293 (bottom), 294 Junghans Uhren GmbH 73 (bottom), 74 Kai International 275 (top), 275 (bottom) Kartell 34 (bottom), 149 (top), 149 (bottom), 150, 298 (top), 298 (bottom), 299, 535 Kawasaki Motors UK 300 (top), 300 (bottom) Kompan A/S 307 (top), 307 (bottom) Leica Camera AG 314 (all images) Lockheed Martin 315 (top), 315 (bottom), 316 (top), 316 (bottom), 317 London Transport Museum 326 (top), 326 (bottom), 327 (top), 327 (bottom), 437 (top) Louis Poulsen 348 (top), 328 (bottom), 329 (top), 329 (bottom) MAG Instruments Inc. 334 (all images) Matsushita 341 (all images) Minale Tattersfield 352 (top), 352 (bottom) Jasper Morrison 35 (bottom) National Motor Museum (Beaulieu) 44 (top), 44 (bottom), 113 (top), 113 (bottom), 116 (top), 116 (bottom), 142 (bottom), 279 (top), 279 (bottom), 283, 259 (bottom), 311 (bottom) Necchi 373 (bottom) Die Neue Sammlung 38, 70 (top), 117 (bottom – photo: Angela Bröhan), 146 (bottom – photo: Angela Bröhan), 256 (bottom left and bottom right), 292 (bottom), 341 (bottom – photo: Angela Bröhan), 395 (bottom – photo: Angela Bröhan), 429, 433 (bottom), 441 (photo: Angela Bröhan), 476 (bottom), 487 (photo: Angela Bröhan), 511, 520 (top) Nike UK Ltd. 366 (top), 366 (bottom), 367 (top), 367 (bottom) Nikon UK Ltd. 368 (top), 368 (bottom), 369 (all images) Nissan Motor Co. 372 (top), 372 (bottom) NPK Industrial Design 370 (top), 370 (bottom), 371 (top), 371 (bottom) Antti Nurmesniemi 378 (top), 378 (bottom) Oakley Inc. 379 (top), 379 (bottom) Opel AG 385 (all images) Olympus 384 (bottom) Parker Pen Company 386 (top), 386 (bottom) Pentagram Design 249 (top), 249 (bottom left, 385 (bottom right), 250, 305, 7 (top), 389 (top), 389 (bottom), 390 Philips NV 394 Pi Design (Bodum) 83 (all images), 84 (all images) Pininfarina SpA 210 (bottom), 396 (top), 396 (bottom), 397 (top), 397 (bottom) Polaroid Corporation (Archives) 399 (top), 399 (bottom), 400, 401 (top) **Porsche Cars UK Ltd.** 306 (top), 306 (bottom), 405 (top), 405 (bottom), 406 (top), 406 (bottom), 407 **Porsche Design** 408 (top), 408 (bottom), 409 (all images) **Quadrant Picture Library** 448 (bottom) **Queensberry Hunt Levien** 410 (top), 410 (bottom) **Radio Corporation of America (Courtesy of Fred Barnum, "His Master's Voice" in America)** 414, 415 (top), 415 (bottom), 416 (top), 416 (bottom) **Raymond Loewy International** 318 (all images), 322 (all images), 323 (top), 323 (bottom), 324 (all images), 325 (top), 325 (bottom) **Renault Presse** 420 (top), 420 (bottom), 421 (top), 421 (bottom) **Revere Ware Corp.** 521 (top), 521 (bottom), 522 **Robert Bosch GmbH** 91 (top), 91 (bottom), 92, 93 (top), 93 (bottom), 94 **Robert Krups GmbH** 308 (top), 308 (bottom), 309 **Rosenthal AG** 255, 422 (top), 422 (bottom), 423 **Rubbermaid Inc.** 424 (top), 424 (bottom), 425 **Saab UK Ltd.** 426 (top), 426 (bottom), 427 (top), 427 (bottom) **Sadler Design** 428 (top), 428 (bottom) **Sanyo UK Ltd.** 430 (top), 430 (bottom) **Science & Society Picture Library** 36 (bottom), 64 (bottom), 65, 111, 112, 113 (bottom), 114, 126 (bottom), 145, 147 (top), 147 (bottom), 156 (bottom left), 156 (bottom right), 160 (top), 160 (bottom), 195 (right), 253, 257 (top), 268, 269 (bottom), 302 (bottom), 303, 304 (top), 304 (bottom), 338 (bottom), 359, 343 (bottom), 344 (top), 356 (top), 360 (top), 360 (bottom), 361, 362, 377 (top), 383 (bottom), 392 (top), 392 (bottom), 393 (top), 411 (bottom), 473, 494 (bottom left), 494 (bottom right), 513 (top), 515, 526 (top – both portrait images), 528, 532, 538 (top) **Sears, Roebuck & Co.**, 439 (bottom), 440, 449 (bottom), 457, 458, 459 (top) **Seymour Powell** 442 (top), 442 (bottom), 443 (top), 443 (bottom) **Sharp.** 444 (top), 444 (bottom), 445 (top), 445 (bottom) **Siemens AG** 446 (top), 446 (bottom), 447 **Sinclair Electronics** 449 (top) **Singer Ltd. Archives** 451 (top), 451 (bottom), 452 (top), 452 (bottom) **Snap-on Inc.** 454 (top), 454 (bottom), 455 (top), 455 (bottom) **Solari** 596 (top) **Sony Corporation.** 456 (top), 456 (bottom), 459 (bottom) **Springfield Armory Archive** 461 (top), 461 (bottom) **Stanley Tools** 462 (top), 462 (bottom – both images) **Steelcase Inc.** 466 (top), 466 (bottom), 467, 468 (top), 468 (bottom), 469 (top), 469 (bottom) **Stelton** 287 (bottom), 335 (top), 335 (bottom), 470 (top), 470/471 (bottom), 471 (top) **Studio X** (photo: John Ross) 330 (top), 330 (bottom), 331 **Swatch International** 474 (top), 474 (bottom), 475 **TASCHEN GmbH Archiv** 107, 281 **Team Design** 453 (top), 453 (bottom) **Tetra Pak Ltd.** 482 (top), 482 (bottom), 483 **Texas Instruments** 484 (top), 484 (bottom), 485 **Torsten Bröhan GmbH** 61, 168 (bottom left) **Toshiba Corp.** 490 (all images), 491 **Toyota Motor Corp.** 492 (top), 492 (bottom), 493 (top), 493 (bottom) **Tupperware Europe** 4–5, 494 (top), 495 **Victorinox** 500 (top), 500 (bottom) **Vignelli Associates** 501 (all images) **Vin-Mag** 67 (bottom), 152, 153 (bottom left), 153 (bottom right), 195 (left), 282 (bottom), 289 (bottom), 290, 311 (top), 418 (bottom), 523 (bottom), 524 **Vitra GmbH** 346 **Volkswagen AG/Audi** 502 (top), 503 (top), 503 (bottom), 504 (top), 504 (bottom), 505 **Volvo Corp.** 506 (top), 506 (bottom), 507 (top), 507 (bottom) **The Wedgwood Museum** 516 (top), 516 (bottom), 517 (top), 517 (bottom), 518 **Werkbund-Archiv** 174 **The Wolfsonian – Florida International University, The Mitchell Wolfson Jr. Collection** 35 (bottom), 173, 193, 196 (Sammlung Finlay B. Matheson), 229 (bottom), 337, 359 (bottom), 364, 393 (bottom), 403, 417, 419, 450, 477 (bottom), 481, 499, 502 (bottom), 509 (bottom), 510, 519, 539 **Xerox Corp.** 529 (top), 529 (bottom), 530, 531 **Yamaha** 533 (top) **Zanussi Design Center** 536 (top), 536 (bottom), 537 (left), 537 (right)